Water Science and Technology Library

Volume 123

The aim of the *Water Science and Technology Library* is to provide a forum for dissemination of the state-of-the-art of topics of current interest in the area of water science and technology. This is accomplished through publication of reference books and monographs, authored or edited. Occasionally also proceedings volumes are accepted for publication in the series. *Water Science and Technology Library* encompasses a wide range of topics dealing with science as well as socio-economic aspects of water, environment, and ecology. Both the water quantity and quality issues are relevant and are embraced by *Water Science and Technology Library*. The emphasis may be on either the scientific content, or techniques of solution, or both. There is increasing emphasis these days on processes and *Water Science and Technology Library* is committed to promoting this emphasis by publishing books emphasizing scientific discussions of physical, chemical, and/or biological aspects of water resources. Likewise, current or emerging solution techniques receive high priority. Interdisciplinary coverage is encouraged. Case studies contributing to our knowledge of water science and technology are also embraced by the series. Innovative ideas and novel techniques are of particular interest.

Comments or suggestions for future volumes are welcomed.

Vijay P. Singh, Department of Biological and Agricultural Engineering & Zachry Department of Civil and Environment Engineering, Texas A&M University, USA Email: vsingh@tamu.edu

All contributions to an edited volume should undergo standard peer review to ensure high scientific quality, while monographs should also be reviewed by at least two experts in the field.

Manuscripts that have undergone successful review should then be prepared according to the Publisher's guidelines manuscripts: https://www.springer.com/gp/authors-editors/book-authors-editors/book-manuscript-guidelines

Rajendra Singh

Soil and Water Conservation Structures Design

 Springer

Rajendra Singh
Agricultural and Food Engineering
Department
Indian Institute of Technology Kharagpur
Kharagpur, India

ISSN 0921-092X   ISSN 1872-4663 (electronic)
Water Science and Technology Library
ISBN 978-981-19-8667-3   ISBN 978-981-19-8665-9 (eBook)
https://doi.org/10.1007/978-981-19-8665-9

This Springer imprint is published by the registered company Springer Nature Singapore Pte Ltd.
The registered company address is: 152 Beach Road, #21-01/04 Gateway East, Singapore 189721,
Singapore

Dedicated
to
My Late Parents
My Wife Archana

Foreword

Soil and water are the fundamental components of the environmental continuum and hence of the ecological continuum. Their conservation is therefore vital which is accomplished by the employment of conservation measures. These measures can be either structural or non-structural or a combination thereof. Development and design of these measures depend on physiographic, hydrometeorological, and field conditions (soil, land use and land cover, and crops) as well as the availability of quality data. Complicating the design of these measures is climate change whose impact needs to be incorporated in the design.

Soil and water conservation is a vast discipline combining related aspects of soil and crop sciences, forestry, hydrology, hydraulics, hydrometeorology, irrigation, drainage, geotechnical engineering, environmental engineering, structural engineering, and decision science. There is a multitude of books and technical reports on soil and water conservation in both the field of soil science and agricultural engineering. Most of the books, however, focus more on concepts, principles, theories, and methods, but much less on design of measures and their implementation in field, and on their effectiveness and methods of their improvements in field. One consequence, especially in developing countries, is that field engineers or soil conservationists base design of these measures on methods which are somewhat outdated and which do not take into account modern methods of design and the impact of global warming and climate change, land use land cover changes, changes in soil and crop management, and environmental impact assessment.

The book by Prof. Rajendra Singh is one of the few books that emphasise design of soil and water conservation structures and is very timely. Spanning 14 chapters it encompasses basic principles and procedures of soil and water conservation, mechanics of erosion, development of measures for the control of erosion, techniques for the measurement and modelling of soil loss, applications of remote sensing (RS) and Geographic Information Systems (GIS) to soil conservation, assessment of the impact of climate and land use land cover (LULC) changes on soil erosion, and design of soil and water conservation structures.

The book will be useful for graduate students, college faculty, field engineers, and watershed managers. I applaud Prof. Singh for writing this book which has long been much needed. The book reflects his long experience in teaching, research, consultancy, and extension.

Vijay P. Singh, Ph.D., D.Sc., P.E., P.H., Hon.D. WRE
Academician (GFA)
Distinguished Professor
Regents Professor
Caroline and William Lehrer Distinguished Chair in Water Engineering
Department of Biological and Agricultural Engineering
Zachry Department of Civil and Environmental Engineering
Texas A&M University
College Station, TX, USA

Preface

As an undergraduate student of *Agricultural Engineering* and a postgraduate student of *Soil and Water Conservation Engineering*, I found the design of soil and water conservation structures fascinating as it allowed me to apply engineering principles to the field conditions. However, I found the design process challenging as it required referring to different books and manuals to get the required information. Later on, while teaching *Soil Conservation Structures Design* at the Indian Institute of Technology Kharagpur, I collected adequate literature on the design concepts and processes of soil and water conservation structures. Subsequently, while offering the course entitled *Soil and Water Conservation Engineering* under National Programme on Technology Enhanced Learning (NPTEL), funded by the Ministry of Education, Government of India, I reckoned it would be a good idea to extend the material with essential details, including the design examples, into a full-fledged textbook on *Soil and Water Conservation Structures Design*. I did so eventually, and while writing, I strived to elucidate design principles and procedures in a simple and unified way so that subject teachers, students, and practising engineers could access them effortlessly.

Although soil and water have supported the development and sustenance of civilisations by facilitating agricultural production in an enabling environment, these natural resources are facing degradation due to uncontrolled anthropogenic activities. Consequently, about 52% of the world's total productive land is subject to different degradation processes. Soil erosion has degraded about 45% of the total land area in India, 32% the USA, 16% in Africa, 31% in China, and 17% in Europe, with an estimated annual average soil erosion rate in the range of 2.87–30 t ha^{-1} yr^{-1}. Globally, water erosion causes sediment flux of 28 Gt yr^{-1}, while wind erosion impacts about 28% of the global land. Therefore, soil and water conservation is essential for maintaining or enhancing land productivity by preventing soil erosion, conserving water, and improving soil fertility. Various soil and water conservation measures, developed focusing on improved management of soil, water, and vegetation, must be vigorously adopted to address global concerns like water scarcity and biodiversity conservation, and ensure sustainable agricultural development and rural livelihood generation. Soil and water conservation structures are often employed to tackle gully erosion, a global phenomenon recognised as the extreme stage of soil erosion. However, the successful design and execution of gully erosion control measures require the services of qualified engineers having sound knowledge of hydrology and soil science, besides a broad understanding of the soil–water–plant–atmosphere relationship.

Soil and Water Conservation Structures Design, focusing on the engineering design of soil and water conservation structures, is intended for undergraduate and graduate students pursuing degrees in agricultural engineering and soil and water conservation engineering, practising engineers engaged in the soil and water conservation or watershed management programmes and environmental consultants. In addition, it will be helpful to those studying soil and water conservation as part of their agriculture, environmental science, soil science, watershed management, and civil engineering courses.

Though several textbooks on soil and water conservation exist, most of these include the entire gamut of land and water management with numerous chapters dedicated to hydrology, irrigation engineering, drainage engineering, and groundwater. Consequently, soil and water conservation engineering coverage is diluted. In particular, the design aspects of the permanent gully control structures, which undergraduate and graduate students and practising engineers often find most challenging, are ignored.

The foremost objective of the book is to present the design principles lucidly. Therefore, every theoretical concept is followed by a worked example for easy understanding of the students. In addition, a set of ten practice questions and ten multi-objective questions are included in each chapter. Faculty members using this textbook in their courses can approach the author for the solution to the practice questions if needed.

The textbook contains 14 chapters. Following the introduction of soil and water conservation principles in Chap. 1, Chap. 2 describes the mechanics and types of water erosion. Chapter 3 presents various soil erosion modelling and measurement techniques for soil loss assessment. Chapters 4 and 5 are dedicated to designing terraces and bunds, two widely practised mechanical erosion control measures. The design of vegetated waterways, including their layout, construction and maintenance, is presented in Chap. 6. Chapter 7 elaborates on the design principles used in designing temporary gully control structures and introduces the permanent structures' hydrologic, hydraulic, and structural design principles. Chapters 8–10 present the comprehensive design of the drop spillway, drop-inlet spillway, and chute spillway. Chapter 11 describes the mechanics of wind erosion and introduces various empirical and process-based wind erosion models. Chapter 12 presents the design of two popular water storage and diversion structures, the earthen embankments, and the farm ponds. Chapters 13 and 14 highlight the recent advances in the field by introducing the applications of remote sensing (RS) and Geographic Information System (GIS) in soil conservation and analysing the impacts of climate and land use land cover (LULC) changes on soil erosion.

I hope this book will be a valuable reference for the faculty, students, and practising engineers and help promote the robust design of soil and water conservation structures for implementation in the integrated watershed management programmes in different parts of the world.

Kharagpur, India   Rajendra Singh

Acknowledgements

I am indebted to my teachers Late Prof. N. N. Sirothia of the University of Allahabad and Late Prof. T. Satyanarayana of the Indian Institute of Technology Kharagpur, for introducing me to the exciting world of Soil and Water Conservation Engineering. I am grateful to Prof. V. P. Singh of Texas A&M University and Prof. S. R. Singh, former Vice Chancellor of Rajendra Agricultural University, Pusa, for their encouragement and constant support. A special mention is due to Prof. H. A. Kazmi, who inspired me to join the academic world.

I want to acknowledge with gratitude the blessings of my late parents and elder sister, and the support and love of my elder brother and sisters and their family. This book would not have been possible without the constant support and encouragement of my sons, Gaurav and Anubhav. I also appreciate the encouragement of my daughter-in-law, Heena. My most significant appreciation is for my wife, Archana, who kept encouraging me to complete the book.

My friends and colleagues, Prof. Nirupama Mallick, Prof. Chandranath Chatterjee, Prof. Ashok Mishra, Dr. Damodhara Mailapally, and Dr. Piyush Singh, offered moral support during the testing times of COVID-19 and encouraged me to complete my dream project of writing a textbook.

Many have helped me prepare this book. I sincerely thank Dr. Kritika Kothari, Dr. Pranesh Paul, Dr. Deepak Singh Bisht, and Dr. Srishti Gaur for critically reviewing several chapters. I am also grateful to Prof. Mahendra Nagdeve, Prof. Arnab Bandyopadhyay, and Prof. Aditi Bhadra for their comments on an earlier chapter. I wish to acknowledge the support of Dr. Manorajan Kumar and Dr. D. R. Sena for helping me locate a few specific technical documents. A big thanks to my Ph.D. students Krishna and Sushanth, and project staff Pramodh for their help in preparing the book's final draft and Partha Pratim Samanta for creating most of the figures in the book.

I wish to place on record the support of several academic and research organisations, including FAO, ICAR-IISWC, ICRISAT, IPCC, University of Nebraska, USBR, and USDA, and publishers like Elsevier and Springer Nature, for permitting the use of the copyright materials, without which this book would not have seen the light of the day. A special word of thanks to Creative Commons for providing open access to the technical reports and images.

I am incredibly grateful to Ms. Swati Meherishi and her team, including Ms. Shalini Selvam and Ms. Aparajita Singh, for all the assistance and technical support.

I have strived hard to keep this book error-free, but still, there may be mistakes. For that, I apologise and invite readers to offer constructive criticism and suggestions to further improve the book in the subsequent editions.

Lastly, I thank the Almighty for giving me the strength and patience to complete this book.

Rajendra Singh

Contents

About the Author

Dr. Rajendra Singh has been a faculty in the Agricultural and Food Engineering Department, Indian Institute of Technology (IIT) Kharagpur since 1989. He has been associated with the Soil Water Conservation Engineering discipline of the Department (presently known as Land and Water Resources Engineering), and has taught "Soil and Water Conservation Engineering" at both undergraduate and postgraduate levels. He has contributed significantly to the Agricultural Engineering profession through teaching, research and technology transfer.

He has developed and popularised five software packages, including the Soil Conservation Structure Designer (SCS_Designer) to improve the agricultural engineering education and research. Dr. Singh has guided 18 Ph.D. and 75 M.Tech. students, with an h-index of 34 and i-10-index of 61. He has received several national and international honours and awards including BOYSCAST Fellowship, AICTE Career Award for Young Teachers, ICAR Young Scientist Award, DAAD Research Fellowship, DAAD Visiting Professorship; and held important positions like Member, Board of Governors, Dean (Undergraduate Studies) and Head of Department at IIT Kharagpur, and Member of Research and Technical Advisory Committees of several National bodies. Dr. Singh has also won the "Recognition Award" in the area of 'Soil and Water Conservation Engineering' from the National Academy of Agricultural Sciences.

Chapter 1
Introduction

Abstract Soil and water conservation is essential to tackle the global challenge of soil erosion, negatively impacting food productivity, water security, and environmental quality. This chapter traces the history of soil erosion and introduces the principles of soil and water conservation. The types of soil erosion and their causes are discussed. The chapter highlights the on-site and off-site effects of soil erosion, and the importance of soil and water conservation measures. It presents the soil and water conservation measures developed and practised under various initiatives worldwide and in India. The chapter introduces various agronomic and biological, and engineering or mechanical measures adopted for soil and water conservation. The chapter also includes the structure of the book to guide prospective readers.

1.1 Overview

Soil and water are the earth's critical natural resources that support the existence of humankind by providing food, feed, fuel, and fibre. Integrated soil–water systems have supported the development and sustenance of civilisations by facilitating agricultural production in an enabling environment. However, during the last century, uncontrolled anthropogenic activities, including intensive cultivation, soil mismanagement, overgrazing, deforestation, and urbanisation, have led to the degradation of these natural resources. Consequently, about 52% of the world's total productive land is subject to different degradation processes (Yousuf and Singh 2019). In India, around 146.8 million hectares out of 328 million hectares of geographical area are degraded (NBSS & LUP 2005). Figure 1.1 presents the world map of human-induced soil degradation (ISRIC 2017; Gupta 2019). Since this widespread land degradation threatens global food security and sustainability, appropriate measures must be adopted to conserve precious soil and water resources.

R. Singh, *Soil and Water Conservation Structures Design*, Water Science and Technology Library 123, https://doi.org/10.1007/978-981-19-8665-9_1

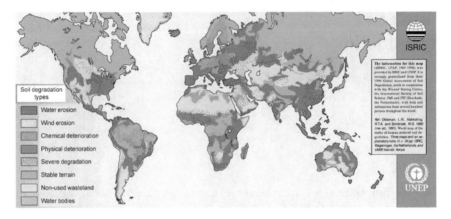

Fig. 1.1 World map showing human-induced soil degradation (After ISRIC 2017; Gupta 2019; Creative Commons Attribution-Share Alike 4.0 International license)

1.2 Soil Erosion

Globally, land degradation may be attributed to soil erosion (Lal 2010; Guo et al. 2019). Soil erosion has been estimated to degrade about 32% of the total land area in the USA, 16% in Africa, 31% in China, 17% in Europe, and 45% in India (Singh et al. 2020). The annual average soil erosion rates reported in the literature vary drastically. For example, the global soil partnership led by the Food and Agriculture Organisation of the United Nations (FAO) reported an annual average soil erosion of 75 gigatonnes per year (Gt year^{-1}) in 1993 (Myers 1993; Montanarella 2015). On the contrary, Borrelli et al. (2013) predicted the annual average soil erosion as 35.9 Gt year^{-1} or 2.87 t ha^{-1} year^{-1} in 2012. Recently, Gachene et al. (2019) reported an annual average soil erosion rate of 30 t ha^{-1} year^{-1}.

Figure 1.2 presents the spatial pattern of soil erosion (Borrelli et al. 2013) and shows that about 13.8 million km^2 suffers from high to severe soil erosion, with 7.5 million km^2 of the area exceeding the generic tolerable soil erosion threshold (T-value). Borrelli et al. (2013) considered a T-value of 10 t ha^{-1} year^{-1}, contrary to 11.2 t ha^{-1} year^{-1} recommended by Wishmeier and Smith (1978). Therefore, there is an urgent need to control soil erosion and check land degradation for sustainable agricultural production and environmental well-being.

Soil erosion is a three-phase process involving the detachment of soil particles from soil mass, their transportation from one place to another by the action of water, wind, ice or gravity, and deposition. Deposition occurs when the transportation agent does not have sufficient energy to carry the soil particles (Fig. 1.3).

Soil erosion existed in many parts of the world, even in Neolithic times, primarily due to land use changes (Vanwalleghem et al. 2017). The deforestation and soil erosion problems also occurred in the Sumerian civilisation in 2700 BCE and the Trojan regime in 1200 BCE (Brown 2003). In modern times, Reverend Jared Eliot of Killingworth, Connecticut, was the first to mention soil erosion, in terms of soil

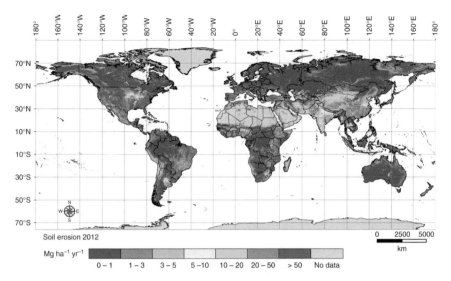

Fig. 1.2 Spatial pattern of soil erosion in 2012 (After Borrelli et al. 2013; Creative Commons Attribution License 4.0 (CC BY))

Fig. 1.3 Three phases of soil erosion (After Sarvade et al. 2019; Reused with permission from Springer Nature)

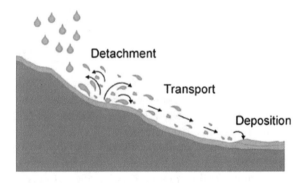

washing and its dangers, in a series of six essays written between 1748 and 1759 (Eliot 1760). Eliot's work brought attention to agriculture, in general, and erosion control, particularly in the USA.

Soil erosion may be of two types: natural or geological and human-induced or accelerated. Natural or geological erosion occurs in the natural environment due to the action of water, wind, gravity, and glaciers. It occurs silently and often continues unnoticed. However, it is usually of little concern because its rate is low, and soil loss is naturally compensated by soil formation. Thus, geological erosion includes soil-forming and soil-eroding processes, which maintain the soil in a favourable balance, suitable for plant growth. Geological erosion is responsible for forming the present topographical features such as canyons, stream channels, and valleys. The Grand Canyon in Arizona, USA, one of the world's geological wonders and the

Marble Rocks, a gorge on the Narmada River in Jabalpur, India, are examples of the topographical features formed by geological erosion.

The human-induced or accelerated erosion occurs due to alterations in vegetative cover and soil conditions caused by human interventions for resource exploitation.

The soil removal, primarily by water and wind, is much faster than the natural soil formation. Accelerated erosion leads to loss of soil productivity and severe environmental damages. It is one of the most important agricultural and natural resource management problems globally. Based on the transportation agent, soil erosion may be further classified as water and wind erosion.

1.2.1 Water Erosion

Globally, water erosion is the most severe type of soil erosion as it causes sediment flux of 28 Gt year^{-1} (Quinton et al. 2010). In India, water erosion affects 68.4% of the total land area (Mandal and Sharda 2013) and results in an annual loss of 13.4 Mt in cereal, pulse, and oil seeds production (Sharda et al. 2010). In water erosion, soil detachment is caused by the raindrops or flowing water and transport is by runoff.

1.2.2 Wind Erosion

Wind erosion is a widespread phenomenon impacting about 28% of the global land (Borrelli et al. 2015). In wind erosion, soil detachment and transportation are caused by wind. Wind erosion is a severe environmental hazard as it causes air pollution and adversely affects human health. It also influences the climate by altering the global radiation budget due to the dust particle concentration in the atmosphere. It occurs extensively in the arid and semiarid regions worldwide, which have inadequate precipitation and little or no vegetation cover. The agricultural lands subject to wind erosion are in Asia, North Africa, Australia, and parts of North and South America (UNEP WMO UNCCD 2016). Rajasthan, Gujrat, Punjab, and Haryana suffer from wind erosion in India.

1.3 Soil and Water Conservation

Soil erosion influences soil productivity and environmental quality through its on-site and off-site effects. The critical on-site effect of erosion is the removal of the nutrient-rich topsoil that leads to reduced crop yield and land degradation (Lal 2009, 2015; Wang 2016). On the contrary, the critical off-site effect of erosion includes sedimentation of streams and reservoirs, pollution of water bodies, disruption of lake ecosystems, and landscape alterations (Ouyang et al. 2010; Gao et al. 2012;

Boardman et al. 2019). Therefore, soil and water conservation is of utmost importance for agricultural and environmental sustainability and the economic well-being of society.

Soil and water conservation is essential for maintaining or enhancing land productivity through (a) prevention or reduction of soil erosion, (b) conservation or drainage of water, and (c) maintenance or improvement of soil fertility (Tideman 1996; WOCAT 2007).

1.3.1 Soil and Water Conservation Measures

Various soil and water conservation measures have been developed and practised since the early twentieth century. The Dust Bowl during the 1930s in the USA prompted extensive research for developing soil conservation measures. It led to the creation of the Soil Conservation Service (now the Natural Resources Conservation Service) in the US Department of Agriculture in 1935 (Kassam et al. 2014). Initially, the soil conservation efforts in the USA were targeted at the on-site effects of soil erosion, i.e., on protecting the nutrient-rich topsoil through reduced tillage. However, the off-site effects of soil erosion, e.g., water quality, have been included in the soil and water conservation plans since the 1980s. Soil and water conservation programmes were also initiated in the 1930s in the British-ruled territories in Africa. However, post-independence, most African nations initiated soil and water conservation programmes in the 1970s (IFAD 1992). In most European countries, scientific interest in soil erosion arose in the 1970s. In India, soil conservation was initiated in 1900 through the enactment of the Land Preservation Act in Punjab to prevent the menace of Cho (mountain torrents). However, post-independence, a Central Soil Conservation Board was set up in 1953 to formulate and implement soil conservation programmes in the country.

Over the last decades, various well-established soil and water conservation measures have been developed. Although these measures focus primarily on the improved management of natural resources, i.e., soil, water and vegetation, they may also simultaneously address other global concerns like water scarcity, poverty alleviation, biodiversity conservation and climate change (through carbon sequestration) (WOCAT 2007). Integrated watershed management programmes adopted worldwide also propagate soil and water conservation measures for economic well-being through sustainable agricultural development and rural livelihood generation (FAO 2006; Joshi et al. 2008; Wani et al. 2008). The holistic management of soil and water resources by adopting modern technological tools for creating inventory, developing participatory soil and water conservation programmes, and introducing policy changes dealing with natural resources management are the future challenges for the agencies and policymakers dealing with the soil and water conservation programmes (Manivannan et al. 2017; Sarvade et al. 2019).

In India, the Mahatma Gandhi National Rural Employment Guarantee Scheme (MGNREGS), Indian legislation enacted in 2005, guaranteeing one hundred days of employment in a year to adult rural community members, focuses on watershed management, including water conservation (Ministry of Rural Development 2022). MGNREGS activities include a comprehensive treatment of a watershed, with a particular focus on recharging groundwater and other drinking water sources through the construction of contour trenches, terraces, bunds, boulder checks, gabion structures, earthen dams, and check dams.

Soil and water conservation measures may be classified as agronomic and biological measures and engineering or mechanical measures.

1.3.1.1 Agronomic and Biological Measures

Agronomic and biological measures focus on soil and crop management to control soil erosion. These measures are typically adopted as an integral part of the farming system, including soil cover, crop rotation and crop mixes. They are relatively inexpensive but quite effective. Crop rotation, cover cropping, contour farming, strip cropping, and mulch tillage or stubble mulching are a few popular agronomic and biological measures. Besides, conservation agriculture, which includes zero-tillage or minimum tillage while maintaining soil cover with crop residues, and a diversified cropping system, is also gaining popularity (Kassam et al. 2014). Agroforestry is another popular measure, particularly in humid, tropical conditions (Udawatta and Gantzer 2022). Agroforestry may also be used in drier conditions as a shelterbelt for reducing wind speeds and minimising wind erosion.

1.3.1.2 Engineering or Mechanical Measures

Engineering or mechanical measures of soil and water conservation include various structures constructed to control runoff and soil erosion. They primarily focus on manipulating the slope gradient and the slope length. The construction typically includes slope cutting and filling and involves substantial expenditure. The structure choice depends on the catchment characteristics, soil physical characteristics, the necessity of retaining or disposing of the runoff, and the availability of construction material. It may also involve engineering inputs depending on the size of the structure.

Engineering or mechanical measures are also preferred to tackle gully erosion, a global phenomenon recognised as the extreme stage of soil erosion (Duvert et al. 2010; Klaus et al. 2014). Gully erosion control measures primarily address the problem of concentrated runoff and focus on gully rehabilitation by stabilising the gully head, bed, and banks. Various types of temporary and permanent gully control structures, including check dams, are constructed across the gully to reduce the channel gradient and stabilise the gully by facilitating vegetative growth over time. The choice of permanent or temporary structures depends on catchment size, expected flow, and the stage of the gully development. The successful execution of gully erosion control measures requires the services of qualified engineers having sound knowledge of hydrology and soil science, besides a broad understanding of the soil–water–plant–atmosphere relationship.

The popular engineering or mechanical measures include terraces, bunds, vegetated waterways, various types of temporary check dams, and permanent structures like drop, drop-inlet, and chute spillways.

1.4 Structure of the Book

The book *Soil and Water Conservation Structures Design* focuses on the engineering design of soil and water conservation structures. The textbook contains 14 chapters. Following the introduction of soil and water conservation principles in *Chapter 1*, *Chapter* 2 describes the mechanics and types of water erosion. It introduces agronomical, biological, and engineering or mechanical measures of erosion control. The Chapter 3 presents various soil erosion modelling and measurement techniques for soil loss assessment. Besides the popular empirical models, the distributed, physically based soil erosion models are discussed. It also highlights the soil loss measurement techniques. Chapters 4 and 5 are dedicated to designing terraces and bunds, two widely practised mechanical erosion control measures. The design of vegetated waterways, including their layout, construction, and maintenance, is presented in Chapter 6. Chapter 7 elaborates on the design principles used in designing temporary gully control structures and introduces the permanent structures' hydrologic, hydraulic, and structural design principles. Chapters 8–10 present the comprehensive design of the drop spillway, drop-inlet spillway and chute spillway. The hydrologic design approaches for estimating the peak flow rate, the hydraulic design for determining various component dimensions, and the structural design for analysing various forces acting on the structures are elaborated. A detailed procedure to analyse the stability of the structures is also included. Chapter 11 describes the mechanics of wind erosion and introduces various empirical and process-based wind erosion models. Chapter 12 presents the design of two popular water storage and diversion structures, the earthen embankments, and the farm ponds. Chapters 13 and 14 highlight the recent advances in the field by introducing the applications of remote sensing (RS) and Geographic Information System (GIS) in soil conservation and analysing the impacts of climate and land use land cover (LULC) changes on soil erosion.

References

Boardman J, Vandaele K, Evans R, Foster IDL (2019) Off-site impacts of soil erosion and runoff: why connectivity is more important than erosion rates. Soil Use Manage 35:245–256

Borrelli P, Robinson DA, Fleischer, LR, Lugato, E et al. (2013) An assessment of the global impact of 21st century land use change on soil erosion. Nat Commun 8. https://doi.org/10.1038/s41467-017-02142-7

Borrelli P, Panagos P, Hessel R, Riksen M, Stolte J (2015) Soil erosion by wind. In: Stolte J, Tesfai M, Øygarden L, Sigrun K, Keizer J, Verheijen F, Panagos P, Ballabio C, Hessel R (eds). Soil threats in Europe, EUR 27607 EN. https://doi.org/10.2788/488054

Brown LR (2003) Eco-Economy: building an economy for the earth, 1st edn. Routledge, Abingdon, Oxfordshire

Duvert C, Gratiot N, Evrard O, Navratil O, Némery J, Prat C, Esteves M (2010) Drivers of erosion and suspended sediment transport in three headwater catchments of the Mexican Central Highlands. Geomorphology 123:243–256

Eliot J (1760) An essay upon field-husbandry in New England: as it is or may be ordered. Edes and Gill, Boston

Food and Agriculture Organization of the United Nations (FAO) (2006) The new generation of watershed management programmes and projects. Food and Agriculture Organization of the United Nations, Rome

Gachene CK, Nyawade SO, Karanja NN (2019) Soil and water conservation: an overview. In: Filho WL, Azul, AM, Brandli L, Özuyar, PG, Wall, T (eds) Zero Hunger. Encyclopedia of the UN sustainable development goals. Springer, Cham

Gao H, Li Z, Li P, Jia L, Zhang X (2012) Quantitative study on influences of terraced field construction and check-dam siltation on soil erosion. J Geograph Sci 22:946–960

Global Soil Partnership (GSP) (2017) Global soil partnership endorses guidelines on sustainable soil management http://www.fao.org/global-soil-partnership/resources/highlights/detail/en/c/416516/

Guo M, Zhang T, Li Z, Xu G (2019) Investigation of runoff and sediment yields under different crop and tillage conditions by field artificial rainfall experiments. Water 11:1019

Gupta GS (2019) Land degradation and challenges of food security. Rev Eur Stud 11:64–72

International Fund for Agricultural Development (IFAD) (1992) Soil and Water Conservation in Sub-Saharan Africa: Issues and Options. Centre for Development Cooperation Services (CDCS), Free University, Amsterdam

International Soil Reference and Information Centre (ISRIC) (2017) http://www.isric.org/projects/global-assessment-human-induced-soil-degradation-glasod

Joshi PK, Jha AK, Wani SP, Sreedevi TK, Shaheen FA (2008) Impact of watershed program and conditions for success: a meta-analysis approach. Global Theme on Agroecosystems Report no. 46. International Crops Research Institute for the Semi-Arid Tropics, Hyderabad

Kassam A, Derpsch R, Friedrich T (2014) Global achievements in soil and water conservation: the case of conservation agriculture. Int Soil Water Conserv Res 2:5–13

Klaus DP, Sebastian O, Johannes BR, Irene M, Ali AH (2014) Soil erosion in gully catchments affected by land-leveling measures in the Souss Basin, Morocco, analyzed by rainfall simulation and UAV remote sensing data. CATENA 113:24–40

Lal R (2015) Restoring soil quality to mitigate soil degradation. Sustainability 7:5875–5895

Lal R (2010) Managing soils and ecosystems for mitigating anthropogenic carbon emissions and advancing global food security. Bioscience 60:708–721

Lal R (2009) Soil degradation as a reason for inadequate human nutrition. Food Security 1:45–57

Mandal D, Sharda VN (2013) Appraisal of soil erosion risk in the eastern Himalayan region of India for soil conservation planning. Land Degrad Dev 24:430–437

Manivannan S, Thillagam VK, Khola OPS (2017) Soil and water conservation in India: strategies and research challenges. J Soil Water Conserv 16:312–319

Ministry of Rural Development (2022) Mahatma Gandhi National Rural Employment Guarantee Act 2005. https://nrega.nic.in/Nregahome/MGNREGA_new/Nrega_home.aspx Accessed 10 August 2022

Montanarella L (2015) Agricultural policy: govern our soils. Nature 528:32–33

Myers N (1993) Gaia: an atlas of planet management. J Acad Librariansh 19:200

National Bureau of Soil Survey and Land Use Planning (NBSS & LUP) (2005) Annual Report 2005, National Bureau of Soil Survey and Land Use Planning, Nagpur

Ouyang W, Hao F, Skidmore AK, Toxopeus AG (2010) Soil erosion and sediment yield and their relationships with vegetation cover in upper stream of the Yellow River. Sci Total Environ 409:396–403

Quinton JN, Govers G, van Oost K, Bardgett RD (2010) The impact of agricultural soil erosion on biogeochemical cycling. Nat Geosci 3:311–314

Sarvade S, Upadhyay VB, Kumar M, Imran Khan M (2019) Soil and water conservation techniques for sustainable agriculture. In: Jhariya M, Banerjee A, Meena R, Yadav D (eds) Sustainable agriculture, forest and environmental management. Springer, Singapore

Sharda VN, Dogra P, Prakash C (2010) Assessment of production losses due to water erosion in rain fed areas of India. J Soil Water Conserv 65:79–91

Singh RK, Chaudhary RS, Somasundaram J, Sinha NK, Mohanty M, Hati KM et al (2020) Soil and nutrients losses under different crop covers in vertisols of Central India. J Soils Sediments 20:609–620

Tideman EM (1996) Watershed management: guidelines for Indian conditions. Omega Scientific Publishers, New Delhi

Udawatta RP, Gantzer CJ (2022) Soil and water ecosystem services of agroforestry. J Soil Water Conserv 77:5A-11A

United Nations Environment Programme; World Meteorological Organisation; United Nations Convention to Combat Desertification (UNEP WMO UNCCD) (2016) Global Assessment of Sand and Dust Storms. United Nations Environment Programme, Nairobi

Vanwalleghem T, Gómez JA, Infante Amate J, González de Molina M, Vanderlinden K, Guzmán G, Laguna A, Giraldez JV (2017) Impact of historical land use and soil management change on soil erosion and agricultural sustainability during the Anthropocene. Anthropocene 17:13–29

Wang Y, Fan J, Cao L, Liang Y (2016) Infiltration and runoff generation under various cropping patterns in the red soil region of China. Land Degrad Dev 27:83–91

Wani SP, Sreedevi TK, Reddy TSV, Venkateswarlu B, Prasad CS (2008) Community watersheds for improved livelihoods through consortium approach in drought prone rainfed areas. J Hydrol Res Dev 23:55–77

Wischmeier WH, Smith DD (1978) Predicting rainfall erosion losses. Agriculture Handbook No. 537, US Department of Agriculture, Science and Education Administration, Washington DC

The World Overview of Conservation Approaches and Technologies (WOCAT) (2007) Where the land is greener—case studies and analysis of soil and water conservation initiatives worldwide. In: Liniger H, Critchley W (eds)

Yousuf A, Singh M (2019) In: Watershed hydrology, management and modeling. CRC Press, Florida

Chapter 2
Water Erosion

Abstract Water erosion encompasses the detachment of soil particles primarily by raindrops and flowing water and their transport by runoff. Comprehending the mechanics of water erosion is essential to developing measures to control erosion. This chapter describes the principal types of water erosion, viz., raindrop splash erosion, sheet erosion, interrill erosion, rill erosion, gully erosion, tunnel erosion, and streambank erosion, and explores the mechanics of each type. The chapter also describes various agronomical and biological measures employed to control water erosion. It also introduces popular engineering or mechanical erosion control measures like terracing, bunding, vegetated waterways, and gully control structures.

2.1 Water Erosion

Water erosion is the detachment and transport of soil particles by water. The soil detachment is primarily caused by raindrops and flowing water, while transport is by runoff. The transported soil may be deposited in the lower landscape or drained into rivers/streams.

Water erosion causes a gradual reduction of topsoil, leading to loss of organic matter and nutrients and a decline in soil fertility and productivity. Also, it is the primary source of sediment that pollutes streams and fills reservoirs. Therefore, understanding the mechanism and knowing various forms of water erosion is essential to developing erosion control practices and minimising its ill effects.

2.2 Types of Water Erosion

Water erosion is broadly classified into the following five types:

1. Raindrop splash erosion
2. Sheet erosion
3. Rill erosion
4. Gully erosion

R. Singh, *Soil and Water Conservation Structures Design*, Water Science and Technology Library 123, https://doi.org/10.1007/978-981-19-8665-9_2

Fig. 2.1 Raindrop splash erosion (Photo courtesy of USDA, NRCS)

5. Streambank erosion

In addition to the above, two more types of water erosion are proposed by researchers. These are interrill erosion and tunnel erosion.

The various water erosion types are discussed in the following sections.

2.2.1 Raindrop Splash Erosion

The impact of free-falling raindrops breaks down the bare soil surface and splashes the individual soil particles, thus, causing their displacement from their original position (Fig. 2.1). Also, the impact loosens the topsoil making it vulnerable to displacement by surface runoff. The loose soil particles subsequently clog the soil pores, reducing infiltration and increasing runoff. Furthermore, the impact of raindrops on thin flowing water layers causes turbulence and enhances soil transportation. Raindrop splash erosion is the primary cause of soil detachment and disintegration; thus, it is considered the first stage of the soil erosion process. The raindrop impact depends on the drop size, velocity, and direction.

2.2.2 Sheet Erosion

Sheet erosion is the removal of a more or less uniform thin layer of the soil due to overland flow (Fig. 2.2). It follows the raindrop splash erosion, resulting in loose soil particles on the surface. Once the rainfall rate exceeds the soil's infiltration capacity, overland flow begins and washes away the loose soil particles. Sheet erosion occurs gradually and almost goes unnoticed until most of the productive topsoil has been lost. It results in the loss of the most fertile soil that contains most of the available nutrients and organic matter in the soil. Cultivated and overgrazed soils are most

Fig. 2.2 Sheet erosion (Photo courtesy of Creative Commons Attribution-Share Alike 3.0 Unported)

vulnerable to sheet erosion. The ability of the overland flow to erode and transport soil particles depends on the rainfall intensity, soil infiltration rate, slope steepness, and vegetative cover.

2.2.3 Rill Erosion

If the overland flow process, causing sheet flow, continues, its erosive action results in the formation of shallow channels, called *rills*. Rill erosion is the removal of soil by concentrated flow occurring through rills (Fig. 2.3). Though rills do not interfere with tillage operations and may be easily levelled by primary tillage implements, these result in significant soil loss. Rill erosion usually occurs in bare agricultural lands, overgrazed, and freshly cultivated soils. It is most profound when intense storms occur on soils having high runoff-producing characteristics and loose, shallow topsoil. Since rill erosion occurs due to concentrated flow, unlike shallow overland flow during sheet flow, the detachability and transportability are higher in rill erosion than in sheet erosion.

Fig. 2.3 Rill erosion (Photo courtesy of USDA, NRCS)

2.2.4   Interrill Erosion

The concept of interrill erosion emerged because tiny rills develop as soon as the sheet erosion begins. The region between rills is called the interrill area, and the erosion due to raindrop splash and sheet erosions occurring concurrently in the interrill area is called interrill erosion (Nearing et al. 1989) (Fig. 2.4). During interrill erosion, raindrop impact detaches the soil particles, as in raindrop splash erosion, and overland flow transports the loose soil particles to rills, as in sheet erosion. Though raindrop splash and sheet erosion coincide, splash erosion dominates during the initial phases, but sheet erosion starts dominating with increasing overland depth. Interrill erosion is a function of rainfall intensity, soil properties, and land slope.

2.2.5   Gully Erosion

Since gully erosion is recognised as a critical global environmental problem (Vanmaercke et al. 2016), it has been one of the widely researched topics world-wide (Castillo and Gomez 2016). However, gullies have been defined confusingly, often mixing them with other landforms like rills, stream channels, ravines, and Badlands. Similarly, gullies have been classified ambiguously based on size, shape, and morphology.

Based on a thorough review covering more than a century of work, Thwaites et al. (2022) provided an objective definition of a gully, based on the definition proposed by Brooks et al. (2019) and modified by Thwaites and Brooks (2021). As per the definition, 'A gully is a persistent, incised erosion feature with walls and/or head

Fig. 2.4 Rill and interrill erosion (Image by M. Mamo, University of Nebraska-Lincoln, https://passel2.unl.edu/)

scarp on average > 0.5 m deep (or that cannot generally be removed by conventional tillage methods for Field Gullies) and cannot be defined as a stream channel (first order or greater) (Fig. 2.5)'.

It further adds, 'The feature erodes residual soils, unconsolidated materials, and saprolite, but not solid bedrock. Mass movement dominates the erosion process within the void, and it has multiple modes of expansion. It always includes extension

Fig. 2.5 Gully erosion (Photo courtesy of Creative Commons Attribution-Share Alike 2.0 Generic)

by head ward retreat into an otherwise undissected land surface by way of a clearly defined head scarp or head wall.

It has erosional side walls of at least moderately steep gradient (~30°; ~60%), and gully walls are dominantly bare soil materials with a distinct break of slope from the uneroded land surface, that exhibit erosion by means of mass movement, with or without sapping by seepage.

The feature overall is typically driven by ephemeral flows associated with direct rainfall into the gully and in the gully catchment. Alluvial gullies, however, are commonly/affected by backwater—by floodwaters from the stream/river they adjoin. The active sediment source is predominantly from within the gully as a clear, bounded, internal erosional zone".

Thus, the essential features of a gully are as follows:

1. It is a continuing erosional feature, with an active head or walls on average > 0.5 m deep.
2. It has an active head that expands upstream and active erosional walls, having a moderately steep gradient (~30°; ~60%).
3. It typically occurs due to ephemeral flows caused by rainfall in the catchment and directly in the gully.
4. It expands and extends due to erosion of the walls, thus, having sediment sources predominantly within the gully.
5. It has a cross-sectional shape: U-shaped, V-shaped, trapezoidal, or rectangular that is permanently recognisable without flow.

2.2.5.1 Classification of Gullies

Thwaites et al. (2022) proposed an Elementary Gully Classification (EGC) that includes classifying gullies at the primary level, based on planform, type-domain, and family; secondary level, based on continuity, cross-section, soil features and soil type; and tertiary level: erosion activity, dominant process, and head morphology.

At the primary level, a gully is broadly classified into the *family* of *Field Gullies* or *Landscape Gullies*. A field gully is primarily formed by cultivation practices in the agricultural fields, while a landscape gully is formed naturally. Under *type-domain*, a field gully may be ephemeral, permanent or classical, or field bank gully. An ephemeral gully is a temporary feature, usually larger than rills, which can be obscured by tillage. A field bank gully is a discontinuous permanent feature controlled by piping and mass movement from upslope runoff rather than gully catchment size. Under *planform shape*, a gully may be linear or dendritic. Thus, a gully may be classified as a linear, ephemeral, and field gully at the primary level.

A gully may be continuous or discontinuous at the secondary level based on *continuity*. The *cross-section* may be V-shaped, U-shaped, or trapezoidal. The soil features may include piping, crust, dispersive, or limiting layer, while the soil type may be sandy, silty, clayey, or loess. Thus, a gully may be continuous, V-shaped, dispersive, and silty soil at the secondary level.

At the tertiary level, a gully is classified based on activities in the head or banks/walls. Subsequently, a gully may be active, semi-active, or stable based on erosion activity. The dominant process may be a failure, seepage, or overfall. The head morphology may be gradual, stepped, or head cut, while banks/walls morphology may be vertical, sloping, or faceted. Thus, a gully may be active, failure, head cut, and faceted walls at the tertiary level.

2.2.6 Tunnel Erosion

Tunnel erosion refers to subsoil removal. It is initiated by water infiltrating through a crack or a hole into dispersible soil layers and carrying soil with the flow to create a small tunnel. In the beginning, the surface soil remains stabilised, but slowly the tunnel grows big and leads to the collapse of the surface soil. Thus, tunnel erosion ultimately leads to gully formation. Water seepage, interflow, or lateral flow are the indicators of tunnel erosion. Tunnel erosion may alter the geomorphological and hydrological characteristics of the region.

2.2.7 Streambank Erosion

Streambank erosion refers to soil removal from stream banks or the bed (Fig. 2.6). Stream channels erode as water flows over the side of the stream banks or scours below the water surface, especially during extreme floods. Streambank erosion is usually both the cause and effect of stream meandering. It is influenced by the removal of protective vegetation from stream banks, cultivation near the banks, excessive sand mining, velocity and direction of flow, channel cross-section and soil texture.

2.3 Water Erosion Processes

Water erosion involves the processes of detachment, transport and deposition of soil. These processes are driven by raindrop impact or flowing water. For example, raindrop impact and flowing water drive the detachment, while flowing water drives transportation. On the contrary, deposition occurs due to a lack of carrying capacity of water. The severity of erosion depends on the quantity of detached soil material and the capacity of the water to transport it. Therefore, erosion may be detachment-limited when the capacity of the water to transport exceeds the eroded soil material, or transport-limited when the eroded soil material exceeds the transportation capacity of water. The three water erosion processes are discussed in the following sections.

Fig. 2.6 Streambank erosion (Photo courtesy of Creative Commons Attribution-Share Alike 2.0 Generic)

2.3.1   Detachment

The kinetic energy of the raindrop impact is the primary cause of the soil particle detachment. The raindrop impact breaks down the soil aggregates and detaches the soil particles. Due to alternate wetting and drying or tillage operations, soil aggregates may also be broken. The detached soil particles seal the soil surface and reduce infiltration. Subsequently, if the rainfall rate exceeds the soil infiltration capacity, overland flow occurs and carries away the detached soil particles. In general, the soil detachability ∞ size of soil particles. Consequently, sand is easily detached compared to clay.

2.3.2   Transportation

The overland flow is the dominant transportation agent, which carries the soil particles down the slope. Water flowing in the rills and gullies also transports the soil particles. The flow velocity and volume determine the quantity of sediment and the distance they are carried away. The smaller soil particles, e.g., clay, are easier to transport than large soil particles like sand.

Soil particles move by suspension, traction, or saltation. Suspension represents the floating of smaller soil particles in the water, while saltation represents the bouncing movement of the smaller soil particles. Traction includes large soil particles rolling and pushing other particles along the soil surface.

2.3.3 Deposition

As the flow velocity reduces, the water loses the energy needed to continue carrying the soil particles. Consequently, the soil particles are deposited at a new location. Deposition mainly occurs in low landscape positions or at the downslope end of the fields. The deposition also occurs in downstream water bodies and causes sedimentation and pollution. Sediments are also deposited in deltas along streams.

2.4 Factors Affecting Erosion by Water

Climate, topography, vegetation, and soil characteristics are the dominant natural factors affecting soil erosion. Also, the interactions among these factors influence the soil erosion rate and magnitude.

Climate Precipitation, temperature, wind, humidity, and solar radiation are the critical climatic factors affecting erosion. Rainfall intensity and magnitude impact erosion. Rainfall intensity influences the raindrop splash and runoff rate, while rainfall magnitude governs the runoff volume. Snowmelt also affects erosion by increasing runoff. Temperature and wind impact erosion through their effects on evaporation and transpiration. Wind also influences the velocity and the angle of impact of raindrops. Humidity and solar radiation influence erosion through their impact on temperature.

Topography Slope length and steepness are the important topographic characteristics influencing erosion. On longer slopes, the overland flow gets accumulated and influences erosion, while on steep slopes, high flow velocity influences erosion by increasing scour.

Vegetation Vegetation plays a vital role in water erosion control. Vegetation cover intercepts rainfall, thus, absorbing the kinetic energy of raindrops and minimising the raindrop splash. Vegetative cover on the soil surface enhances the surface roughness, thus, hindering the overland flow and minimising its soil transportation capacity. Vegetation also increases transpiration, thus, enhancing the soil water storage and reducing surface runoff. Plant roots improve the soil aggregate stability, thus minimising their erodibility. Over the long term, vegetation improves the soil physical properties, including the organic matter content and porosity, and enhances the water holding capacity.

Soil Physical properties of soils, like texture, structure, organic matter content, and infiltration capacity influence soil erosion. Usually, soil detachability is high for coarse-textured soils, but their transportability is low. Thus, sand particles are easy to detach, but difficult to transport. The organic matter content and infiltration capacity affect soil erosion through their impact on surface runoff.

2.5   Control of Water Erosion

Soil conservation measures aim to obtain the maximum sustained level of production from cultivated lands and maintain the soil loss below the natural rate of soil formation. Based on the concept that soil erosion rate should be equal to or below the natural rate of soil formation, *soil loss tolerance* value or T-value is in vogue.

Soil loss tolerance is the maximum permissible rate of erosion at which soil fertility can be maintained over 20–25 years (Morgan 2005). Soil loss tolerance is usually recommended in designing soil conservation measures. The maximum acceptable limit of soil loss or T-value is 11.2 t ha^{-1} year^{-1} (Wischmeier and Smith 1978), while the average rate of natural soil formation is about 0.02–0.5 mm per year (Schumm and Harvey 1982). However, Borrelli et al. (2013) considered a T-value of 10 t ha^{-1} year^{-1} while assessing the global impact of twenty-first century land use change on soil erosion.

2.5.1   Strategies for Controlling Water Erosion

Control of soil erosion by water is based on the following strategies.

1. Reducing raindrop impact on the soil surface
2. Increasing infiltration capacity of soil and reducing runoff
3. Improving aggregate stability of soil
4. Increasing surface roughness to reduce flow velocity

Reducing raindrop impact on soil surface Raindrop impact can be minimised by covering the soil surface with crops or mulches. The protecting plant cover absorbs the raindrop impact and protects the soil surface from raindrop impact.

Increasing infiltration capacity of the soil and reducing runoff Soil infiltration capacity can be increased by providing mulches and increasing the infiltration opportunity time. Consequently, the surface runoff would reduce, and so will the erosion. The land slope may also be manipulated to reduce the flow velocity, thus, minimising the soil detachment and transportation.

Improving aggregate stability of soil The aggregate stability of soil can be improved by enhancing the organic matter content of the soil. Stable soil aggregates are resistant to detachment.

Increasing surface roughness to reduce flow velocity Surface roughness due to soil cloudiness or the topographical variations due to tillage operations enhances the water storage on the soil surface and infiltration and reduces the flow velocity. Consequently, soil erosion reduces.

2.5.2   Water Erosion Control Measures

Water erosion control measures may be broadly classified into the following two categories:

1. Agronomical and biological measures
2. Engineering or mechanical measures

2.5.2.1   Agronomical and Biological Measures

Agronomical and biological measures focus on utilising crops or vegetation to control soil erosion. A vegetative or surface cover is provided for as long as possible to reduce runoff, increase infiltration, and reduce soil loss. The popular measures include,

- Crop rotation
- Cover cropping
- Contour farming
- Strip cropping
- Mulch tillage or stubble mulching

Crop Rotation

Crop rotation is a simple form of multicropping, which involves growing different crops in the same land in regular succession. A multicropping system is called intercropping when multiple crops are grown on the land in a single season. In agroforestry, perennial trees or shrubs occupy a part of the land over growing seasons.

Crop rotation aims to improve soil fertility by alternating shallow-rooted and deep-rooted crops. Besides replenishing nitrogen, crop rotation increases soil infiltration capacity and reduces runoff and soil erosion. Typically, leguminous crops like gram, peas, pulses, beans, groundnut, and alfalfa are used with cereal crops like wheat and rice. Leguminous plants have root nodules that help fix soil nitrogen.

Cover Cropping

Cover crops are grown as a conservation measure either during the off-season or in between rows as companion crops. Legumes like gram, peas, pulses, beans, groundnut and alfalfa, and grasses are preferred as cover crops. Cover crops protect soil against erosion while fixing nitrogen and improving soil fertility. These are also grown under tall trees like rubber to protect the soil against raindrop impact. Figure 2.7 shows rye cover crop with soybean.

Contour Farming

Contour farming refers to performing all farming operations like ploughing, planting and harvesting on the contour, i.e., across the slope rather than up and down the slope (Fig. 2.8). Contour farming reduces runoff due to water impoundment in between contours and leads to the deposition of loose soil particles along the contours. Also,

Fig. 2.7 Rye cover crop with soybean (Photo courtesy of USDA, NRCS)

the impounded water gets more time to infiltrate, resulting in water conservation. Thus, contour farming is most suitable for low rainfall areas with a 2–10% slope. When properly managed, contour farming may reduce soil loss by up to 50% while enhancing the crop yield by 10%. However, the effectiveness depends on slope steepness and slope length. On steep slopes or in high rainfall areas, contour farming may fail as row breaks release impounded stored water and cause gullying. Therefore, contour farming may be combined with strip cropping in high rainfall areas or on steep slopes.

Strip Cropping

It is the practice of growing alternate strips of crops across the field slope. Crops of varying root and shoot characteristics are chosen for the alternate strips. For example, cereal crops and grasses or legumes may be grown in alternate strips. The strips break the land slope in several segments, intercept runoff and sediment and enhance infiltration. Narrow strips are preferred as they may reduce the flow lengths significantly. However, the strip width depends on the land slope, erosion potential, cropping system, and farm implements used. The minimum strip width must permit smooth farm equipment operations.

Based on the topography, erosion hazards or the prevalent cropping system at a location, strip cropping may be of three types: (i) contour strip cropping, (ii) field strip cropping, and (iii) buffer strip cropping.

Contour Strip Cropping

Contour strip cropping has strips of row crops, alternating with strips of grass, legumes, or small grains, planted on the contours (Fig. 2.9). The strips of grass, legumes, or small grains reduce the flow velocity and trap sediments, leaving row crop strips. Contour strip cropping is more effective than contour farming due to diverse crop strips and is preferred in erosion-prone areas.

Fig. 2.8 Contour farming (Photo courtesy of USDA, NRCS)

Fig. 2.9 Contour strip cropping (Photo courtesy of USDA, NRCS)

Fig. 2.10   Field strip cropping (Photo courtesy of USDA, NRCS)

Field Strip Cropping

Field strip cropping has crop strips placed across the slope without following the contours (Fig. 2.10). The strips follow a similar pattern as in contour strip cropping, i.e., strips of grass, legumes, or small grains alternate with the strips of row crops. It is preferred in flat terrains or where the topography is too irregular to adopt the contour strip cropping.

Buffer Strip Cropping

Buffer strip cropping has strips of grasses or legumes grown between contour strips of crops and can be perpetual or seasonal (Fig. 2.11). They are used to protect soils on erosion-prone terrains. Buffer strips may have varying widths and distances depending on the land slope. These strips capture sediments and nutrients, thus preventing soil erosion and water pollution downstream.

Mulch Tillage or Stubble Mulching

A mulch is a cover spread over the surface to protect the soil. Mulch tillage is a conservation tillage practice that utilises the residual mulches of preceding crops (Fig. 2.12). Mulches reduce the raindrop impact and protect soil aggregates from breaking down. They increase infiltration and reduce runoff magnitude and velocity. In stubble mulching, the stalks of previous crops are retained during the tillage operations for the next crop.

The conservation tillage practices like zero-tillage or minimum tillage are also gaining popularity for controlling soil erosion and improving soil quality. In zero-tillage, also called no-tillage, the soil is left undisturbed, i.e., the no-tillage operation is performed, and the seeds are planted directly in the previous crop's residues. On the contrary, minimum tillage involves the minimum disturbance of the soil.

Fig. 2.11 Buffer strip cropping (Photo courtesy of USDA, NRCS)

Fig. 2.12 Mulch tillage (Photo courtesy of USDA, NRCS)

The scope of conservation tillage has been expanded into *Conservation Agriculture (CA)*, which includes no-till seeding, maintaining a continuous soil cover by crop residue, stubble or cover crops, and growing diverse plant species in the cropping system (Kassam et al. 2014). CA is being promoted by FAO as the sustainable agricultural intensification strategy (FAO 2011).

Agroforestry practices that include crop mixes, multistorey tree plantations, alley cropping, windbreaks, and shelterbelts are also being used extensively for erosion control worldwide. Agroforestry systems contribute to soil conservation through the effects of canopy cover, ground cover by litter, and the soil stabilising by plant roots (Atangana et al. 2014).

2.5.2.2   Engineering or Mechanical Measures

Engineering measures, also called mechanical measures, are adopted to control soil erosion in fields where agronomical and biological measures cannot limit the soil erosion within the tolerance value. Engineering measures include terraces, bunds, and vegetative waterways. Moreover, temporary and permanent gully control structures are used in gullies facing severe erosion. Temporary structures are preferred in areas where labour is inexpensive, and construction materials like planks, logs, stones, brush, woven wire, or earth are readily available. On the other hand, permanent structures need technical skills for their design and construction and are built using concrete or masonry.

Terraces

Terraces are the most widely practised soil erosion control measure worldwide (Fig. 2.13). Terraces consist of earth embankments constructed across the steep slopes to intercept surface runoff. Subsequently, water may be diverted at a non-erosive velocity to a safe outlet or stored to enhance soil infiltration. Terraces reduce the slope length, control runoff, conserve soil moisture, and facilitate cultivation on steep slopes. Terraces are broadly classified as common or normal terraces or bench terraces. Common (or normal) terraces, consisting of a ridge and a channel, are classified as narrow-, medium-, and broad-base terraces. Bench terraces consist of a series of level or sloping benches constructed across the slope and are classified according to the slope as level or tabletop, inward-sloping or outward-sloping terraces.

Bunds

Bunds are small embankment type structures constructed across the land slope, either with soil or stone (Fig. 2.14). These reduce the slope length, runoff, and soil erosion and enhance soil infiltration. These are similar to the narrow-base terraces, but cultivation is not practised on bunds, though they may be kept under permanent grass cover for protection. Bunds are also constructed along field boundaries and are called peripheral bunds. Bunds are broadly classified as contour and graded bunds.

Vegetated waterways

Vegetated waterways are natural or constructed channels lined with suitable vegetation to dispose of runoff safely without causing erosion. These are constructed along the natural slope and have a carrying capacity to handle the peak runoff rate generated from the area. Also, the *permissible velocity approach* is adopted in their design to carry the peak flow without damaging the vegetal lining. Besides conveying runoff, vegetated waterways prevent gully formation, act as outlets for terraces or graded bunds, and protect and improve water quality.

Fig. 2.13 Terrace system (Photo courtesy of USDA, NRCS)

Fig. 2.14 Contour bunding in a farm (Photo courtesy of ICRISAT)

Farm Ponds

Though primarily used as water harvesting structures, farm ponds also conserve soil and nutrients (Fig. 2.15). These are adopted in semi-arid regions to serve multiple purposes. Farm ponds provide food and financial security in dryland agriculture

Fig. 2.15 Farm pond (Photo courtesy of <u>Creative Commons Attribution-Share Alike 2.0 Generic</u> license)

areas by augmenting irrigation to boost crop productivity besides generating additional income through fish and duck farming. Farm ponds are broadly classified as embankment or excavated types.

Gully Control Structures

Gully control structures are used in gullies facing severe erosion that cannot be controlled with agronomical, biological, or engineering erosion control measures.

These structures are built across the gully to reduce the channel gradient and stabilise the gully to prevent further erosion. The structures could be either temporary or permanent.

Temporary structures are preferred in areas where labour is inexpensive, and construction materials like planks, logs, stones, brush, woven wire, or earth are readily available. The temporary gully control structures are simple to construct and maintain. These include woven-wire check dams, brushwood check dams, loose rock dams, log check dams, and gabion check dams.

On the other hand, permanent structures are used in medium to large gullies having significantly high flows that the temporary structures could not handle. These are typically constructed using masonry or reinforced concrete. The permanent gully control structures are mainly categorised as drop, drop inlet and chute spillways. Figure 2.16 shows a typical drop spillway.

Fig. 2.16 Drop spillway (Photo courtesy of USDA, NRCS)

Questions

1. Describe the geological and accelerated soil erosion. What are the causes and effects of these erosion processes?
2. Describe the types of water erosion. What are the causes and effects of various types of water erosion?
3. Distinguish between the rill and interrill erosion processes.
4. Define gully erosion and enlist the essential features of a gully.
5. Describe the process of classifying gullies at primary, secondary and tertiary levels under the Elementary Gully Classification system.
6. What are the processes involved in soil erosion by water? Describe the factors governing the water erosion processes.
7. Describe the factors affecting water erosion.
8. Describe various strategies that may be adopted to control water erosion. Given an option, which of these strategies you would choose for a watershed in your locality and why?
9. Describe various agronomical and biological measures of water erosion control. What are the specific features of each of these measures?
10. Describe various engineering measures of water erosion control. What are the specific features of each of these measures?

Multiple Choice Questions

Answer the questions by choosing the correct option.

1. The maximum permissible rate of erosion at which soil fertility can be maintained over 20–25 years is called

(A) soil formation confines   (B) soil loss tolerance
(C) geological erosion rate   (D) accelerated erosion rate

2. Which of these is not a form of soil movement in water erosion?

(A) Suspension   (B) Saltation
(C) Surface creep   (D) Traction

3. Which one of the following is not a cover crop?

(A) Wheat   (B) Gram
(C) Peas   (D) Alfalfa

4. For coarse-textured soils, the soil

(A)   detachability is high
(B)   detachability is low
(C)   transportability is high
(D)   detachability and transportability are equal

5. The erosion due to raindrop splash and sheet erosions occurring concurrently in
the region between rills is called

(A) gully erosion   (B) interrill erosion
(C) rill erosion   (D) tunnel erosion

6. As per the Elementary Gully Classification (EGC), a gully may be continuous,
V-shaped, dispersive, silty soil at the

(A)   elementary level
(B)   primary level
(C)   secondary level
(D)   tertiary level

7. Which of the following may not qualify to be designated as a strategy to control
soil erosion by water?

(A)   Reducing raindrop impact on the soil surface
(B)   Increasing infiltration capacity of soil
(C)   Reducing surface roughness
(D)   Improving aggregate stability of soil

8. Which of these is not a method of water erosion control?

(A) Contour farming   (B) Intercropping
(C) Mulching   (D) Shelter belt

9. Growing two or more crops in the same field but in different rows is called

    (A) alternate cropping   (B) crop rotation
    (C) kharif cropping   (D) strip cropping

10. Which of these is not a form of conservation tillage?

    (A) Deep tillage   (B) Minimum tillage
    (C) Mulch tillage   (D) Zero-tillage

References

Atangana A, Khasa D, Chang S, Degrande A (2014) Agroforestry for soil conservation. In: Tropical agroforestry. Springer, Dordrecht

Borrelli P, Robinson DA, Fleischer, LR, Lugato, E et al. (2013) An assessment of the global impact of 21st century land use change on soil erosion. Nat Commun 8. https://doi.org/10.1038/s41467-017-02142-7

Brooks AP, Thwaites RN, Spence J, Pietsch T, Daly J (2019) A gully classification scheme to underpin Great Barrier Reef water quality management: 1st edn. NESP Project 4.9 Final Report. The National Environmental Science Programme, Townsville. http://www.nesptropical.edu.au

Castillo C, Gomez JA (2016) A century of gully erosion research: Urgency, complexity and study approaches. Earth- Sci Rev 160: 300–319

FAO (2011) Save and grow. A policymakers' guide to the sustainable intensification of smallholder crop production. Food and Agriculture Organization of the United Nations, Rome

Kassam A, Derpsch R, Friedrich T (2014) Global achievements in soil and water conservation: the case of conservation agriculture. Int Soil Water Conserv Res 2:5–13

Morgan RPC (2005) Soil erosion and conservation, 3rd edn. Blackwell Publishing, Oxford

Nearing M, Foster GR, Lane LJ, Finkner SC (1989) A process-based soil erosion model for USDA-water erosion prediction project technology. Trans ASAE 32:1587–1593

Schumm SA, Harvey MD (1982) Natural erosion in the USA. In: Determinants of soil loss tolerance. Special publication 45. American Society of Agronomy and Soil Science Society of America, Madison, USA

Thwaites RN, Brooks AP, Pietsch TJ, Spencer JR (2022) What type of gully is that? the need for a classification of gullies. Earth Surf Process Landf 47:109–128

Thwaites RN, Brooks AP (2021) A gully is not a stream channel. How do we tell the difference and why does it matter? Full paper 74446, 10th Australian stream management conference 2021. Kingscliff, NSW. 2–4 August 2021

Vanmaercke M, Poesen J, Van Mele B, Demuzere M, Bruynseels A, Golosov V, Fuseina Y (2016) How fast do gully headcuts retreat? Earth-Sci Rev 154:336–355

Wischmeier WH, Smith DD (1978) Predicting rainfall erosion losses: a guide to conservation planning. USDA Agriculture Handbook No 537: US Department of Agriculture, Washington DC

Chapter 3
Soil Loss Estimation

Abstract The soil loss estimated using soil erosion models is vital in evaluating the existing soil conservation practices and identifying priority areas and appropriate measures to control erosion. This chapter presents various soil erosion modelling and measurement techniques for soil loss assessment. The Universal Soil Loss Equation (USLE), an empirical modelling approach, is introduced along with its factors: rainfall erosivity, soil erodibility, slope length-gradient, land cover and management, and soil conservation practice factor. Also, the Modified USLE (MUSLE), which has a runoff factor in place of the rainfall factor, and the Revised USLE (RUSLE), which includes several process-based concepts, are discussed. The chapter introduces the Water Erosion Prediction Project (WEPP) and the European Soil Erosion Model (EUROSEM), the distributed, physically-based soil erosion models that can simulate soil loss under diverse land uses and hydrologic conditions. Also, the Soil Conservation Service (SCS) curve number method and the rational method used for estimating the runoff volume and peak runoff rate are included. The chapter discusses the soil loss measurements from runoff plots. The different size plots are discussed along with commonly used devices, namely the multislot divisor and Coshocton wheel.

3.1 Background

Soil erosion has been recognised as a typical land degradation problem worldwide due to its unfavourable economic and environmental impacts. It is a well-established fact that soil erosion results in soil nutrients loss, leading to a decline in crop productivity. Also, it results in an increased sedimentation load in rivers and reservoirs, impacting their carrying and storage capacities, respectively, besides impacting the water quality and ecosystem.

The accurate assessment of soil loss is vital for the government agencies, managing the soil and water conservation programmes, and policymakers. The soil loss estimation may help in evaluating the existing soil management practices and in devising better management strategies to check the soil loss. It may also help in identifying priority areas for initiating or strengthening the soil conservation measures.

R. Singh, *Soil and Water Conservation Structures Design*, Water Science and Technology Library 123, https://doi.org/10.1007/978-981-19-8665-9_3

Several modelling and measurement techniques exist for the soil loss assessment. Zingg (1940) may be credited to have published the first equation to relate soil loss to the topographic factors like slope length and slope percentage. Subsequently, Smith and Whitt (1948) supplemented the equation by introducing crop and conservation practice factors. Wischmeier and Smith (1965) first introduced the Universal Soil Loss Equation (USLE) as a comprehensive model and subsequently updated it in 1978 (Wischmeier and Smith 1978). Williams and Berndt (1977) proposed the Modified USLE (MUSLE) by introducing a runoff factor to replace the USLE rainfall factor. USLE was subsequently revised as Revised USLE (RUSLE) by incorporating several process-based concepts for estimating the model inputs (Renard et al. 1997).

USLE, MUSLE, and RUSLE follow an empirical modelling approach and remain the most widely adopted soil loss estimation tools around the world. USLE/MUSLE/RUSLE modelling approach has been adopted by several well-established models including Agricultural Non-point Source Pollution (AGNPS), Chinese Soil Loss Equation (CSLE), Simulation of Production and Utilisation of Rangeland (SPUR), Soil and Water Assessment Tool (SWAT), and Simulator for Water Resources in Rural Basins (SWRRB) (Williams et al. 1985; Wight and Skiles 1987; Arnold et al. 1998; Cronshey and Theurer 1998; Liu et al. 2020).

The empirical nature of the USLE methodology and its inability to handle sediment deposition limits its use in modelling large watersheds, especially for areas under soil conservation measures like contouring and terracing. Consequently, the research community has focussed on developing models that are independent of the USLE. Water Erosion Prediction Project (WEPP) and European Soil Erosion Model (EUROSEM) are examples of the modelling tools that adopt the physically-based approach (Laflen et al. 1997; Morgan et al. 1998).

Soil loss data from laboratory or field measurements are needed for calibration and evaluation of various soil loss estimation models. Soil loss measurements have customarily been carried out from different measurement units under simulated and natural rainfall conditions. The measurement units include laboratory and field runoff plots, and watersheds having a uniform or diverse land use. The following sections present the details of soil loss modelling and measurements.

3.2   Modelling Soil Loss

Modelling soil erosion by water, vital to understanding the soil erosion processes and predicting soil erosion rates, started in the 1930s when the first research study was published. Since then, numerous empirical and process-based models of varying complexities and prediction capabilities have been developed for evaluating soil erosion. Although, process-based models like WEPP and EUROSEM have been developed in the recent past, the empirical models, like USLE and RUSLE, are extensively used globally because of their simplicity and wider acceptability.

3.2.1 *Universal Soil Loss Equation (USLE)*

The Universal Soil Loss Equation (USLE) is the most famous empirical modelling tool that estimates average annual soil loss from an area over a long period. USLE considers the climate, soil, topography, and land use to predict soil loss owing to sheet or rill erosion. It may also be used to assess the efficacy of the soil conservation measures adopted in the area. The USLE is given as

$$A = R\,K\,LS\,C\,P \qquad (3.1)$$

where A = average annual soil loss, Mg ha^{-1}; R = rainfall-runoff erosivity factor; MJ mm h^{-1} ha^{-1}; K = soil erodibility factor, Mg h MJ^{-1} mm^{-1}; LS = slope length-gradient factor, dimensionless; C = land cover and management factor, dimensionless; and P = soil conservation practice factor, dimensionless.

Rainfall-Runoff Erosivity Factor (R)

Rainfall-runoff erosivity factor (R) includes two components: the erosion resulting from the raindrop impact effect and the associated runoff, given by *rainfall erosion index*, *EI* (also referred to as *rainfall erosivity index*), and the erosion due to runoff from the thaw, snowmelt, and rain on frozen soil, R_S.

Rainfall Erosion Index (EI)

Rainfall erosion index or rainfall erosivity index (EI) denotes the inherent ability of rainfall to cause soil erosion. EI is used as an abbreviation for energy-times-intensity and is expressed as

$$EI = KE\,I_{30} \qquad (3.2)$$

where EI = rainfall erosion index (E and I representing energy and intensity), MJ mm ha^{-1} h^{-1}; KE = kinetic energy of the rainstorm, MJ ha^{-1}; and I_{30} = maximum 30-min rainfall intensity during the storm, mm h^{-1}.

Equation (3.2) is based on the fact that the product of KE and I_{30} gives the best estimation of soil loss (Wischmeier and Smith 1965).

The kinetic energy of the rainstorm, KE (in MJ ha^{-1}), is obtained for a particular event using the following relationship:

$$KE = \left(\sum_{r=1}^{m} E_r\,V_r \right) \qquad (3.3)$$

where E_r = kinetic energy per unit of rainfall, MJ ha^{-1} mm^{-1}; and V_r = rainfall depth for the time interval r of the hyetograph, which has been divided into $r = 1, 2, ..., m$ subintervals, mm. Wischmeier and Smith (1958) gave the following relationship between E_r and I:

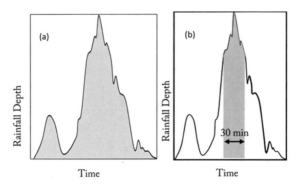

Fig. 3.1 Isolated storm event from the rain gauge chart **a** for estimating the rainfall intensity and the kinetic energy, **b** for estimating the maximum 30-min rainfall intensity during the storm

$$E_r = 0.119 + 0.0873 \log_{10} I \tag{3.4}$$

where I = rainfall intensity of the storm, mm h^{-1}.

KE, E_r, V_r, I, and I_{30} are obtained for a storm from rain gauge charts.

Estimation of Rainfall Erosion Index (*EI*) From Rain Gauge Data

From the rain gauge chart, the storm event is isolated for estimating V_r, I, E_r, I_{30}, and KE. Figure 3.1a, b present the storm event isolated from the recorded rainfall. The shaded portion in Fig. 3.1a highlights the entire isolated event that may be used for estimating the rainfall depth, V_r, and rainfall intensity, I, for different time intervals to estimate the kinetic energy per unit rainfall depth, E_r, using Eq. (3.4), and kinetic energy, KE, using Eq. (3.3). The shaded portion in Fig. 3.1b, on the other hand, highlights the 30-min duration for estimating the I_{30} for the storm. The values of KE and I_{30} are subsequently multiplied to obtain the rainfall erosion index, EI, as per Eq. (3.2).

Wischmeier and Smith (1965) developed *isoerodent maps* by plotting lines joining the points of equal rainfall erosivity, which was later extended by Wischmeier (1976) for almost the entire USA. A global erosivity map has also been developed using the Global Rainfall Erosivity Database (GloREDa) (Panagos et al. 2017).

Example 3.1 The following rainfall event is isolated from a rain gauge chart. Estimate the rainfall erosivity index for the event.

Date	Time (h)	09:55	10:15	10:20	10:25	10:45	10:55	11:05	11:15
27 August 2020	Rainfall (mm)	0	1.8	3.3	23.3	30.0	35.5	37.1	37.4

Solution The calculations are performed using the isolated storm event data, as shown in Table 3.1.

Soil Erodibility Factor (*K*)

Erodibility denotes the soil resistance to detachment and transportation. It may be estimated by measuring soil loss from the standard USLE runoff plot. The USLE runoff plot is 22.13 m long and has a 9% slope, and it is kept fallow with periodic tillage up and down the slope. Under these standard conditions, USLE factors, *LS*, *C*, and *P*, are 1.0. Hence, the soil loss depends only on *R* and *K*. Thus, from the measured soil loss data for known *R*, *K* is estimated.

Since the plot experiments are resource-intensive, i.e., these require time and money, researchers have put efforts to relate the soil erodibility factor to the readily accessible soil physical properties. Wischmeier et al. (1971) introduced a nomograph relating *K* to four soil properties, viz., particle-size, soil permeability, organic matter content, and the soil structure code used for classifying soils. The following equations, representing the nomograph, may be used to estimate *K* (Wischmeier and Smith 1978; Foster et al. 1981).

$$K = 2.77 \times 10^{-7}\left(M^{1.14}\right)(12 - a) + 0.043(b - 2) + 0.033(c - 3) \tag{3.5}$$

$$M = (\% \text{ silt} + \% \text{ very fine sand}) \times (100 - \% \text{ clay}) \tag{3.6}$$

where M = a parameter representing particle-size; a = organic matter content of the soil, %; b = soil structure code; and c = profile-permeability class, defined based on saturated hydraulic conductivity.

The soil particles size is < 0.002 mm for clay, 0.002–0.05 mm for silt, and 0.05–0.10 mm for very fine sand. The organic matter content of the soil, a, is considered equal to 1.72 times the % organic carbon content of the soil. The soil structure code, b, is considered as 1 for very fine granular, 2 for fine granular, 3 for medium or coarse granular, and 4 for block, platy, or massive structure. The profile-permeability class, c, is defined based on saturated hydraulic conductivity as 1 for rapid (150 mm h^{-1}), 2 for moderate to rapid (50–150 mm h^{-1}), 3 for moderate (15–50 mm h^{-1}), 4 for slow to moderate (5–15 mm h^{-1}), 5 for slow (1–5 mm h^{-1}), and 6 for very slow (< 1 mm h^{-1}) (Wischmeier and Smith 1978; Blanco and Lal 2010).

Equation (3.5), however, should be used only if the measured soil loss data are not available or if no relationship is available for estimating *K*-factor under given local conditions. In recent past, researchers have attempted to develop soil erodibility maps utilising the global database like SoilGrids 1 km, SoilGrids 250 m, and remote sensing techniques (Panagos et al. 2014; Wang et al. 2016). Table 3.2 presents the values of the *K*-factor determined based on the runoff plot studies for different soils in India (Sharda et al. 2007).

Slope Length-Gradient Factor (*LS*)

USLE considers topography in the form of slope length (*L*) and slope steepness (*S*).

Table 3.1 Calculations using the isolated storm data (Example 3.1)

Time (h)	Rainfall (mm)	Time duration (min)	Rainfall during the period, V_r (mm)	Rainfall intensity during the period, I (mm h^{-1})	Kinetic energy per unit rainfall, E_r (MJ ha^{-1} mm^{-1})	Kinetic energy, KE (MJ ha^{-1})	Consecutive 30-min rainfall (mm)	I_{30} (mm h^{-1})	Rainfall erosion index, EI (MJ mm ha^{-1} h^{-1})
(1)	(2)	(3)	(4)	(5)	(6)	(7)	(8)	(9)	(10)
09:55	0							56.4	593.39
10:15	1.8	20	1.8	5.4	0.183	0.329			
10:20	3.3	5	1.5	18	0.229	0.343	23.3 (09:55–10:25)		
10:25	23.3	5	20	240	0.313	6.536	28.2 (10:15–10:45)		
10:45	30.0	20	6.7	20.1	0.233	1.560	12.2 (10:25–10:55)		
10:55	35.5	10	5.5	33	0.252	1.384	7.4 (10:45–11:15)		
11:05	37.1	10	1.6	9.6	0.205	0.328			
11:15	37.4	10	0.3	1.8	0.141	0.042			
Sum KE						10.521			

Columns (1) and (2): given rainfall data; columns (3), (4), and (5): by manipulating columns (1) and (2); column (6): by applying Eq. (3.4); column (7): by multiplying columns (4) and (6) and obtaining the sum as per Eq. (3.3); column (8): by adding column (4) values for consecutive 30-min slots; column (9): maximum value obtained in column (8) multiplied by 2; column (10): by multiplying the sum of column (7) with column (9) as per Eq. (3.2)

Table 3.2 K-factor values developed in India for various soils (after Sharda et al. 2007)

Soil type	Loam	Loamy sand	Sandy loam	Silt loam	Clay loam	Laterite
K-factor (Mg h MJ^{-1} mm^{-1})	0.17	0.07	0.08	0.15	0.11	0.04

The slope length (L) represents the ratio of soil loss from a field of a given length to that from a plot of 22.13 m length. It is given as

$$L = \left(\frac{l}{22.13}\right)^m \tag{3.7}$$

where l = length of the field, m. The exponent 'm' depends on the field slope and is expressed as

$$m = 0.6\left[1 - \exp(-35.835 \times s)\right] \tag{3.8}$$

where s = field slope, %.

The value of factor m may be considered as 0.5 for $s \geq 5\%$, 0.4 when s ranges over 3.5–4.5%, 0.3 when s ranges over 1–3%, and 0.2 for $s < 1\%$.

The slope steepness (S) is given as

$$S = \lfloor 65.41 \, \text{Sin}^2\theta + 4.56 \, \text{Sin}\theta + 0.065 \rfloor \tag{3.9}$$

where θ = slope steepness in degrees.

The slope length-gradient factor (LS), also called *topographic factor*, may be expressed by combining Eqs. (3.7) and (3.9) as

$$LS = \left(\frac{l}{22.13}\right)^m \lfloor 65.41 \, \text{Sin}^2\theta + 4.56 \, \text{Sin}\theta + 0.065 \rfloor \tag{3.10}$$

As evident from Eq. (3.10), soil loss would be higher for longer and steeper slopes. However, as compared to the slope length, the slope steepness has a more significant influence on the soil loss.

Land Cover and Management Factor (C)

The land cover and management factor (C) accounts for the combined effect of the interrelated land cover and management practices like crop cover, crop sequence, tillage practices, and residue management. It is the *soil loss ratio* (SLR), given as the ratio of soil loss in a field cropped with specific management to the corresponding loss from a field having up-and-down cultural practice. The information on the erosion causing rainfall distribution, the cover provided by the growing crops, crop residues and the management practices followed during the high rainfall periods is needed to determine the value of C for an area.

Table 3.3 *C* factor values for different crops in various regions of India (after Sharda et al. 2007)

Region	Crop	Value of *C*
Agra	Pearl millet	0.61
	Marvel grass (Dichanthium annulatum)	0.13
Dehradun	Lemongrass (Cymbopogon grass)	0.13
	Strawberry	0.27
Hyderabad	Grass	0.12
	Pearl millet	0.40
	Pearl millet-cowpea	0.38
Rehmankhera	Sorghum-red gramme	0.28
	Sesame-gramme	0.45

In recent times, GIS and remote sensing techniques, combined with the crop statistics and vegetation-based indices for land use classification, are widely being used to develop *C* factor for soil loss estimation at macro-regional scale (Zhang et al. 2011; Borrelli et al. 2016; Schmidt et al. 2018). Table 3.3 presents the values of the *C* factor for different cropping conditions in various regions of India (Sharda et al. 2007).

Soil Conservation Practice Factor (P)

The soil conservation practice factor (*P*) represents the ratio of soil loss from a field with specific conservation practices like contouring, strip cropping on the contour, and terracing to that from a field having no conservation practices. The value of *P* may range from 0.25 for contour strip cropping to 1.0 for conditions where no conservation practices are adopted. In fields where multiple conservation practices have been adopted, the *P* factor is considered as the product of *P* values for different conservation practices. Table 3.4 presents the *P* factor values for different conservation practices in various regions of India (Sharda et al. 2007).

Table 3.4 *P* factor values for different conservation practices in various regions of India (after Sharda et al. 2007)

Region	Conservation practice	Value of *P*
Dehradun	Contour cultivation	0.74
	Channel terraces + Graded furrows	0.35
	Channel terraces + Contour farming	0.38
	Terracing + Bunding	0.03
	Brushwood check dams	0.52
Ootacamund	Contour cultivation	0.51
Hazaribagh	Contour cultivation	0.31
Kanpur	Contour cultivation	0.39
Chandigarh	Contour bunding	0.28

Example 3.2 A 150 m long field having 4% slope in Dehradun is under Cymbopogon grass. The conservation practices adopted by the management are contour cultivation and strip cropping. The soil is loam (having 30% clay, 40% silt, 10% coarse and medium sand, and 20% very fine sand) and has a fine granular structure. The saturated hydraulic conductivity of the soil is 50 mm h^{-1}, and it contains 3% organic matter. Determine the soil loss under a rainstorm having $R = 200$ MJ mm ha^{-1} h^{-1}.

Solution Given: $R = 200$ MJ mm ha^{-1} h^{-1}; length of the field, $l = 150$ m; field slope, $s = 4\%$; location = Dehradun; crop management: Cymbopogon grass; conservation practice = contour cultivation and strip cropping; soil type and texture = loam, 30% clay, 40% silt, 10% coarse and medium sand, and 20% very fine sand; soil structure: fine granular; organic matter content = 3%; saturated hydraulic conductivity = 50 mm h^{-1}.

For determining the average soil loss, we need to use USLE (Eq. (3.1)),

$$A = R\,K\,LS\,C\,P$$

We need to have the values of various USLE factors to use in the equation.
$R = 200$ MJ mm ha^{-1} h^{-1} (known); K can be determined using Eqs. (3.5) and (3.6).

Putting the known values of % silt, % very fine sand and % clay in Eq. (3.6),

$$M = (\%\ \text{silt} + \%\ \text{very fine sand}) \times (100 - \%\ \text{clay})$$
$$= (40 + 20) \times (100 - 30) = 4200$$

Equation (3.5) is

$$K = 2.77 \times 10^{-7}(M^{1.14})(12 - a) + 0.043(b - 2) + 0.033(c - 3)$$

From the given information, the soil organic matter content, $a = 3\%$, soil structure code, $b = 2$ (for fine granular structure), and profile-permeability class, $c = 3$ (for saturated hydraulic conductivity of 50 mm h^{-1}). Hence, putting known values of M, a, b, and c in Eq. (3.5),

$$\begin{aligned}
K &= 2.77 \times 10^{-7}(M^{1.14})(12 - a) + 0.043(b - 2) + 0.033(c - 3) \\
&= 2.77 \times 10^{-7}(4200^{1.14})(12 - 3) + 0.043 \times (2 - 2) + 0.033 \times (3 - 3) \\
&= 0.03\ \text{Mg h MJ}^{-1}\ \text{mm}^{-1}
\end{aligned}$$

LS may be determined using Eqs. (3.8) and (3.10),

$$m = 0.6\big[1 - \exp(-35.835 \times s)\big] = 0.6\big[1 - \exp(-35.835 \times 4)\big] = 0.6;$$
$$\theta = \tan^{-1}(4/100) = 0.04$$

$$LS = \left(\frac{l}{22.13}\right)^m \left[65.41 \, \text{Sin}^2\theta + 4.56 \, \text{Sin}\theta + 0.065\right]$$

$$= \left(\frac{150}{22.13}\right)^{0.6} \left[65.41 \, \text{Sin}^2(0.04) + 4.56 \, \text{Sin}(0.04) + 0.065\right]$$

$$= 1.11$$

C may be referred from Table 3.3; $C = 0.13$ (for Dehradun; Cymbopogon grass).

P may be referred from Table 3.4. For Dehradun, P values for contour cultivation and strip cropping are 0.74 and 0.51, respectively. Thus, the P value for use in Eq. (3.1) will be, $P = 0.74 \times 0.51 = 0.38$.

Putting R, K, LS, C, and P in Eq. (3.1),

$$A = R \, K \, LS \, C \, P = 200 \times 0.03 \times 1.11 \times 0.13 \times 0.38 = 0.33 \, \text{Mg ha}^{-1}$$

Answer: Average annual soil loss $= 0.33 \, \text{Mg ha}^{-1}$.

3.2.2 Modified Universal Soil Loss Equation (MUSLE)

Williams (1975) modified USLE and developed the *Modified Universal Soil Loss Equation* (MUSLE) by introducing *runoff factor* to replace the rainfall-runoff erosivity factor (R). The modification was based on the concept that though rainfall may cause soil detachment, the soil transport or sediment yield depends mainly on the runoff yield. The modification made it possible to determine the soil loss from individual storm events and thus, overcome the most significant drawback of the USLE.

The following equation gives MUSLE:

$$A = 11.8\left(V_Q Q_P\right)^{0.56} K \, LS \, C \, P \tag{3.11}$$

where A = sediment yield for a single event, Mg; V_Q = volume of runoff, m^3; and Q_p = peak flow rate, m^3 s^{-1}. K, LS, C, and P remain the same as in USLE.

The volume of runoff, V_Q, and the peak flow, Q_P, is determined using the *SCS curve number method* and the *rational method*, respectively.

Example 3.3 A storm produced a total runoff volume of 100 m^3 with a peak discharge of 5 m^3 s^{-1}. Compute the sediment yield for a storm using soil erodibility factor $= 0.33$ Mg h MJ^{-1} mm^{-1}, slope length-gradient factor $= 0.68$, land cover and management factor $= 0.14$, and soil conservation practice factor $= 0.4$.

Solution Given: Volume of runoff, $V_Q = 100$ m^3; peak discharge, $Q_P = 5$ m^3 s^{-1}; soil erodibility factor, $K = 0.33$ Mg h MJ^{-1} mm^{-1}; slope length-gradient factor, LS

= 0.68; land cover and management factor, $C = 0.14$; and soil conservation practice factor, $P = 0.4$.

Putting the known values of V_Q, Q_P, K, LS, C and P in Eq. (3.11),

$$A = 11.8(V_Q Q_P)^{0.56} K \, LS \, C \, P = 11.8 \times (100 \times 5)^{0.56}$$
$$\times 0.33 \times 0.68 \times 0.14 \times 0.4 = 4.8 \, \text{Mg}$$

Answer: Sediment yield for the storm $= 4.8$ Mg.

3.2.3 SCS Curve Number Method

The SCS curve number method (Soil Conservation Service 1972) is based on the hypothesis that when a rainfall event occurs, the *initial abstraction*, which includes interception, infiltration (before the runoff begins), and depression storage, takes place. Subsequently, runoff begins, but even then, the infiltration process continues in the form of *actual retention* (Fig. 3.2). If the rainfall continues, the actual retention may attain its maximum value, i.e., the *potential maximum retention*.

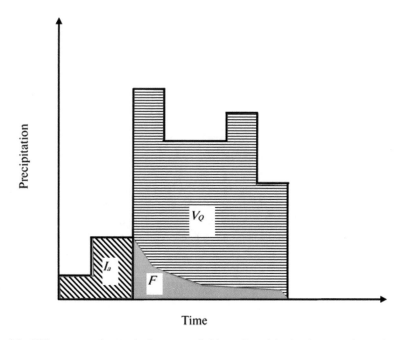

Fig. 3.2 SCS curve number method concept: division of precipitation into actual retention (F), initial abstraction (I_a), and runoff (V_Q)

SCS assumed that the ratio of the actual retention (F) to the potential maximum retention (S) is equal to the ratio of the actual runoff (V_Q) to the potential maximum runoff ($P - I_a$). Mathematically, this relationship is expressed as

$$\frac{F}{S} = \frac{V_Q}{P - I_a} \tag{3.12}$$

where P = accumulated rainfall depth; and I_a = initial abstraction.

From Fig. 3.2, it is evident that once runoff begins, then the total rainfall is divided between V_Q and F. Thus,

$$F = P - I_a - V_Q \tag{3.13}$$

Substituting F from Eq. (3.13) into (3.12), we get

$$\frac{P - I_a - V_Q}{S} = \frac{V_Q}{P - I_a} \tag{3.14}$$

Equation (3.14) may be solved to yield,

$$V_Q = \frac{(P - I_a)^2}{(P - I_a) + S} \tag{3.15}$$

The measured rainfall-runoff data of several events were analysed, and subsequently, SCS recommended the following relationship:

$$I_a = 0.2\,S \tag{3.16}$$

Substituting I_a from Eq. (3.16) into (3.15), we get

$$V_Q = \frac{(P - 0.2S)^2}{(P + 0.8S)} \quad \text{for} \quad P > 0.2S \tag{3.17}$$

Equation (3.17) represents the fundamental relationship for the SCS curve number method. For a rainfall event, the runoff depth could be computed if the potential maximum retention, S, of the area is known. The potential maximum retention or the potential infiltration after runoff begins is related to the dimensionless curve number, CN, as follows:

$$S = \frac{25{,}400}{CN} - 254 \quad (S \text{ is in mm}) \tag{3.18}$$

The CN ranges from 0 to 100 and can be determined from a table based on the SVL (soil-vegetation-land use) complex of the area. It is evident from Eq. (3.18) that if CN is high, S will be low, indicating lower infiltration and higher magnitude of runoff. For example, when *CN* is 100, $S = 0$, and the total rainfall gets transformed to runoff.

SCS (1972) tabulated the *CN* values for antecedent moisture condition (AMC) Class II and $I_a = 0.2S$ as a function of land use or cover, land treatment or practice, hydrological condition, and hydrological soil group. The antecedent moisture condition (AMC) reflects the soil moisture status of the study basin preceding the rainfall event. It is defined based on the 5-day accumulated rainfall before the runoff event, and variations in the evapotranspiration are considered by segregating the dormant season and the growing season. Three AMC classes are identified: AMC I (reflects dry condition), AMC II (average condition), and AMC III (saturated condition).

The land use or cover is categorised as fallow (bare land having the highest runoff potential), row crops (crops grown in distant rows, exposing the maximum soil surface to the rainfall), small grain (crops grown in close rows, protecting the maximum soil surface from rainfall), and close-seeded legumes or rotational meadow (that protect the soil surface from rainfall throughout the year).

Land treatment or practice represents agronomical practices like crop residue cover and straight row and soil conservation treatments like contouring and terracing. Hydrological condition is categorised as poor or good based on the factors governing the infiltration and runoff characteristics of the surface (good if the surface cover and roughness enhance infiltration and lessen runoff).

Hydrological soil group represents the classification of soils into four groups based on infiltration rate or runoff potential. Hydrological soil *Group A* represents soils that have high infiltration rate (7.62–11.43 mm h^{-1}) and low runoff potential, e.g., sand, sandy loam, or loamy sand. *Group B* soils have moderate infiltration rate (3.81–7.62 mm h^{-1}) and moderately low runoff potential, e.g., silty loam or loam. *Group C* soils have low infiltration rate (1.27–3.81 mm h^{-1}) and moderately high runoff potential, e.g., sandy clay loam. *Group D* soils have very low infiltration rate (0–1.27 mm h^{-1}) and high runoff potential, e.g., clay, silty clay, or clay loam.

Table 3.5 presents the typical *CN* values for agricultural lands for AMC II, and Table 3.6 presents the 5-day accumulated rainfall values for different AMC classes. SCS has also given a table for converting the AMC II *CN* values obtained from Table 3.5 to their corresponding AMC I or AMC III values (Table 3.7).

To apply the SCS curve number method for a study area, the applicable *CN* values are determined from Table 3.5 for different combinations of land use and cover, land treatment or practice, hydrological condition, and hydrological soil group. Subsequently, a weighted *CN* is determined based on the area represented by the individual combinations. Since the value determined from Table 3.5 is for AMC II class, the 5-day accumulated rainfall before the runoff event is determined, and the applicable AMC class is ascertained using Table 3.6. If the AMC class obtained from Table 3.6 is different from AMC II, then Table 3.7 is used to determine the equivalent *CN* value for the applicable AMC class. Once *CN* for the study area is known, then Eqs. (3.18) and (3.17) are used to determine the depth of runoff volume.

Example 3.4 Determine the runoff amount from a 4.0 km^2 agricultural watershed due to a storm generating 40 mm of rainfall during the growing season. The cumulative

Table 3.5 Curve numbers (CN) for agricultural lands for AMC II and $I_a = 0.2S$ (after NRCS 2008)

Land use or cover	Treatment or practice	Hydrological condition	Curve numbers for hydrological soil group			
			A	B	C	D
Fallow	Bare soil		77	86	91	94
	Crop residue (CR)	Poor	76	85	90	93
		Good	74	83	88	90
Row crops	Straight row (SR)	Poor	72	81	88	91
		Good	67	78	85	89
	SR + CR	Poor	71	80	87	90
		Good	64	75	82	85
	Contoured (C)	Poor	70	79	84	88
		Good	65	75	82	86
	C + CR	Poor	69	75	82	86
		Good	64	74	81	85
	C + Terraced (T)	Poor	66	74	80	82
		Good	62	71	78	81
	C + T + CR	Poor	65	73	79	81
		Good	61	70	77	80
Small grain	Straight row (SR)	Poor	65	76	84	88
		Good	63	75	83	87
	SR + CR	Poor	64	75	83	86
		Good	60	72	80	84
	Contoured (C)	Poor	63	74	82	85
		Good	61	73	81	84
	C + T	Poor	61	72	79	82
		Good	59	70	78	81
	C + T + CR	Poor	60	71	78	81
		Good	58	69	77	80
Close-seeded legumes or rotation meadow	Straight row (SR)	Poor	66	77	85	89
		Good	58	72	81	85
	Contoured (C)	Poor	64	75	83	85
		Good	55	68	78	83
	C + T	Poor	63	73	80	83
		Good	51	67	76	60

Table 3.6 5-day antecedent rainfall for different AMC classes (after NRCS 2008)

AMC class	5-day antecedent rainfall (mm)	
	Dormant season	Growing season
I	<13	<36
II	13–28	36–53
III	>28	>53

Table 3.7 Conversion of AMC II *CN* values to AMC I and AMC III (after NRCS 2008)

CN AMC II	*CN* AMC I	*CN* AMC III	*CN* AMC II	*CN* AMC I	*CN* AMC III
100	100	100	58	38	76
98	94	99	56	36	75
96	89	99	54	34	73
94	85	98	52	32	71
92	81	97	50	31	70
90	78	96	48	29	68
88	75	95	46	27	66
86	72	93	44	25	64
84	68	92	42	24	62
82	66	91	40	22	60
80	63	90	38	21	58
78	60	89	36	19	56
76	58	88	34	18	54
74	55	87	32	16	52
72	53	86	30	15	50
70	51	85	25	12	43
68	48	84	20	9	37
66	46	82	15	6	30
64	44	81	10	4	22
62	42	79	5	2	13
60	40	78	0	0	0

rainfall during the last five days before the storm is 70 mm. The watershed has three SVL combinations: row crops, contoured under a good condition in 50% of the watershed, close-seeded legumes planted in straight rows under a good condition in 25% of the watershed, and fallow land with crop residue under a good condition in the remainder of the watershed. The soils in the watershed belong to hydrological soil group *C.*.

Solution Given: Watershed area $= 4.0 \text{ km}^2$; rainfall, $P = 40$ mm; 5-day antecedent rainfall for growing season $= 70$ mm; area of the watershed under row crops, contoured under a good condition $= 2.0 \text{ km}^2$; area of the watershed under close-seeded legumes planted in straight rows under a good condition $= 1.0 \text{ km}^2$; area of the watershed kept as fallow with crop residue under a good condition $= 1.0 \text{ km}^2$; hydrological soil group $= C$.

First, we need to find the weighted CN for the watershed under AMC II. From Table 3.5, for row crops, contoured under a good condition, hydrological soil group C, $CN = 82$; watershed area $= 2.0 \text{ km}^2$. For close-seeded legumes planted in straight rows under a good condition, hydrological soil group C, $CN = 81$; watershed area $= 1.0 \text{ km}^2$ and for fallow with crop residue under a good condition, hydrological soil group C, $CN = 88$; Watershed area $= 1.0 \text{ km}^2$.

Hence, the weighted CN for the watershed under AMC II class $= \frac{(82 \times 2 + 81 \times 1 + 88 \times 1)}{4} = 83.25$.

The cumulative rainfall during the last five days before the storm is given as 70 mm. Therefore, from Table 3.6, the AMC class is found as AMC III.

Thus, we need to determine the equivalent weighted CN for the watershed from Table 3.7. We find that for AMC II CN values of 82 and 84, the corresponding values for AMC III CN are 91 and 92. Therefore, by interpolation, the weighted CN for AMC III is 91.875.

$$S = \frac{25,400}{CN} - 254 = \frac{25,400}{91.875} - 254 = 22.5 \, \text{mm}$$

Putting the value of weighted CN in Eq. (3.18),

$$V_Q = \frac{(P - 0.2S)^2}{P + 0.8S} = \frac{(40 - 0.2 \times 22.5)^2}{(40 + 0.8 \times 22.5)} = 21.7 \, \text{mm}$$

Answer: Runoff volume generated from the watershed $= 21.7$ mm.

3.2.4 Rational Method

The rational method, developed by Mulvany (1851), is popularly used for estimating the peak runoff rate due to relatively frequent storms having a return period of 2–10 years. It is useful in designing soil and water conservation structures in small watersheds. The method assumes that the runoff producing rainfall occurs at a constant intensity for a duration equal to the time of concentration of the watershed, and it is uniformly distributed over the entire watershed. It is expressed as

$$Q = c \, i \, A \tag{3.19}$$

Table 3.8 Runoff coefficient c for various soils under different land use conditions (after Venkateswarlu et al. 2013)

Soil type and infiltration	Land use		
	Cultivated	Pasture	Forest
Above average infiltration, usually sandy or gravelly	0.29	0.15	0.10
Average infiltration, no clay pans, loams, or similar soils	0.40	0.35	0.30
Below average infiltration, heavy clay soils, or soils with a dry pan near the surface, shallow soils having impervious rock	0.50	0.45	0.40

where Q = peak runoff rate, m^3 s^{-1}; c = runoff coefficient, dimensionless; i = average rainfall intensity for a duration equal to the time of concentration of the watershed, m s^{-1}; and A = watershed area, m^2.

The runoff coefficient c represents the portion of rainfall that gets converted into the runoff and reflects the impact of land use and cover, land slope, soil, and management practices on the rainfall-runoff transformation process. Table 3.8 presents the values of c for various soil types with varying infiltration characteristics and different land use (Venkateswarlu et al. 2013). The runoff coefficient c for various hydrological soil groups under different land use and rainfall conditions may also be referred from Horn and Schwab (1963).

The average rainfall intensity, i, can be obtained from the intensity–duration–frequency curve developed for the area. The storm duration is considered equal to the time of concentration, T_c, which is defined as the time taken by water to travel from the hydraulically remotest point of the watershed to the outlet. It is expected that at T_c, the entire watershed would contribute runoff to the outlet. Though several methods have been developed for the estimation of T_c, for small watersheds, the following equation, developed by Kirpich (1940), is preferred.

$$T_c = 0.0195L^{0.77}S^{-0.385} \qquad (3.20)$$

where T_c = time of concentration, minutes; L = maximum length of flow, m; and S = average slope of the watershed, fraction.

To apply the rational method for determining the peak runoff rate for a watershed, the value of runoff coefficient, c, is determined, based on the land use and cover conditions. For a watershed having multiple land uses or covers, the weighted value of c is determined based on the area represented by different land uses or covers. Alternatively, the peak runoff rates may be determined for individual land use/cover subareas and summed up to get the peak runoff rate for the watershed.

The latter approach is recommended for watersheds having distinct land use/cover areas. The application of the rational method, however, is limited to small watersheds, having area up to 1200 ha.

Example 3.5 A 0.70 km^2 agricultural watershed has a slope of 0.4% and the maximum length of flow 1.0 km. The 10-year maximum depth of rainfall is tabulated. Half of the watershed has row crops grown, whilst the other half of the watershed has pasture. The watershed has loam soils having average infiltration rate. Estimate the peak flow rate of runoff for the watershed for the 10-year return period.

Duration (minutes)	5	10	20	30	40	60
Maximum depth of rainfall (mm)	15	25	45	53	60	65

Solution Given: Watershed area, $A = 0.70 \, km^2 = 7 \times 10^5 \, m^2$; average slope of the watershed $S = 0.004$; maximum length of flow $L = 1000 \, m$; land use = row crops $(0.35 \, km^2)$, pasture $(0.35 \, km^2)$; Soil = loam having average infiltration rate.

Using Kirpich formula to calculate T_c,

$$T_c = 0.0195 L^{0.77} S^{-0.385} = 0.0195 \times (1000)^{0.77}(0.004)^{-0.385} = 33.4 \, min$$

On interpolating the given time versus maximum rainfall depth,

Depth of rainfall for 33.4 min $= 53 + \frac{(60-53)}{10} \times 3.4 = 55.4 \, mm$.

Storm intensity, $i = \frac{55.4}{33.4} \times 60 = 99.5 \, mm \, h^{-1} = \frac{99.5 \times 10^{-3}}{3600} = 2.76 \times 10^{-5} \, m \, s^{-1}$.

Next, we need to determine the weighted runoff coefficient c for the watershed. From Table 3.8, for cultivate land on loam soil having average infiltration rate, $c = 0.40$, and for pasture, $c = 0.35$.

Hence, the weighted c for the watershed $= \frac{(0.4 \times 0.35 + 0.35 \times 0.35)}{0.7} = 0.375$.

Putting the values of c, i and A in Eq. (3.19),

$$Q = ci A = 0.375 \times (2.76 \times 10^{-5}) \times \left(7 \times 10^5\right) = 7.25 \, m^3 \, s^{-1}$$

Answer: The peak rate of runoff from the watershed $= 7.25 \, m^3 \, s^{-1}$.

3.2.5   Revised Universal Soil Loss Equation (RUSLE)

The USLE was revised during the 1990s to the RUSLE by including process-based components and improved procedures to estimate various factors involved (Renard et al. 1997). The basic multiplicative structure of USLE is, however, preserved in RUSLE.

RUSLE provides an updated procedure for estimating the erosivity factor (R) using the following relationship:

$$R = \frac{1}{n} \sum_{J=1}^{n} \sum_{k=1}^{m_j} (EI)_k \qquad (3.21)$$

where, n = number of years of record (usually more than 20 years); m_j = number of events causing erosion during year j; and EI = rainfall erosion index for rainfall event k, MJ mm ha^{-1} h^{-1}.

Thus, the modification paves the way for the estimation of rainfall erosion index (EI), even for an individual event, using the relationship:

$$EI = \left(\sum_{r=1}^{m} e_r v_r \right) I_{30} \qquad (3.22)$$

where e_r = unit rainfall energy, MJ ha^{-1} mm^{-1}; v_r = rainfall depth for the time interval r of the hyetograph, which has been divided into $r = 1, 2, \ldots, m$ subintervals, mm.

The unit rainfall energy is given by the following relationship (McGregor and Mutchler 1976):

$$e_r = 0.29\left(1 - 0.72\, e^{-0.082\, I}\right) \qquad (3.23)$$

RUSLE typically uses the erodibility nomograph, developed following Eq. (3.5), for estimating the soil erodibility factor, K. It guides the user in identifying the soils for which the nomograph is not applicable and then suggests alternative methods for estimating K. RUSLE also includes variability of K over different seasons by considering freezing and thawing. Equations relating K to the annual R for the frost-free period are used to estimate K, and the seasonal variability is taken into account by considering weights proportional to EI for a 15-day interval.

RUSLE provides improved relationships to estimate the slope length-gradient factor, LS, for complex topographies by considering the upslope contributing area and flow accumulation. RUSLE removes the dominance of slope steepness and gives more even weightage to length and steepness. It considers ponding over the surface area and adjusts soil loss accordingly.

RUSLE includes computer routines to estimate the above subfactors for various tillage operations and crops and is especially helpful for conditions for which experimental data are not available to obtain SLR.

RUSLE provides improved estimates of the soil conservation practice factor (P). For example, P values for contouring and terracing are re-evaluated and presented by considering the furrow grade and ridge height, and terrace grades, respectively. Similarly, a broader range of options is provided for strip cropping. P factors are also included for rangelands based on surface slope, roughness, and runoff reduction.

RUSLE is available as a computer model, with RUSLE1.06c being its last version (USDA-ARS 2020). RUSLE2 (Foster et al. 2003) is the advanced version of RUSLE that provides a user-friendly computer interface and uses a hybrid approach to model soil loss. It includes empirical equations to compute sheet and rill erosion due to rainfall erosivity and process-based equations to compute the detachment, transport, and deposition of sediments due to runoff. The input data of RUSLE2 may be referred from the existing databases to compute soil loss at a daily time step.

3.2.6 Water Erosion Prediction Project (WEPP) Model

The Water Erosion Prediction Project (WEPP) is a distributed, physically-based soil erosion model (Flanagan et al. 1995). It integrates hydrology, hydraulics, plant science, and soil erosion mechanics for predicting erosion. It is capable of evaluating diverse land uses and hydrologic conditions whilst simulating soil erosion at the watershed scale and hillslope. Its modular structure includes modules for the climate to generate weather information; hydrology to compute infiltration and flow; erosion to compute soil loss and sediment deposition; daily water balance; irrigation; plant development; and residue decomposition.

WEPP operates in two modes, one meant for individual hillslopes and the other for multiple hillslopes located in a watershed. The hydrology module includes the Green-Ampt equation for infiltration estimation and a kinematic wave equation for flow routing. The erosion module includes a steady-state mass balance equation to estimate the sediment generation from rill and interrill areas. The bedload equation of Yalin (1963) is used to calculate the transport capacity.

WEPP requires four inputs dealing with the topography of the model area, soil characteristics, land use and management, and climate. It offers climate databases for USA, and soil and land use databases with a large variety of options. The model outputs include water balance, soil detachment and deposition on a hillslope, and sediment yield from the hillslope. The application of WEPP, however, is limited to small watersheds, having area up to 2.6 km^2.

3.2.7 EUROSEM

The European Soil Erosion Model (EUROSEM) is a dynamic distributed model that simulates sediment transport, erosion, and deposition under different environmental conditions (Morgan et al. 1998). It is designed to model rill and interrill erosion processes within a storm, i.e., it is an event-based model, unlike WEPP, which is a continuous simulation model.

EUROSEM has a modular structure having modules dealing with rainfall interception, infiltration, soil surface condition, surface runoff, soil detachment due to rainfall and runoff, and the sediment transport. The outputs include runoff, soil loss, the flow hydrograph, and the sediment graph. The interrill and rill flow processes are modelled explicitly.

The erosion model EUROSEM is coupled with the hydrology model, kinematic runoff and erosion model (KINEROS) (Woolhiser et al. 1990) for determining the values of Q and A. KINEROS is a distributed, physically-based, deterministic model that describes the processes like rainfall, interception, and infiltration during a storm event. KINEROS considers only the Hortonian overland flow and neglects the saturation-excess overland flow.

3.3 Measuring Soil Loss

As mentioned earlier, soil loss data from laboratory or field measurements are essential to have a fair idea on the magnitudes of soil loss under various agronomic and management practices. Besides, measured data are also useful for calibrating and evaluating various soil loss estimation models. The soil loss measurement is usually carried out in erosion plots.

3.3.1 Erosion Plots

Typically, two types of plots are used for measuring soil loss, bounded, and unbounded.

Bounded plots, usually rectangular in shape, are constructed by separating a portion of the land using metal sheets or wood, sufficiently inserted in the soil, and extended above the ground surface (Fig. 3.3). These plots of known size, shape, and slope are used to monitor the soil loss with the assumption that the soil loss takes place uniformly from the plot area and is independent of the shape and size of the plot. Since the runoff and sediment from the upslope cannot enter the plot, these plots are not suitable for measuring soil loss under natural conditions. Instead, these are preferred in research studies focussing on analysing the factors affecting soil erosion.

A *Gerlach Trough*, named after its inventor (Gerlach 1967), is typically used as an unbounded plot or a plot without boundaries. The trough has a 0.5 m long and 0.2 m wide metal gutter with a movable lid, which is set into the soil surface perpendicular

Fig. 3.3 Bounded runoff plot

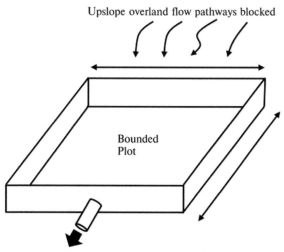

Upslope overland flow pathways blocked

Bounded Plot

Runoff generated from within the plot only

to the land slope and connected to a collecting tank for water and sediment collection. The contributing area to the trough may be computed using the topographic survey data, or by locating flow path using a dye.

Both bounded and unbounded plots are employed for sheet and rill erosion measurement. In both cases, runoff and sediment are directed to collecting tanks for retaining all or a part of the water and sediment. For large plots, two collecting tanks are used. The runoff and sediment get collected in the first tank, and the over-flow from this tank is passed through either a *multislot divisor* or *Coshocton wheel*, which divides the flow into several equal portions and diverts one part into the second tank.

The erosion plots may have provisions of automatic flow measurement and sediment sampling. For flow measurement, usually, an H-flume along with water level gauges or a flow metre is preferred. Further, an automatic sediment sampler may be installed to extract runoff samples at regular time intervals for analysing the sediment concentration. The collected data may also be used to develop a *sediment rate versus time graph*, commonly known as the *sediment graph*. It may, however, be noted that the automation would make the system expensive. Thus, in most cases, a composite sampler such as *Coshocton wheel* may serve the purpose. The total weight of the soil collected may be determined by first drying and weighing sediment samples of known volume to find the mean sample weight and then multiplying it by the total sediment volume collected in the tank. In case a divisor or Coshocton wheel is used, then the total weight of the sediment is adjusted based on the percentage of the runoff and sediment collected in the tank.

3.3.2 Multislot Divisor

A multislot divisor (Fig. 3.4) consists of several slots of equal dimensions fitted at the end of a rectangular divisor box (Brakensiek et al. 1979). It functions on the assumption that the flow will pass through the divisor box with a uniform horizontal velocity, and the flow and sediment would be divided equally amongst the multiple slots. Consequently, the flow passing through one of the slots is led into a collecting tank, and the remainder of the flow is drained away. The water accumulated in the tank is subsequently analysed to determine the runoff and sediment volume.

The application of the multislot divisor is limited to low flow rates, and it provides only the accumulated flow and sediment volumes. It is, however, preferred due to simple design and ease in operation. The number of slots and their dimension of slots and their dimension governs the capacity of the divisor. The number of slots determines the aliquot size and is referred to as the *divisor ratio*, e.g., a divisor having seven slots has a divisor ratio of 1:7. Based on the expected flow from the plot, usually, 3, 5, 7, 9, 11, or 13 slots of sizes 125 mm × 100 mm (width × height) are used. The expected flow and the divisor ratio used govern the capacity of the collecting tank.

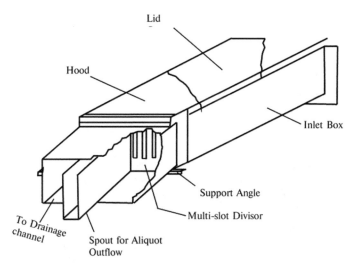

Lid

Hood

Inlet Box

Support Angle

Multi-slot Divisor

To Drainage channel

Spout for Aliquot Outflow

Fig. 3.4 Multislot divisor (after Brakensiek et al. 1979)

3.3.3 *Coshocton Wheel*

The Coshocton wheel sampler is named after the place where it was developed, i.e., Coshocton, Ohio, USA (Parsons 1954). It consists of a slightly inclined wheel (usually called a water wheel) having eight vanes and a radially mounted sampling slot (Fig. 3.5). Water from the H-flume drops on the water wheel and rotates it. The sampling slot collects an aliquot sample as it passes through the flow jet during each revolution of the wheel. The aliquot sample is routed to a sample storage tank through the base plate of the wheel. The sample is subsequently analysed to determine the sediment volume.

3.3.4 *Size of Plots*

The nature of the study governs the selection of the plot size. For example, microplots of 1 m × 1 m or 1 m × 2 m may be used to analyse the impact of rain splash but may be too small for overland flow studies. The microplots are typically used to analyse the impact of experimental treatments like the variations in the rainfall intensity and the per cent of the soil surface under residue. These plots are usually equipped with a rainfall simulator to study the treatment impact under a controlled environment. Small-scale plots of about 100 m^2 are most commonly used for analysing the impact of agronomical and management practices and for deriving the input data for models based on USLE methodology.

The standard plots used to derive the inputs for USLE, usually called USLE runoff plots, are 22.13 m long and have 9% slope. Though the width of USLE plots

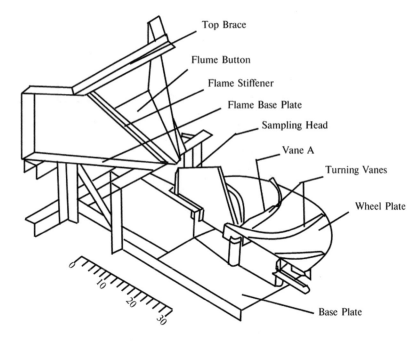

Fig. 3.5 Coshocton wheel (after Parsons 1954)

is flexible, a plot of 4.1 m or larger width is preferred for rill and interrill erosion studies. These plots may be with or without a rainfall simulation system; however, a rainfall simulation system is desirable for accurate analysis of the impacts of various treatments.

Field plots of about 1 ha are preferred for analysing the impact of treatments like bunding and terracing, which cannot be studied on small-scale plots. It may, however, be noted that due to the lack of soil deposition considerations, erosion plots usually result in the overestimation of soil loss. Also, plot studies may not be suitable for analysing the impact of human settlement or land use changes or the impact of tree plantation programmes. Such impacts can best be analysed only at the watershed or catchment scale. Therefore, many a time, it may be advisable to monitor the flow and sediment movement from the entire watershed at the outlet to draw meaningful conclusions regarding soil erosion treatments.

Questions

1.   Enumerate the strengths and weaknesses of the USLE methodology.
2.   What are the changes introduced in RUSLE to improve the USLE? Discuss the advantages of the changes introduced.
3.   Assume that a watershed near your hometown has $R = 250$ MJ mm ha^{-1} h^{-1}, $K = 0.25$ Mg h MJ^{-1} mm^{-1}, length of the field, $l = 100$ m, field slope, $s = 6\%$, $C = 0.2$, and $P = 1.0$. Determine the soil loss from the watershed. Suppose you

wish to reduce the soil loss by 25%, then what conservation practices would you recommend to the watershed authorities?

4. Assume that the soil loss from the watershed near your hometown, for known values of R, K, C, and P, is 5.0 Mg ha^{-1} for a 75-m slope length with 4% slope. What would be the soil loss if the slope length is doubled?

5. A 4-h storm, having a peak discharge of 4 m^3 s^{-1}, results in a total runoff volume of 80 m^3. Determine the sediment yield due to the storm. Use soil erodibility factor = 0.20 Mg h MJ^{-1} mm^{-1}, slope length-gradient factor = 0.60, land cover and management factor = 0.24, and soil conservation practice factor = 0.5.

6. Determine the runoff from a 200-ha agricultural watershed due to a storm having 50 mm of rainfall. It is growing season, and the previous five days had a cumulative rainfall of 40 mm. The watershed has three SVL combinations, distributed equally over the watershed: row crops, contoured, and terraced under good condition; small grains planted in straight rows with crop residue under good condition; and close-seeded legumes planted in straight rows under good condition. The soils in the watershed belong to hydrological soil group B.

7. Determine the peak runoff rate from 20 ha watershed for a 10-year storm resulting in a magnitude of 50 mm. The storm duration equals the time of concentration of the watershed. The average slope of the watershed is 0.6%, and the maximum length of flow is 500 m. The watershed has heavy clay soils having below average infiltration, and small grains are grown following good practice.

8. A 100-ha watershed has soils belonging to the soil hydrologic group C. The watershed has an average slope of 2.0% and the maximum flow length of 800 m. Compute the time of concentration for the watershed.

9. Describe the exclusive features of various physically-based soil erosion models. What are their strengths and weaknesses?

10. Describe the working principle of the Coshocton wheel.

Multiple Choice Questions
Answer the questions by choosing the correct option.

1. For an agricultural watershed, the rainfall erosivity index is estimated as 250 MJ mm ha^{-1} h^{-1}. The soil erodibility index is 0.05 Mg h MJ^{-1} mm^{-1}, crop management factor is 0.64, conservation practice factor is 1.0, and slope length factor is 0.25. If the conservation practices adopted subsequently in the watershed reduce the conservation practice factor to 0.75, then the reduction in the soil loss in Mg ha^{-1} would be

(A) 0.5 (B) 1.0
(C) 1.5 (D) 2.0

2. An experimental plot needs to be chosen for obtaining the soil erodibility factor value for applying the universal soil loss equation. The length (in m) and slope (in %) of the plot, respectively, should be

(A) 9, 22 (B) 11, 22
(C) 22, 9 (D) 22, 11

3. Whilst using the SCS curve number method, if $CN = 60$, then the runoff volume (in mm) for a rainfall event having 80 mm depth would be

(A) 10 (B) 27
(C) 48 (D) 80

4. The rational formula is used for determining the peak flow Q_p whilst designing a structure. If a storm having the same intensity as used in the design but of twice larger duration occurs, then the resulting peak discharge would be

(A) Q_p (B) $2Q_p$
(C) $Q_p/2$ (D) Q_p^2

5. A soil belongs to hydrological soil Group C. The infiltration rate of the soil is

(A) high   (B) moderate
(C) low   (D) very low

6. The value of curve number, CN, for concrete surface could be taken as

(A) 0   (B) 20
(C) 50 (D) 100

7. The Coshocton wheel is used for

(A) measuring rainfall
(B) measuring peak runoff
(C) measuring sediment load
(D) partitioning direct runoff and baseflow

8. The sediment collected at the watershed outlet is expected to be

(A) same as the gross erosion from the watershed
(B) less than the gross erosion from the watershed
(C) more than the gross erosion from the watershed
(D) difficult to judge

9. A multislot divisor is used to study

(A) depth of rainfall
(B) spatial distribution of rainfall
(C) particle-size analysis of sediments
(D) volume of sediment from an erosion plot

10. Match column I and column II and identify the most appropriate combinations:

Column I		Column II	
P	Rational formula	1	Time of concentration
Q	Curve number	2	Physically-based soil erosion model
R	Kirpich formula	3	Hybrid soil erosion model
S	Gerlach trough	4	Runoff
T	RUSLE2	5	Peak flow rate
U	WEPP	6	Erosion plot

(A) P-5, Q-4, R-1, S-2, T-6, U-3
(B) P-5, Q-4, R-1, S-6, T-3, U-2
(C) P-5, Q-6, R-3, S-4, T-1, U-2
(D) P-5, Q-6, R-1, S-3, T-2, U-4

References

Arnold JG, Srinivasan R, Muttiah RS, Williams JR (1998) Large-area hydrologic modeling and assessment: Part I. Model development. J Am Water Resour Assoc 34:73–89

Blanco H, Lal R (2010) Principles of soil conservation and management. Springer Science and Business Media BV, Dordrecht

Borrelli P, Panagos P, Langhammer J, Apostol B, Schütt B (2016) Assessment of the cover changes and the soil loss potential in European forestland: first approach to derive indicators to capture the ecological impacts on soil-related forest ecosystems. Ecol Indic 60:1208–1220

Brakensiek DL, Osborn HB, Rawls WJ (Coordinators) (1979) Field manual for research in agricultural hydrology. Agriculture Handbook 224, US Department of Agriculture, Washington DC

Cronshey RG, Theurer FD (1998) AnnAGNPS—non-point pollutant loading model. In: Proceedings of the first federal interagency hydrologic modelling conference, Las Vegas, 19–23 April 1998

Flanagan DC, Ascough II JC, Nicks AD, Nearing MA, Laflen JM (1995) Overview of the WEPP erosion prediction model. In: Flanagan DC, Nearing MA (eds) USDA—water erosion prediction project hillslope profile and watershed model documentation. NSERL Report No. 10, USDA-ARS National Soil Erosion Research Laboratory, West Lafayette

Foster GR, McCool DK, Renard KG, Moldenhauer WC (1981) Conversion of the universal soil loss equation to SI metric units. J Soil Water Conserv 36:355–359

Foster GR, Toy TJ, Renard KG (2003) Comparison of the USLE, RUSLE1.06c, and RUSLE2 for application to highly disturbed land. In: First interagency conference on research in the watershed. USDA-Agricultural Research Service, Washington DC

Gerlach T (1967) Hillslope troughs for measuring sediment movement. Revue Geomorphologie Dynamique 4:173

Horn DL, Schwab GO (1963) Evaluation of rational runoff coefficients for small agricultural watersheds. Trans ASAE 6:195–198

Kirpich ZP (1940) Time of concentration of small agricultural watersheds. Civil Eng 10:362

Laflen JM, Elliot WJ, Flanagan DC, Meyer CR, Nearing MA (1997) WEPP—predicting water erosion using a process-based model. J Soil Water Conserv 52:96–102

Liu B, Xie Y, Li Z, Liang Y, Zhang W, Fu S, Yin S, Wei X, Zhang K, Wang Z, Liu Y Zhao Y, Guo Q (2020) The assessment of soil loss by water erosion in China. Int Soil Water Conserv Res 8:430–439

McGregor KC, Mutchler CK (1976) Status of the R factor in northern Mississippi. In: Soil erosion: prediction and control. Soil Conservation Society of America

Morgan RPC, Quinton JN, Smith RE, Govers G, Poesen JWA, Auerswald K, Chisci G, Torri D, Styczen ME (1998) The European soil erosion model (EUROSEM): a dynamic approach for predicting sediment transport from fields and small catchments. Earth Surf Process Landf 23:527–544

Mulvany T (1851) On the use of self-registering rain and flood gauges in making observations of the relation of rainfall and flood discharges in given catchment. Trans Inst Civil Eng 4:18–33

Natural Resources Conservation Service (NRCS) (2008) National engineering handbook, Section 2: Estimating runoff volume and peak discharge. USDA Natural Resources Conservation Service, Washington DC

Panagos P, Meusburger K, Ballabio C, Borrelli P, Alewell C (2014) Soil erodibility in Europe: a high-resolution dataset based on LUCAS. Sci Total Environ 479–480:189–200

Panagos P, Borrelli P, Meusburger K, Yu B, Klik A, Jae Lim K et al (2017) Global rainfall erosivity assessment based on high-temporal resolution rainfall records. Sci Rep 7:4175

Parsons DA (1954) Coshocton-type runoff samplers: laboratory investigations. United States Soil Conservation Service, US Department of Agriculture, Washington DC

Renard KG, Foster GR, Weesies GA, Porter JP (1997) RUSLE: revised universal soil loss equation. J Soil Water Conserv 46:30–33

Schmidt S, Alewell C, Mausburger K (2018) Mapping spatio-temporal dynamics of the cover and management factor (C-factor) for grasslands in Switzerland. Rem Sens Environ 211:89–104

Sharda VN, Juyal GP, Prakash C, Joshi BP (2007) Training Manual: Soil Conservation And Watershed management, Vol-II (soil water conservation engineering). Central Soil and Water Conservation Research and Training Institute (CSWCRTI), Dehradun

Smith DD, Whitt DM (1948) Evaluating soil losses from field areas. Agric Eng 29:394–396

Soil Conservation Service (1972) National engineering handbook. Section 4—Hydrology. USDA Natural Resources Conservation Service, Washington DC

US Department of Agriculture—Agricultural Research Service (USDA-ARS): Revised Universal Soil Loss Equation. Available at https://www.ars.usda.gov/southeast--area/oxford-ms/national-sedimentation-laboratory/watershed-physical-processes-research/docs/revised-universal-soil-loss-equation-rusle-welcome-to-rusle-1-and-rusle-2/1. Accessed 12 October 2020

Venkateswarlu B, Osman M, Padmanabhan MV, Kareemulla K, Mishra PK, Korwar GR, Rao KV (2013) Field manual on watershed management. Central Research Institute for Dryland Agriculture, Hyderabad

Wang GQ, Fang QQ, Teng YG, Yu JS (2016) Determination of the factors governing soil erodibility using hyperspectral visible and near-infrared reflectance spectroscopy. Int J Appl Earth Obs Geoinf 53:48–63

Wight JR, Skiles JW (eds) (1987) SPUR: simulation of production and utilisation of rangelands. Document and user guide. ARS 63, US Department of Agriculture, Washington DC

Williams JR (1975) Sediment-yield prediction with universal equation using runoff energy factor. In: Present and prospective technology for predicting sediment yield and sources. ARS S-40, US Department of Agriculture, Washington DC

Williams JR, Berndt HD (1977) Sediment yield prediction based on watershed hydrology. Trans ASAE 20:1100–1104

Williams JR, Nicks AD, Arnold JG (1985) Simulator for water resources in rural basins. J Hydraul Eng ASCE 111:970–986

Wischmeier WH, Smith DD (1958) Rainfall energy and its relationship to soil loss. Trans Am Geophys Uni 39:285–291

Wischmeier WH, Smith DD (1965) Predicting rainfall-erosion losses from cropland east of the Rocky Mountains. Agriculture Handbook No. 282, US Department of Agriculture, Washington DC

Wischmeier WH, Johnson CB, Cross BV (1971) A soil erodibility nomograph for farmland and construction sites. J Soil Water Conserv 26:189–192

Wischmeier WH (1976) Use and misuse of the universal soil loss equation. J Soil Water Conserv 31:5–9

Wischmeier WH, Smith DD (1978) Predicting rainfall erosion losses. Agriculture Handbook No. 537, US Department of Agriculture, Science and Education Administration, Washington DC

Woolhiser DA, Smith RE, Goodrich DC (1990) KINEROS: a kinematic runoff and erosion model: documentation and user manual, ARS 77. US Department of Agriculture, Washington DC

Yalin MS (1963) An expression for bed-load transportation. J Hyd Div ASCE 98:221–250

Zhang W, Zhang Z, Liu F, Qiao Z, Hu S (2011) Estimation of the USLE cover and management factor C using satellite remote sensing: a review. In: 19th International conference on geoinformatics. IEEE, Shanghai, China

Zingg AW (1940) Degree and length of land slope as it affects soil loss in runoff. Agric Eng 21:59–64

Chapter 4
Terrace

Abstract Terraces are the most widely practised soil erosion control measure around the world. The practice consists of earth embankments constructed across the steep slopes to intercept surface runoff and divert it at a non-erosive velocity to a safe outlet or store it to enhance soil infiltration. This chapter broadly categorises terraces into common (or normal) terraces and bench terraces. The chapter presents the design of common (or normal) terraces in terms of their spacing, grades, length, and cross-section. The design of bench terraces includes spacing, bench width, cross-section, and length, besides the volume of cut and fill or earthwork and area lost under them. The chapter also contains the terrace system planning, including its location, layout, and maintenance. The design procedures are demonstrated through solved examples.

4.1 Definition

The English word *terrace* has its root in the Latin (L) *terra*, which means earth or ground. *Terra* was adopted as *terrassa* by *Old Occitan* (a French Dialect), which was later adopted as *terrasse* by French.

Terraces are an ancient erosion control practice that consists of earth embankments constructed across the steep slopes. The earth embankments intercept surface runoff and either divert it at a non-erosive velocity to a safe outlet or store it to enhance soil infiltration.

Terracing is the most widely practised soil erosion control measure around the world. The earliest record of terracing dates back to the Iron Age when it was practised in the Middle East, sub-Mediterranean, and Alpine regions. Subsequently, in the late fourteenth century, terracing was adopted in the Mediterranean region. Terracing was also practised in Asia (China, Indonesia, Palestine, Philippines, Thailand, Vietnam, and Yemen) about 5000 years ago; in Americas (Guatemala, Mexico and Peru) for about 2000 years; and in Africa (Tanzania) for about 300–500 years. In India, terracing was included as a measure in the Land Preservation Act of Punjab in 1900 for protecting land from the menace of *Cho* (mountain torrents).

4.2 Functions

The primary functions of the terraces are:

1. To decrease the soil erosion by cutting down the slope length and carrying the runoff at a non-erosive velocity to a safe outlet
2. To conserve moisture by increasing soil infiltration
3. To facilitate cultivation and tillage operation on steep slopes

Besides the above primary functions, terraces also help in improving soil fertility and land productivity by improving the soil nutrient status, mainly when used for irrigated agriculture. Terraces ensure food security by enhancing crop yield in the hilly regions. Terraces play a vital role in afforestation or reforestation programmes and add to biodiversity. Terraces also augment recreational options by providing appealing landscapes.

4.3 Classification

Terraces are broadly classified into two categories:

1. Common or normal terraces
2. Bench terraces

The classification of terraces is given below (Fig. 4.1).

Common (or normal) terraces consist of a ridge and a channel. The channel may be level or graded. These terraces are preferred in areas having less than 20% of the land slope. Common terraces may further be classified according to their base width as *narrow-*, *medium-*, and *broad-base terraces* (Fig. 4.2). For narrow-, medium-, and broad-base terraces, the earthwork is carried out within 3.0 m, 3.0 to 6.0 m, and 6.0 to 12.0 m, respectively.

Fig. 4.1 Classification of terraces

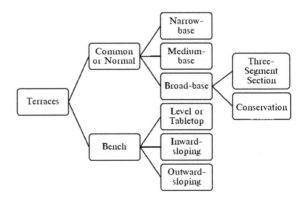

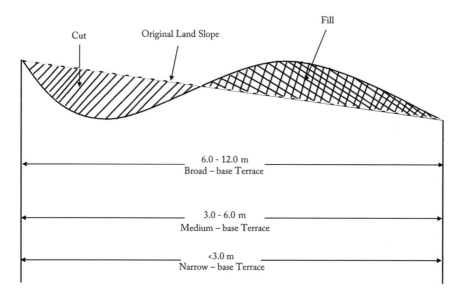

Fig. 4.2 Common terrace with different variants according to the base sizes (After FAO 2000; Reproduced with permission from FAO)

Narrow-base terraces are preferred in high rainfall areas having 10 to 16% land slope and soils with low to moderate permeability. The front and back slopes are 2:1 or flatter and kept under permanent grass cover. The terrace channel may have a triangular, parabolic, or trapezoidal shape with the water depth maintained at the existing ground line (Fig. 4.3). A freeboard of 20% of the depth of flow or 0.10 m may be provided to avoid the overtopping.

Sometimes a back-cut slope may be provided to supplement the earth fill for the terrace ridge. The provision of access roads is needed for entering the terrace interval as farm equipment cannot cross the terrace ridge due to steep slopes on the front and the back.

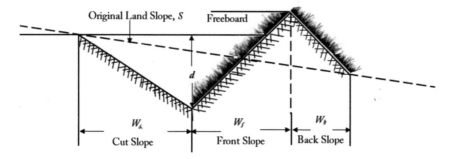

Fig. 4.3 Narrow-base terrace cross-section with nomenclature

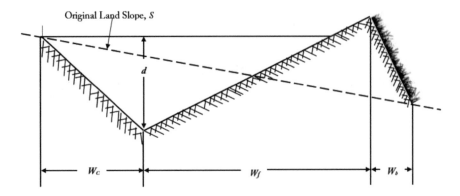

Fig. 4.4 Steep- or grassed-back slope terrace

Medium-base terraces are a combination of narrow-base and broad-base terraces. The medium-base terraces are preferred in medium rainfall areas having 6 to 12% land slope and soils with moderate permeability. The back slope is 2:1, and the front slope is 5:1 or flatter. The back slope is not cultivated and is kept under permanent grass cover. The medium-base terraces are also named as *steep-back slope terraces* or *grassed-back slope terraces* (Fig. 4.4). Similar to the narrow-base terraces, the terrace channel may be triangular, parabolic, or trapezoidal shaped with the maximum water depth limited to the existing ground surface. A back-cut slope may be provided to supplement the earth fill for the terrace ridge. The farm equipment cannot cross the terrace ridge due to a steep-back slope.

Broad-base terraces are the old-style terraces that have been used for many years. These terraces are preferred in low rainfall areas having less than 8% land slope and soils with high permeability. The broad-base terraces may have two variations: the *three-segment section broad-base* and the *conservation bench broad-base*.

All slopes (the cut, front, and back slopes) of the *three-segment section broad-base terrace* are flatter than 5:1 (Fig. 5a). The flat slopes allow the operation of field equipment over the whole terrace cross-section. Thus, the entire terrace is cultivated. The width of the equipment governs the length of the side slopes.

The *conservation bench* variation of the broad-base terrace is provided with a wide, flat channel to retain runoff and enhance the infiltration (Fig. 5b). The conservation bench is preferred in drier areas to enhance the water availability for crop production.

Bench terraces may further be classified according to the slope as *level or tabletop, inward-sloping,* or *outward-sloping* terraces. The level or tabletop bench terraces are favoured in areas having medium rainfall and deep soils with high permeability (Fig. 4.6). The primary function of the level or tabletop bench terraces is moisture conservation.

The inward-sloping bench terraces are preferred in high rainfall areas having soils with low permeability (Fig. 7a). The primary function of the inward-sloping bench

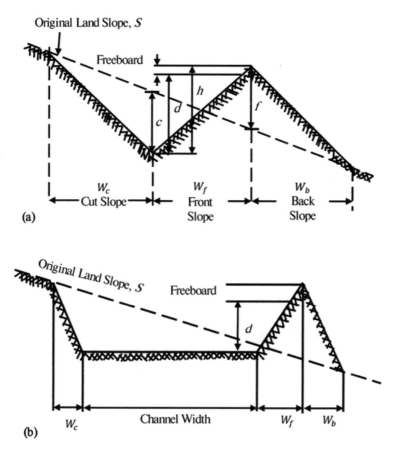

Fig. 4.5 Broad-base terrace cross-sections **a** three-segment section broad-base, **b** conservation bench broad-base

terraces is to prevent the ponding of water. These terraces are provided with a suitable drain at the inward-end for the safe disposal of the excess runoff. These terraces are useful for crops that are susceptible to waterlogging. The outward-sloping bench terraces are preferred in areas having low rainfall and medium deep soils with high permeability (Fig. 7b). These terraces are usually provided with a shoulder bund at the outward-end to prevent damages to the downstream terraces.

4.4 Design of Common (or Normal) Terraces

The design of a common or normal terrace system includes specifying the terrace spacing and length with a suitable cross-section and channel grade. The conservation bench broad-base terraces are designed to conserve the runoff, while graded terraces

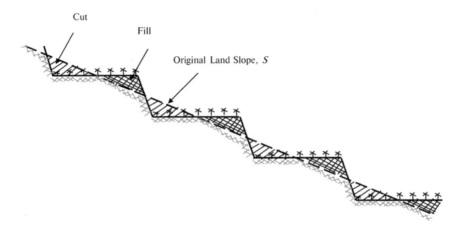

Fig. 4.6 Typical level or tabletop bench terrace

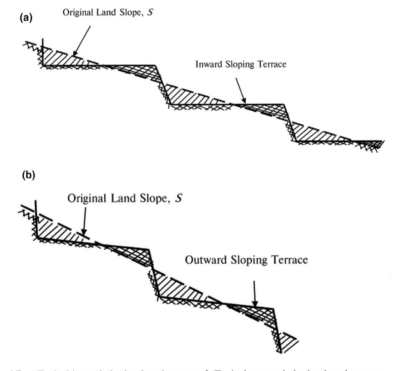

Fig. 4.7 **a** Typical inward-sloping bench terrace, **b** Typical outward-sloping bench terrace

are designed to dispose of runoff. A 24-h-duration storm having 10 year frequency is usually adopted for designing terraces. Land slope, climatic conditions, soil characteristics like soil type and depth, and soil and crop management practices are the common factors that govern the design of the terraces.

4.4.1 Terrace Spacing

Terrace spacing is expressed in terms of vertical or horizontal distance between two adjacent terraces, measured at corresponding points of the neighbouring terraces (Fig. 4.8). The vertical and horizontal distances are referred to as the *vertical interval* (*VI*) and the *horizontal interval* (*HI*).

Terrace spacing may be estimated using the Revised Universal Soil Loss Equation (RUSLE) or the Universal Soil Loss Equation (USLE), provided the values of the various factors used are available. The USLE (Eq. (3.1), Chap. 3) or RUSLE is used to estimate the topographic factor, *LS*, or the length of the slope factor, *L*, for a target mean annual or seasonal soil loss. Subsequently, the horizontal interval (*HI*) is estimated in terms of the slope length.

Alternatively, the terrace spacing may be estimated using the empirical *VI* equation, first proposed by Ramser (1931). The equation is given as,

$$VI = XS + Y \qquad (4.1)$$

where VI = vertical interval; X = a constant that depends on the geographical location; S = average land slope in %; and Y = a constant that depends on the soil characteristics and crop conditions.

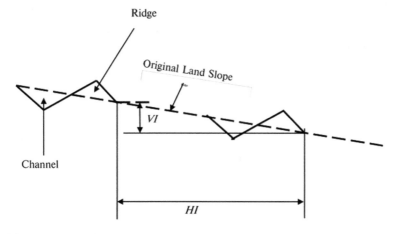

Fig. 4.8 Vertical Interval (*VI*) and Horizontal Interval (*HI*)

The value of X varies from 0.4 to 0.8 in the FPS system for different locations in the USA (0.12–0.24 in SI units). The values of Y originally recommended by Ramser have been revised from time to time, and the current values, published in the Engineering Field Handbook (NRCS 2011), are:

- 1.0 foot (0.3 m) for soils having high erodibility and below-average infiltration rates and cropping systems providing sparse cover during high rainfall events
- 4.0 feet (1.2 m) for soils having erosion-resistant characteristics and cropping systems providing equal to or more than 30% cover at the time of planting,
- 2.5 feet (0.75 m) when one of the factors, i.e., either soil erodibility or crop cover, is favourable and the other is unfavourable
- User may choose other values between 1.0 and 4.0 feet (0.3–1.2 m) based on soil and crop cover conditions.

A version of the VI equation, credited to MP Cox, a researcher with the United States Agency for International Development (USAID), is as follows:

$$VI = (xS + y) \times 0.3 \tag{4.2}$$

where VI = vertical interval, m; x = rainfall factor; and y = infiltration and crop cover factor.

Tables 4.1 and 4.2 present the values of x and y, respectively.

FAO has also suggested the following VI equations for common terraces for specific regions and soils (FAO 1986):

For erodible soils in humid regions,

$$VI = \frac{(S + 4)}{10} \tag{4.3}$$

For normal soils in semi-arid regions,

Table 4.1 Rainfall factor (x)

Rainfall condition	Annual rainfall, mm	x
Scanty	<640	0.8
Moderate	640–900	0.6
Heavy	>900	0.4

Table 4.2 Infiltration and crop cover factor (y)

Intake rate	Crop cover during the critical period	y
Below average	Low coverage	1.0
Average or above average	Good coverage	2.0
One of the above two factors is favourable, and the other is unfavourable		1.5

$$VI = \frac{(S + 6)}{10} \qquad (4.4)$$

where VI is in m and S is in %.

The horizontal interval (HI) may be estimated using the following equation.

$$HI = VI \times (100/S) \qquad (4.5)$$

The computed values of VI and HI may be varied up to 25% to account for varying rainfall, soil, and crop cover conditions.

Example 4.1 In an area, the average annual soil loss is 19 t ha^{-1}. The values of rainfall erosivity factor, R, and soil erodibility factor, K, are 200 MJ mm ha^{-1} h^{-1} and 0.34 t ha h ha^{-1} MJ^{-1} mm^{-1}, respectively. The other USLE factors are: crop management factor, $C = 0.3$; conservation practice factor, $P = 0.5$; slope length factor, $L = 60$ m; and slope steepness factor, $S = 6\%$. What should be the length of the slope and the vertical interval of the terrace if the soil loss is to be reduced to 10 t ha^{-1}?

Solution

Given, $A = 19$ t ha^{-1}, $R = 200$ MJ mm ha^{-1} h^{-1}, $K = 0.34$ t ha h ha^{-1} MJ^{-1} mm^{-1}, $C = 0.3$, $L = 60$ m, $S = 6\%$, and $P = 0.5$.

Using USLE, $A = R\,K\,LS\,C\,P$

$$LS = \frac{A}{RKCP} = \frac{19}{(200 \times 0.34 \times 0.3 \times 0.5)} = 1.86$$

From Eq. (3.10) (Chap. 3),

$$LS = \left(\frac{L}{22.13}\right)^{m} 0.065 + 0.0456S + 0.006541S^2$$

Putting the values of LS, L and S in the equation, we get $m = 1.1786$.
Now for limiting the soil loss to 10 t ha^{-1},

$$LS = \frac{A}{RKCP} = \frac{10}{(200 \times 0.34 \times 0.3 \times 0.5)} = 0.98$$

Now, putting $LS = 0.98$, $S = 6\%$, and m $= 1.1786$ in Eq. (4.10), we get $L = 34.8$ m. The horizontal interval, HI, is equal to L.

Thus, terrace vertical interval,

$$VI = \left(\frac{S}{100}\right) \times L = \left(\frac{6}{100}\right) \times 34.8 = 2.1 \text{ m}$$

Answer: Length of slope $= 34.8$ m; vertical interval $= 2.1$ m.

4.4.2 Terrace Grades

The graded terraces are designed to collect and transport the runoff at non-erosive velocities to a safe outlet. The terrace channel should have sufficient grade to avoid prolonged ponding, thus preventing the damages to crops or delay in the farming operations. The graded terraces may be constructed with uniform or variable grades.

The uniform-graded terraces are usually short and have constant slope throughout their length. Though the grade may range from 0.1 to 0.6%, a grade lower than 0.5% is usually recommended. The steeper grades, however, may be opted if the terraces are short and the soil is impervious.

The variable-graded terraces are constructed with increasing grades towards the outlet, i.e., the grade is kept minimum in the upstream end and maximum at the outlet end. The enhanced grade towards the outlet increases the channel capacity and helps in accommodating the continually augmented discharge from the growing area contributing to the channel. The lower grade in the upper portion results in increased absorption of runoff.

While designing the terrace channel, the flow velocities are kept lower than 0.46 m s^{-1} for soils with high erodibility and 0.61 m s^{-1} for other soils. The Manning equation is usually adopted for the channel design with the roughness coefficient value of 0.035 or less.

The minimum grade recommended for the variable-grade terraces is 0% for highly permeable soils and 0.2% for soils having low permeability. The maximum grade may vary from 0.3% to 2.0%. However, for terraces having a length of 150 m or more, the maximum grade is limited in between 0.3% and 0.5%. The maximum permissible grade may be determined using the following equation (Beasley et al. 1984):

$$S_T = \frac{213.7 V^{S/3} \left(\frac{Z^2+1}{Z}\right)^{2/3}}{(W_s L_T)^{2/3}} \tag{4.6}$$

where S_T = maximum permissible terrace grade in %; Z = channel side slope; Ws = spacing between terraces which may be taken equal to HI; L_T = distance of the location from the upstream end (measured along the length of the terrace towards the outlet); and V = maximum velocity (0.61 m s^{-1}).

The conservation bench broad-base terraces are designed to conserve the runoff. Hence, these are constructed with a flat channel bed.

4.4.3 Terrace Length

The length of the terrace channel depends on the field size and shape, outlet position, flow rate, soil infiltration, and channel capacity. The length should be such that the flow velocity remains non-erosive. The length could be longer for permeable soils as compared to the impermeable soils.

The graded terrace length should be within 500 m in soils with high permeability and 600 m in soils with low permeability to avoid overtopping during high rainfall events. In cases where it is necessary to construct a terrace longer than these limits, the flow may be divided at the midway of the terrace length and runoff may be drained through outlets at both ends. Alternatively, the terrace length may be divided into six sloping sections having successive gradients of 0, 0.1, 0.2, 0.3, 0.4, and 0.5%.

For conservation bench broad-base terraces, the channel length governs the volume of water stored. The terrace channel length should not exceed 1000 m for reducing the potential risk of failure. These terraces, however, could be more extended, provided the channel has small dams at intermittent locations to protect it against failure.

4.4.4 Terrace Cross-Section

The terrace cross-section includes specification of channel shape and size to have adequate flow capacity and side slopes to facilitate farming operations. The factors affecting the terrace cross-section are the land slope and farm implement used.

While designing a broad-base terrace, the cross-section may include a triangular channel, as shown in Fig. 5a. The width of the front slope (W_f) is usually kept the same as the machinery width. A storm having a 10-year return period is used to find the peak rate of runoff and the depth of flow. A freeboard of 20% of the depth of flow or 0.10 m is then considered to get the ridge height, h. The cut slope and back slope are usually kept flatter than 5:1 to facilitate the farm operations.

With equal side slope widths $[W_c$ (cut slope) $= W_f$ (front slope) $= W_b$ (back slope) $= W$ (width of side slope or width of the terrace)], cut (c) and fill (f) are:

$$c + f = h + S \times W \tag{4.7}$$

where $h =$ height of the ridge (equal to the sum of the depth of the channel and the freeboard), m; $S =$ original land slope, fraction.

The cut is equal to the fill when a balanced cross-section is used. The terrace cross-section is usually kept the same throughout its length, though it leads to over-design of the upstream channel.

In case of the conservation bench broad-base terraces, a trapezoidal channel cross-section is usually preferred. The peak runoff volume for a storm having a return period of 10 years is used for determining the flow depth, d, and the channel width. The side slopes are chosen following the recommendations for the three-segment broad-base terraces.

In case of the narrow-base terraces, the front slope and the back slope are usually 2:1 or flatter and kept under permanent grass cover. A triangular channel is usually provided to handle the peak runoff rate for 10-year return period storm.

In case of steep-back slope or grassed-back slope terraces, the back slope is 2:1 and kept under permanent grass cover. The width of the front slope (W_f) is kept the same as the machinery width. The cut slope is usually kept flatter than 5:1 to facilitate the farm operations. A triangular channel is provided to handle the peak runoff rate for 10-year return period storm.

Equations similar to Eq. (4.7) have been derived for grassed-back slope terraces by Larson (1969) (Ghidey et al. 1992).

Example 4.2 For a broad-base terrace made on 8% land slope, the ridge height is 0.3 m and the terrace width is 6.0 m. Assuming a balanced cross-section, determine the heights of cut and fill. What are the front slope and back slope ratios?

Solution

Given: Land slope, $S = 8\%$; ridge height, $h = 0.3$ m; and terrace width, $W = 6.0$ m.

Since the cross-section is balanced, the cut height and the fill height will be the same, i.e., $c = f$.

From Eq. (4.7), $c + f = h + S \times W$.

Substituting the known values of W, S, and h for $c = f$,

$$c = f = \frac{(h + S \times W)}{2} = \frac{(0.3 + 0.08 \times 6)}{2} = 0.39 \text{ m}$$

From Fig. 5a, it is evident that for the front slope, the change in elevation over W is equal to h.

Hence, front slope, $S_f = \frac{W}{h} = \frac{6}{0.3} = 20$; hence, front slope ratio $= 20{:}1$

For the back slope, the change in elevation over W is equal to $(f + W \times S)$.

Hence, back slope, $S_b = \frac{W}{(f + W \times S)} = \frac{6}{(0.39 + 6 \times 0.08)} = 6.90$; hence, back slope ratio $= 6.90{:}1$.

Answers: Height of cut $=$ Height of fill $= 0.39$ m; front slope ratio $= 20{:}1$: Back slope ratio $= 6.9{:}1$.

Example 4.3 Design an 800 m long conservation bench broad-base terrace for a 4% land slope. The extreme rainfall having 10-year frequency is 0.18 m. The soil absorbs 40% of the rainfall in the field. The region has a moderate rainfall condition. The soil has above average infiltration rate, and the cover condition during the critical period is good.

Solution

Given: Length of the terrace, $l = 800$ m; land slope, $S = 4\%$; maximum expected rainfall having 10-year recurrence interval $= 0.18$ m; infiltration capacity of the soil $= 40\%$ of the rainfall; rainfall condition $=$ moderate; soil infiltration rate $=$ above average; and cover condition during the critical period $=$ good.

Step 1: We need to estimate the vertical interval, VI, and based on the information available, we may use Eq. (4.2),

$$VI = (xS + y) \times 0.3$$

From Tables 4.1 and 4.2, x = 0.6 and y = 2.0.

$$\text{Hence,} \quad VI = (0.6 \times 4 + 2) \times 0.3 = 1.32 \text{ m}$$

Using Eq. (4.5), Horizontal Interval, $HI = VI \times (100/S) = 1.32 \times (100/4) =$ 33 m

Step 2: We need to determine the maximum runoff volume to be stored.

Since the maximum expected rainfall having 10-year recurrence interval is 0.18 m, and 40% of the rainfall gets infiltrated, the runoff potential is 60% of the rainfall.

The conservation bench broad-base terrace is designed to handle the maximum runoff potential.

Hence, the maximum runoff to be stored = 60% of rainfall = $0.6 \times 0.18 =$ 0.108 m.

Step 3: Area required to store the maximum runoff from the terrace,
$A =$ Maximum runoff depth $\times HI = 0.108 \times 33 = 3.564 \text{ m}^2$

Step 4: Now, we need to choose the channel shape and determine its dimensions

Let us consider a trapezoidal section with cut slope, front slope, and back slope of 6:1, 8:1, and 8:1, respectively (Fig. 4.9). Hence, the cross-section area of channel,

$$A = a_1 + a_2 + a_3 = \frac{1}{2} \times 6d \times d + bd + \frac{1}{2} \times 8d \times d = bd + 7d^2$$

Since the storage area required = 3.564 m², the cross-section area of the channel should be equal to the required storage area.

Assuming b = 6.0 m, $bd + 7d^2 = 3.564$ or $7d^2 + 6d = 3.564$.

Since it is a quadratic equation, its solution is

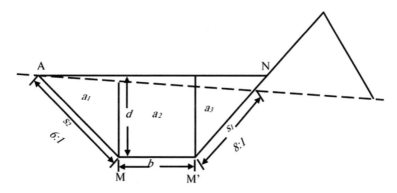

Fig. 4.9 Broad-base terrace with trapezoidal channel section

$$d = \frac{-6 \mp \sqrt{(6)^2 - (4 \times 7 \times (-3.564))}}{2 \times 7} = \frac{-6 \mp 11.65}{14} = 0.40 \text{ or } -1.26 \text{ m}$$

Hence, depth of flow, $d = 0.40$ m.
Step 5: Assuming a freeboard = 20% of the flow depth,
Ridge height = $1.2d = 0.48$ m.
The front slope width = $8 \times 0.48 = 3.84$ m.

Answers: Terrace vertical interval = 1.32 m; terrace horizontal interval = 33 m; flow depth = 0.40 m; channel bottom width = 6.0 m; terrace slope ratios, $S_c = 6{:}1$, $S_f = 8{:}1$ and $S_b = 8{:}1$; terrace ridge height = 0.48 m; and terrace front slope width = 3.84 m.

Example 4.4 Design a 300 m long variable-graded terrace for 6% land slope. The flow velocity in the terrace channel should be within 0.60 m s^{-1}. The intensity of 1-h rainfall with 10-year frequency is 100 mm h^{-1}. The region has a heavy rainfall condition. The soil has below-average infiltration rate, and the cover condition during the critical period is good.

Solution

Given: Length of the variable-graded terrace, $l = 300$ m; land slope, $S = 6\%$; maximum permissible velocity of water = 0.60 m s^{-1}; intensity of 1-h, 10-year rainfall = 100 mm h^{-1}; heavy rainfall condition, soil with below-average infiltration rate, and good cover condition during the critical period.
Step 1: We need to estimate the vertical interval, VI, and based on the information available, we may use Eq. (4.2),

$$VI = (xS + y) \times 0.3$$

From Tables 4.1 and 4.2, rainfall factor, $x = 0.4$, and infiltration and crop cover factor, $y = 1.5$.

$$\text{Hence, } VI = (0.4 \times 6 + 1.5) \times 0.3 = 1.2 \text{ m}$$

Using Eq. (4.5), Horizontal Interval,

$$HI = VI \times (100/S) = 1.2 \times (100/6) = 20 \text{ m}$$

Step 2: Let the grades of the variable-graded terrace be 0.5, 0.4, 0.3, and 0.2 for each quarter length from the outlet, respectively.
Step 3. Next, we need to determine the peak rate of runoff, and we may use the rational method for the purpose (Eq. (3.19), Chap. 3),

$$Q = ciA$$

Assuming runoff coefficient, $c = 0.6$

The drainage area, A, is the terrace area, i.e., $A =$ length of the terrace (l) $\times HI$.

$$\text{Hence, } A = 300 \times 20 = 6000 \text{ m}^2$$

Now we need to determine the time of concentration, and subsequently, the rainfall intensity, i in m s^{-1}, for the same duration.

Determining time of concentration, T_c, using Kirpich formula (Eq. (3.20), Chap. 3),

$$T_c = 0.0195 L^{0.77} S^{-0.385}$$

Maximum length of flow for graded terrace,

$$L = \text{length of the terrace } (l) + HI = 300 + 20 = 320 \text{ m}$$

$$\text{The slope along the flow path, } S = \frac{H}{L}$$

where H is the elevation difference between the outlet and the farthest point, i.e., the sum of the elevation difference between adjacent terraces and the elevation difference over the terrace length.

Elevation difference between adjacent terraces, i.e., over $HI = 0.06 \times 20 = 1.2$ m.

Elevation difference over the channel length,

0.5% slope in lower quarter (75 m) $= 0.005 \times 75 = 0.375$ m.

0.4 % Slope in Third Quarter (75 m) $= 0.004 \times 75 = 0.3$ m.

0.3% slope in second quarter (75 m) $= 0.003 \times 75 = 0.225$ m.

0.2% slope in upper quarter (75 m) $= 0.002 \times 75 = 0.15$ m.

Total elevation difference, $H = 1.2 + 0.375 + 0.3 + 0.225 + 0.15 = 2.25$ m.

$$S = \frac{H}{L} = \frac{2.25}{320} = 0.007$$

Hence, the time of concentration $T_c = 0.0195 \, (320)^{0.77}(0.007)^{-0.385} = 11.2$ min.

Now, we need to obtain the rainfall intensity (equivalent to the given 1-h intensity of 1-h, 10-year rainfall of 100 mm h^{-1}) for a storm having a duration of 11.2 min. Since we do not have the storm data for the region, we may use the standard monograph (Fig. 4.10).

From the figure, the 1-h intensity of 100 mm h^{-1} will be converted approximately into 11.2 min intensity of 168 mm h^{-1}. Hence, $i = 4.7 \times 10^{-5}$ m s^{-1}.

Putting the values of c, i, and A in the rational formula,

$$Q = ci A = 0.6 \times \left(4.7 \times 10^{-5}\right) \times 6000 = 0.17 \, m^3 s^{-1}$$

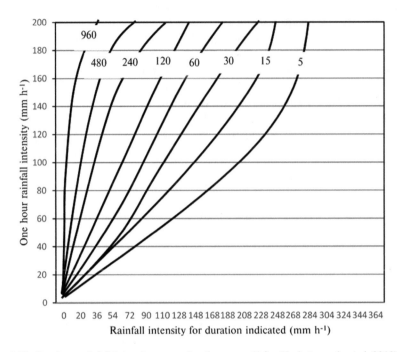

Fig. 4.10 One-hour rainfall intensity versus duration curve (After Venkateswarlu et al. 2013)

Step 4: Assuming a steep-back slope or grassed-back slope terrace, with the cut slope, front slope, and back slope of 6:1, 8:1, and 2:1, i.e., $S_c = 6:1$, $S_f = 8:1$, and $S_b = 2:1$. Also, assuming a triangular channel cross-section (Fig. 4.11).

$$\text{Cross - section area of channel, } A = a_1 + a_2 = \frac{1}{2} \times 6d \times d + \frac{1}{2} \times 8d \times d = 7d^2$$

$$\text{Assuming } d = 0.30 \text{ m}, \ A = 7 \times 0.30 \times 0.30 = 0.63 \text{ m}^2$$

$$\text{Wetted perimeter, } P = S_1 + S_2 = \sqrt{\left(d^2 + (8d)^2\right)} + \sqrt{\left(d^2 + (6d)^2\right)} = 4.25 \text{ m}$$

$$\text{Hydraulic radius, } R = A/P = 0.63/4.25 = 0.15 \text{ m}$$

$$\text{Flow velocity from Manning equation, } V = \frac{1}{n} R^{2/3} S^{1/2}$$

$$= \frac{1}{0.035} (0.15)^{2/3} (0.0026)^{1/2}$$

$$= 0.41 \text{ m s}^{-1}$$

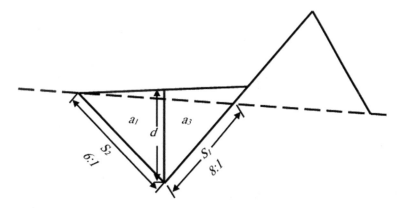

Fig. 4.11 Steep-back slope terrace cross-section (Example 4.4)

Here, Manning n is assumed as 0.035. Also, the average slope along the terrace length = average of slopes determined for the four quarters = $(0.375 + 0.3 + 0.225 + 0.15)/4 = 0.26\%$.

The flow velocity determined is less than 0.61 m s^{-1} (soil being less erodible) and is non-erosive and, hence, acceptable.

The designed flow capacity of the channel, $Q = AV = 0.63 \times 0.41 = 0.26$ m^3 s^{-1}

The designed channel has sufficient capacity, i.e., higher than the estimated peak flow rate of 0.17 m^3 s^{-1}.

Step 6: Assuming a freeboard = 20% of the flow depth,
Ridge height = $d + 0.2\, d = 1.2 \times 0.30 = 0.36$ m

Answers: Terrace vertical interval = 1.2 m; terrace horizontal interval = 20 m; terrace grades = $0.5, 0.4, 0.3$, and 0.2 for each quarter length from outlet; flow depth = 0.30 m; terrace channel flow capacity = 0.26 m^3 s^{-1}; terrace slope ratios, $S_c = 6:1$, $S_f = 8:1$, and $S_b = 2:1$; terrace ridge height = 0.36 m.

4.5 Design of Bench Terraces

The design of a *bench terrace system* involves the specification of the depth of cut or the terrace spacing, grade and length of the terraces, and a suitable cross-section. The selection of the type of bench terrace depends on the rainfall and soil conditions. The *level or tabletop bench terraces* are preferred in areas having deep soils with high permeability and medium rainfall. The *inward-sloping bench terraces* are useful in areas having soils with low permeability and high rainfall. In contrast, the *outward-sloping bench terraces* are useful in areas having medium deep soils with high permeability and low rainfall. The factors governing the design of bench terraces are the same as in the case of the common terraces.

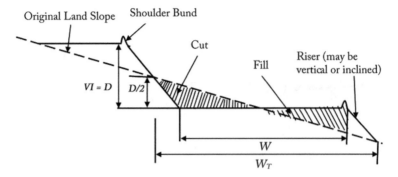

Fig. 4.12 Bench terrace with nomenclature

4.5.1   Terrace Spacing

Like common terraces, the bench terrace spacing is also given in terms of the vertical interval (*VI*), which represents the vertical distance between two successive bench terraces (Fig. 4.12). The *VI* is twice the *depth of cut*. The land slope, the soil depth, and the bench width govern the depth of cut. The prevalent local practices and practical experience guide the decision.

4.5.2   Bench Width

The bench width (*W*) is governed by the tillage implements used for the dominant cropping system in the region. The soil depth is yet another factor that governs the bench width. Wide benches require deep soils and involve higher construction costs. Recommended widths are 2.5–5.0 m for manual cultivation and 3.5–8.0 m for mechanised cultivation.

Once *W* is decided, the depth of cut, *D*/2, and the vertical interval, *VI* (= *D*), can be calculated.

Vertical Interval when the riser is vertical

Considering triangle NMA in Fig. 13a, NM = *S* and MA = 100 (as the land slope is *S* %).

Now considering ABC and NMA as similar triangles,

$$\frac{(D/2)}{(W/2)} = \frac{S}{100} \text{ or } D = \frac{WS}{100}$$

$$\text{Or, } VI = D = \frac{WS}{100} \tag{4.8}$$

(a)

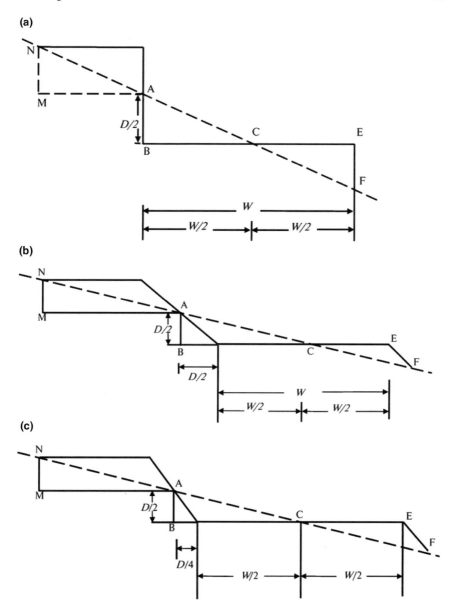

(b)

(c)

Fig. 4.13 **a** Bench terrace with a vertical riser, **b** Bench terrace with a batter slope of 1:1, **c** Bench terrace with a batter slope of ½:1

The vertical riser (or vertical cut), however, is used only if the soil is very stable or when the depth of cut is small (up to 1.0 m). Usually, for the stability of the fill or the embankment, a batter or riser slope of ½:1 or 1:1 is provided. It must, however, be kept in mind that flatter the batter slope, the larger will be the area lost due to bench terracing.

The relationship between W and D for the batter slope of ½:1 or 1:1 can also be found out as follows:

Vertical interval when the batter slope is 1:1

Using the equivalent triangles ABC and NMA (Fig. 13b),

$$\frac{(D/2)}{\left(\frac{W}{2} + \frac{D}{2}\right)} = \frac{S}{100} \quad \text{or, } D = \frac{WS}{(100 - S)}$$

$$\text{Or, } VI = D = \frac{WS}{(100 - S)} \tag{4.9}$$

Vertical interval when the batter slope is 1/2:1

When the batter slope is 1/2:1 (Fig. 13c),

$$\frac{(D/2)}{\left(\frac{W}{2} + \frac{D}{4}\right)} = \frac{S}{100} \quad \text{or, } D = \frac{2WS}{(200 - S)}$$

$$\text{Or, } VI = D = \frac{2WS}{(200 - S)} \tag{4.10}$$

General form of vertical interval equation for different batter slopes

A general form of the vertical interval equation may be derived and written as,

$$VI = D = \frac{WS}{(100 - SS_r)} \tag{4.11}$$

where S_r = batter or riser slope.

By putting the value of the batter slope as 0, 1, or ½ in Eq. (4.11), Eqs. (4.8)–(4.10) could be obtained.

4.5.3 Terrace Cross-Section

The terrace cross-section includes dimensions of the shoulder bund, the inward slope of the terrace, and dimensions of the drainage channel in case of bench terraces sloping inward and outward slope in case of bench terraces sloping outward.

The shoulder bund dimensions depend on the type of the bench terrace. In case of terraces sloping inward, the size of shoulder bund is kept nominal, while in the terraces sloping outward and level top, the shoulder bund cross-section is broader for holding the rainwater on the terrace top.

The typical dimension of shoulder bund used in bench terraces sloping outward is given as under:

$$\text{Top width} = 0.30 \text{ m; bottom width} = 1.20 \text{ m; height} = 0.45 \text{ m}$$

The inward and outward slopes of bench terraces depend on the type of soil and the average rainfall in the area. Usually, the inward slope varies from 2.0 to 10% and the outward slope from 2.0 to 8½%. The inward-sloping terraces include a drainage channel at the inner edge of the terrace to dispose of the excess runoff. The channel dimensions are calculated knowing the excess runoff magnitude and using the Manning equation to ensure non-erosive flow velocity in the channel.

4.5.4 Terrace Length

The following equation can calculate the linear length of terraces per hectare:

$$L = \frac{10000}{W_T} \tag{4.12}$$

where L = linear length of terraces per hectare, m; and W_T = terrace width = (W $+ D \times S_r$).

Thus, when the riser is vertical, the length of the terrace per hectare will be $\left(\frac{10000}{W}\right)$, and when the batter slope is 1:1, the length of the terrace per hectare will be $\left(\frac{10000}{(W+D)}\right)$.

4.5.5 Net Cultivated Area

The net cultivated area is the area of the flat strips (benched) which are used for cultivation. It is calculated by multiplying the terrace length (L) per hectare with the width of the bench (W) and expressed by the following equation:

$$A = L \times W = \frac{10,000W}{W_T} \tag{4.13}$$

When calculating the net area, the shoulder bund width should be subtracted.

4.5.6 Volume of Cut and Fill or Earthwork

The earthwork involved in the construction of bench terraces is needed to compute the cost of earthwork. It may be noted that the volume of cut and fill is kept the same to minimise the cost of construction.

The computation of earthwork involves finding the volume of soil that is cut, which may be obtained by finding the area of the cut section (which is a triangle) and multiplying it by the length of the terrace.

$$\text{Area of cut, } A_c = \frac{1}{2}\left(\frac{D}{2}\right)\left(\frac{W}{2}\right) = \frac{WD}{8} \tag{4.14}$$

Putting D from Eq. (4.11) in Eq. (4.14),

$$\text{Area of cut, } A_c = \frac{WD}{8} = \frac{W}{8}\frac{WS}{(100 - SS_r)} \tag{4.15}$$

The volume of the earthwork, $E_w = A_c \times L$.
Substituting A_c and L from Eqs. (4.15) and (4.12), respectively,

$$\text{The volume of earthwork, } E_w = A_c \times L$$
$$= \frac{W}{8}\frac{WS}{100 - SS_r} \times \frac{10,000}{W_T} \tag{4.16}$$

Equation (4.16) gives the generalised form of the equation for the volume of earthwork in m^3 ha^{-1}.

For a bench terrace with a vertical riser, $S_r = 0$ (and, thus, $W_T = W$). Using Eq. (4.16), the volume of earthwork,

$$E_w = \frac{W}{8}\frac{WS}{(100 - SS_r)} \times \frac{10000}{(W)} = \frac{W}{8}\frac{WS}{(100)} \times \frac{10000}{W}$$

Substituting $\frac{WS}{100} = D$ (from Eq. (4.8)), $E_w = \frac{W}{8} \times D \times \frac{10000}{W} = 1250DE_w = \frac{W}{8} \times D \times \frac{10000}{W} = 1250D$

Thus, for a bench terrace with a vertical riser,

$$E_w = 1250D \text{ m}^3 \text{ ha}^{-1} \tag{4.17}$$

Equation (4.16) may also be used to calculate the volume of earthwork for inward-sloping or outward-sloping bench terraces as follows:

In case of inward-sloping bench terraces, S is replaced by $(S + s)$, and in case of outward-sloping bench terraces, S is replaced by $(S—s)$, where $s =$ inward or outward slope of the bench, %.

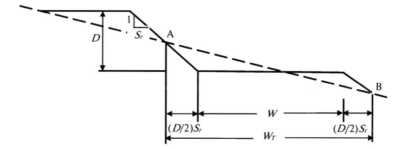

Fig. 4.14 Typical bench terrace with an inclined riser

4.5.7 Area Lost Under Bench Terraces

The cultivable area lost due to bench terraces can be calculated as follows:

Let D be the vertical interval of the benches to be laid out on the land with a slope of S % along AB, and W_T $[= (W + (D \times S_r))]$ be the projected length of AB on the horizontal plane (Fig. 4.14).

From Fig. 4.14,

$$AB = \sqrt{W_T^2 + D^2}$$

Using Newton's generalised binomial theorem,

$$AB = \left(W_T^2 + D^2\right)^{1/2} = \left(W_T + \frac{D^2}{2W_T} - \frac{D^4}{8W_T^3} + \ldots\right)$$

Neglecting higher-level terms,

$$AB = \left(W_T + \frac{D^2}{2W_T}\right)$$

But, $W_T = \dfrac{100D}{S}$ (as the land slope is $S\%$)

Hence, $AB = \left(\dfrac{100D}{S} + \dfrac{D^2}{2} \times \dfrac{S}{100D}\right) = \left(\dfrac{100D}{S} + \dfrac{DS}{200}\right)$

$$AB = \left(\frac{100D}{S} + \frac{DS}{200}\right) \tag{4.18}$$

Terrace width available for cultivation, W, is

$$W = W_T - (D \times S_r)$$

Substituting, $W_T = \frac{100D}{S}$,

$$W = \frac{100D}{S} - DS_r$$

$$W = \frac{D(100 - SS_r)}{S} \tag{4.19}$$

Width not available for cultivation after terracing $= AB-W$
Substituting AB and W from Eqs. (4.18) and (4.19),
Width not available for cultivation after terracing is given as,

$$AB - W = \frac{100D}{S} + \frac{DS}{200} - \frac{100D}{S} + DS_r = \left(\frac{DS}{200} + DS_r\right) \tag{4.20}$$

Width lost expressed as % of original inclined width AB [Eq. (4.20)/Eq. (4.18)]

$$= \frac{\frac{DS}{200} + DS_r}{\frac{100D}{S} + \frac{DS}{200}} \times 100 = \frac{\frac{S}{200} + S_r}{\frac{100}{S} + \frac{S}{200}} \times 100 = \frac{\frac{S+200S_r}{2}}{\frac{20000+S^2}{200S}}$$

Thus, the width lost as % of the original width

$$= \frac{(S + 200S_r)}{\left(\frac{200}{S} + \frac{S}{100}\right)} \tag{4.21}$$

The % width lost may be taken as the % area lost.

Example 4.5 A hillslope having 15% slope is to be provided with bench terraces. The vertical interval of the terrace is 1.5 m. Determine the length of the terrace and the earthwork per hectare. What will be the area lost if the batter slope of the terrace is 1:1?

Solution

Given, land slope, $S = 15\%$; vertical interval, $D = 1.5$ m; batter slope, $S_r = 1$.
   From Eq. (4.11), we have, $D = \frac{WS}{(100 - SS_r)}$
Putting the known values of D, S, and S_r,

$$1.5 = \frac{W \times 15}{(100 - 15 \times 1)}, \text{ we get bench width, } W = 8.5 \text{ m}$$

From Eq. (4.12), the length of terraces per hectare, $L = \frac{10000}{(W_T)}$
The terrace width, $W_T = W + (D \times S_r) = 8.5 + (1.5 \times 1) = 10$ m
Putting the values of W_T in Eq. (4.12), $L = \frac{10,000}{10} = 1000$ m
From Eq. (4.16), the volume of earthwork, $E_w = \frac{W}{8} \frac{WS}{(100 - SS_r)} \times \frac{10,000}{(W_T)}$
Putting the values of W, S, S_r, and W_T in Eq. (4.16),

$$E_w = \frac{W}{8} \times \frac{WS}{(100 - SS_r)} \times \frac{10,000}{(W_T)}$$

$$= \frac{8.5}{8} \times \frac{(8.5 \times 15)}{(100 - 15 \times 1)} \times \frac{10,000}{(10)}$$

$$= 1594 \text{ m}^3 \text{ ha}^{-1}$$

From Eq. (4.21), % area lost $= \frac{(S+200S_r)}{\left(\frac{200}{S}+\frac{S}{100}\right)}$

Putting the values of S and S_r in Eq. (4.21),

$$\% \text{ area lost} = \frac{(15 + 200 \times 1)}{\left(\frac{200}{15} + \frac{15}{100}\right)} = 15.95\%$$

Answers: Length of the terrace = 1000 m; volume of the earthwork per hectare = 1594 m³ ha⁻¹; % area lost = 15.95%.

Example 4.6 Design a bench terrace with a uniform channel grade of 0.4%. The length of the bench terrace is 150 m. The maximum rainfall intensity for a duration same as the time of concentration, and having a 10-year frequency is 200 mm h⁻¹. Assume other information as per standards.

Solution

Given, terrace channel grade = 0.4%; maximum rainfall intensity for 10-year frequency = 200 mm h⁻¹; length of the terrace = 150 m.

For the given rainfall condition, an inward-sloping bench terrace with a slope of 5% is selected. It is further assumed that the bench width is 6 m, and the batter slope is 1:1 (Fig. 4.15).

The area of the terrace that would act as the catchment area, $A = L \times W$
= 150 × 6 = 900 m²

The peak discharge to be handled by the terrace can be estimated using the rational formula:

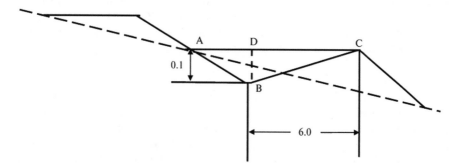

Fig. 4.15 Figure for Example 4.6

$$Q = ci\,A$$

Assuming the runoff coefficient, $c = 0.4$

Also, $i = 200$ mm h^{-1} = $\left(\frac{200}{1000 \times 3600}\right)$ = 5.6×10^{-5} m s^{-1}.

Thus, $Q = 0.4 \times (5.6 \times 10^{-5}) \times 900 = 0.02$ m^3 s^{-1}

Assuming the depth of the channel as 0.10 m (Fig. 4.15).

The area of flow available is equal to the area under triangle ABC.

Total area of ABC = area of ADB + area of BDC

$$= \frac{1}{2} \times 0.10 \times 0.10 + \frac{1}{2} \times 0.10 \times 6 = 0.31 \text{ m}^2$$

The wetted perimeter, $P = AB + BC = \sqrt{0.1^2 + 0.1^2} + \sqrt{0.1^2 + 6^2} = 6.14$ m

The hydraulic radius, $R = \frac{A}{P} = \frac{0.31}{6.14} = 0.05$ m

The flow velocity using Manning's equation (assuming $n = 0.04$):

$$V = \frac{1}{n} R^{2/3} S^{1/2} = \frac{1}{0.04} (0.05)^{2/3} (0.004)^{1/2} = 0.21 \text{ m s}^{-1}$$

The calculated velocity is non-erosive, hence acceptable.

Using continuity equation, flow $Q = AV = 0.31 \times 0.21 = 0.07$ m^3 s^{-1}

The calculated discharge capacity of the channel is higher than the required capacity. Hence, the design is acceptable.

Answer: Terrace type = inward-sloping bench terrace; terrace grade = 5%; bench width = 6 m; batter slope = 1:1; depth of the channel = 0.1 m.

4.6   Terrace System Planning

Planning Considerations

While planning the terrace system for a region, the soils, landscape, and tillage equipment need special consideration (Huffman et al. 2013). The selection of a proper outlet or disposal area, however, is the first step while planning the terrace system, particularly for terrace layout and construction.

Soils

Soil characteristics influence both terrace design and construction. Soil characteristics like permeability and erodibility impact the terrace spacing, grade, and cross-section. Soils having a good structure, i.e., those having a mix of particle sizes (and pore sizes), and stability during wetting and drying cycles are preferred for the terrace construction. Terraces are most useful in areas facing sheet and rill erosion. Terraces may not be practical on sandy soils, soils with shallow subsoil, or stony soils.

Landscape

Common or normal terraces are most effective in areas where the land slope is between 2 and 18%. Terraces may require excessive earthwork, adding to the overall cost, if used for slopes higher than 18%. In areas having higher than 20% land slope, bench terraces are constructed for preserving soil and water and ensuring crop cultivation. For slopes flatter than 2%, terraces may not be economical, and hence, less expensive measures may be chosen. Terraces are not practical in excessively gullied areas as they may prove to be too expensive.

Tillage equipment

Terrace spacing and grading are influenced by tillage equipment popularly used in the region. The proper operation of the tillage implements may be hampered if the terrace spacing is too close or the terrace slopes are too steep.

Terrace Outlets

Natural drainage channels, especially the grassed waterways, are usually preferred as the terrace channel outlet (Fig. 4.16). In the absence of a natural drainage outlet, vegetated waterways may be constructed along field boundaries. The capacity of such an outlet must be sufficient to handle the flow being carried by the terrace channel. The gradient should be low to maintain non-erosive flow velocities. The gradient, however, should be enough to prevent silting. Sod flumes, permanent pastures, stabilised gullies, stable road ditches, concrete channels, or pipe outlets are the other possible alternatives. The cost of construction and maintenance may govern the choice of the outlet. The cultivation of the graded terraces should be avoided in the last five to ten metres of its length. The portion should be permanently kept under the grass to avoid erosion due to the large volume of water that may pass.

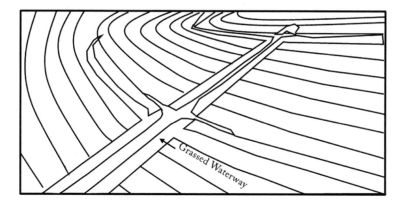

Fig. 4.16 Grassed waterway as terrace outlet

4.7   Terrace Location

The selection of the terrace location follows the selection of the terrace outlet.

Besides minimising the erosion, terrace location aims at minimising the maintenance requirement while facilitating farm operations.

While locating graded terraces, the top terrace must be located near the top of the slope in such a way that the runoff approaching it is within reasonable limits. The drainage area of the top terrace should be similar to the drainage area of any other terrace of equal length. Terrace ridges should be located just above depressions or abrupt changes in slope. Minor adjustments, limited to 15–20% of the design vertical intervals, may be made while locating the terraces. Sharp bends in terraces should be avoided. A short terrace length is preferable for ease of maintenance. Farm roads, located on the contour or the ridge, must be provided for accessing various parts of the farm. For terraces meant for water conservation, usually, the need for outlets is limited. However, on steeper slopes, where the excess flow is a possibility, the principles adopted in locating the graded terraces are followed.

4.8   Terrace System Layout

Terrace system layout design should include locations of terraces, waterways, farm roads, and other farm installation needs. A topographic map or digital elevation model may be used for developing the preliminary plans. The initial plan development is followed by staking out all the contours or graded contours by graded contouring (for graded terraces) or levelling (for conservation terraces) surveys.

The alignment between multiple lines of terraces influences the farmability. Terrace alignment could be parallel (parallel terrace alignment) or non-parallel (gradient terrace alignment). Parallel terrace alignment is preferable for row crop operations. In parallel alignment, two or more terraces are spaced a uniform (horizontal) distance apart. The horizontal distance between terraces is kept lower than the maximum estimated horizontal distance and maintained at a multiple of the farm machinery width. Terraces with parallel alignments usually require a higher amount of cuts or fills (or earthwork) and, thus, are costly to design, layout, and construct.

Non-parallel terraces follow the gradient terrace alignment. Non-parallel terraces are laid on natural contours and have variable spacing. Non-parallel terraces are easy to layout and build and best suited for non-row crop farming, i.e., for small grains or pastures.

4.9   Terrace Maintenance

Regular maintenance and monitoring of terraces are imperative. Physical inspections, especially after severe storms, and swift repair are essential. While inspecting the terraces, special care should be taken of the terrace ridge, channel and cross-sections, and the outlet elevations. Similarly, removal of sediment from the terrace channel bed is essential for maintaining the capacity and grade. The capacity of the grassed waterway outlets must be maintained by clipping the grass and removing the sediment deposited. In the case of underground outlets, sediment or trash deposition should be avoided.

Questions

1. Classify terraces, and describe specific features of different types of terraces.
2. Based on the information from the mountainous regions closer to your home town, enlist the types of terraces popularly used in the area.
3. Determine the vertical and horizontal intervals of terraces for a semi-arid region with normal soils. The land slope is 5%.
4. For a machinery width of 4 m, ridge height of 0.35 m, and a 4% land slope, determine the cut slope, front slope, and back slope ratios for a three-segment broad-base terrace.
5. A 150 m long terrace is constructed on 6% land slope. The terrace cross-section is balanced, and the widths of the cut slope, front slope, and back slope are 4.2 m. The flow depth is 0.30 m, and a freeboard of 0.10 m is provided. Determine the volumes of cut and fill.
6. Design a 150 m long inward-sloping bench terrace for 20% land slope. The area has clay loam soil, and the maximum one-hour rainfall intensity is 100 mm h^{-1}. The terrace channel is provided with 0.5% uniform grade.
7. Bench terraces are planned for land having a 15% slope. The vertical interval is 2.5 m, and the riser has a slope of 1:1. The earthwork in cutting is equal to filling. Determine the volume of earthwork per hectare.
8. In a hilly region having a land slope of 30%, bench terracing is planned. The maximum depth of cut is 0.7 m, and the riser slope is 1:1. Compute the terrace width and the volume of earthwork per hectare.
9. The average annual soil loss at a location is 14 t ha^{-1}. What would be the terrace spacing and the slope length if the soil loss has to be reduced to 10 t ha^{-1}? The USLE factors are: rainfall erosivity factor $= 1800$ MJ mm ha^{-1} h^{-1}, soil erodibility factor $= 0.26$ t ha h ha^{-1} MJ^{-1} mm^{-1}, crop management factor $= 0.25$, conservation practice factor $= 0.6$, slope steepness $= 7\%$, and slope length $= 140$ m.
10. Compute the maximum permissible grade for terraces having a length of 200 m, if the maximum permissible velocity $= 0.61$ m s^{-1}, channel side slope ratio $= 5:1$, the vertical spacing between terraces $= 1.2$ m, and land slope $= 8\%$.

Multiple Choice Questions

Answer the questions by choosing the correct option.

1. Broad-base terraces are useful in areas with

   (A) moderately permeable soils and high rainfall
   (B) highly permeable soils and high rainfall
   (C) moderately permeable soils and low rainfall
   (D) highly permeable soils and low rainfall

2. Tabletop or level bench terraces are useful in

   (A) medium rainfall areas with highly permeable soils
   (B) medium rainfall areas with low permeability soils
   (C) heavy rainfall areas with highly permeable soils
   (D) low rainfall areas with moderately permeable soils

3. Conservation bench broad-base terraces are preferred in

   (A) semi-arid region   (B) arid region
   (C) humid region   (D) cold desert region

4. For a three-segment section broad-base terrace having equal side slopes, the width of the terrace is W, the cut is c, the fill is f, the height of the ridge is h, and the existing land slope is S. For a balanced cross-section, the cut will be

   (A) $(h-S \times W)$ (B) $(h + S \times W)$
   (C) $(h + S \times W)/2$ (D) $(h-S \times (W/2))$

5. For a broad-base terrace on 10% land slope, the width of the terrace is 5.0 m and the channel depth is 0.3 m. The terrace cross-section is balanced, and the freeboard is included in the channel depth. The heights of cut and fill are

   (A) 0.2 m   (B) 0.4 m
   (C) 0.6 m   (D) 0.8 m

6. The function of the shoulder bund is to

   (A) hold runoff water over the terrace bed
   (B) check soil erosion
   (C) demarcate the field boundary
   (D) protect margin of side bund

7. If the existing land slope and the bench width are given as S and W, respectively, then for a batter slope of 150%, the vertical interval will be

   (A) $WS/100$ (B) $WS/(100-S)$
   (C) $2WS/(200-3S)$ (D) $2WS/(300-2S)$

8. Bench terraces are planned on a 10% hill slope. For a riser slope of ½:1, the % area lost will be

(A) 4.62 (B) 5.47
(C) 6.25 (D) 6.78

9. Bench terraces are planned on 15% hill slope. The vertical interval is 2.5 m, and the riser has a slope of 1:1. The earthwork in cutting is equal to filling. The volume of earthwork per hectare for a balanced cross-section in cubic m per ha will be

(A) 2656 (B) 2818
(C) 4248 (D) 5312

10. For an area having a land slope of 16%, a bench terrace is being planned. The recommended depth of cut and batter slopes is 0.40 m and ½:1, respectively. Assuming cut = fill, the bench width and the earthwork per hectare are

(A) 3.2 m, 980 m^3 (B) 4.6 m, 920 m^3
(C) 4.8 m, 980 m^3 (D) 5.0 m, 920 m^3

References

Beasley RP, Gregory JM, McCarthy TM (1984) Erosion and sediment pollution control. Iowa State University Press, Ames

FAO (2000) Manual on Integrated Soil Management and Conservation Practices. FAO Land and Water Bulletin 8, Rome

FAO (1986) Watershed management field manual: gully control. FAO Conservation Guide, Rome

Ghidey F, Gregory JM, Thompson AL (1992) Terrace channel design and evaluation. Trans ASAE 35:1513–1520

Huffman R, Fangmeier D, Elliot W, Workman S, Schwab G (2013) Soil and water conservation engineering, 7th edn. American Society of Agricultural and Biological Engineers, St Joseph

Natural Resources Conservation Service (NRCS) (2011) Terraces. In: National Engineering Handbook, Part 650 Engineering Field Handbook. USDA Natural Resources Conservation Service, Washington DC

Ramser CE (1931) Farm terracing. Farmers' Bulletin. Number 1669. USDA, Washington DC

Venkateswarlu B, Osman M, Padmanabhan MV, Kareemulla K, Mishra PK, Korwar GR, Rao KV (2013) Field manual on watershed management. Central Research Institute for Dryland Agriculture, Hyderabad

Chapter 5
Bunds

Abstract Bunds are among the most common mechanical measures of erosion control. These consist of small embankments constructed across the land slope to reduce the slope length, runoff, and soil erosion and enhance soil infiltration. This chapter presents a broad classification of bunds. It includes the common design considerations for contour and graded bunds like storm frequency, spacing, side slopes, freeboard, and seepage through them. The chapter elaborates on the design of contour and graded bunds in terms of their height, cross-section, length, the volume of earthwork, and area lost under them. The chapter also contains the planning considerations and construction of bunds. The design procedures are demonstrated through solved examples.

5.1 Definition

The English word *bund* has its root in the Persian *band* through Hindi *Bandh*, which means an embankment, dyke, or levee.

Bunds are a widespread soil and water conservation measures, categorised as the mechanical measures of erosion control. These are small embankment type structures, constructed across the land slope, either with soil or stone. Bunds reduce the slope length, resulting in the diminished runoff, reduced soil erosion, and enhanced soil infiltration (Pathak et al. 1989; Singh et al. 1990; Mati 2007). These are similar to the narrow-base terraces, but cultivation is not practised on bunds, though they may be kept under permanent grass cover for protection.

The first known use of the term *bund* may be credited to the levees or bunds along the Tigris River in Baghdad in the early nineteenth century. Soil bunds are widely practised in Ethiopia, Tanzania, Kenya, and Rwanda as a prevalent indigenous technology. Soil bunds are constructed to get them converted into bench terraces in 10–20 years due to deposition of eroded soil in between the bunds. In Kenya, Malawi, and Zambia, construction of soil bunds was compulsorily practised over large areas of communal lands in the 1950s. In India, the term was first used in 1938 to refer to *contour bunds* as a component of the dry farming development scheme.

R. Singh, *Soil and Water Conservation Structures Design*, Water Science and Technology Library 123, https://doi.org/10.1007/978-981-19-8665-9_5

5.2 Functions

The primary functions of the bunds are:

1. To reduce the velocity of overland flow by breaking up a long slope into short segments, and consequently checking the soil erosion
2. To form water storage area on their upslope side for retaining runoff within the field, and consequently improving soil infiltration and groundwater recharge
3. To dispose of excess runoff safely from the agricultural fields and avoid water stagnation.

Besides the above primary functions, bunds also help in improving land productivity through effective utilisation of the fertilisers and soil nutrients.

5.3 Classification

Bunds are mainly classified into two categories:

(i) Contour bunds
(ii) Graded bunds

Contour bunds are narrow-base trapezoidal embankments constructed along contours (Fig. 5.1). These bunds divide the area into strips that act as barriers to runoff and store water on their upslope side. Consequently, the flow rate and flow volume are reduced, leading to decreased soil erosion and enhanced soil infiltration and groundwater recharge.

Contour bunds are preferred in medium to low rainfall areas having less than 700 mm of annual rainfall. The area may have 2–6% land slope and highly permeable soils. Though contour bunds are primarily designed to conserve rainfall, the occasional high rainfall intensities may produce high runoff volume. The bund breaching or overtopping, in such situations, may be avoided by incorporating surplus weirs (also known as waste weirs) in the bund to dispose of excess water (Fig. 5.2).

Surplus weirs are also useful in handling prolonged water stagnation which may otherwise reduce the crop yields. The height of the weir crest governs the depth of impoundment behind the bund. The height is usually maintained at 0.30 m from the ground level.

Graded bunds are similar to contour bunds in cross-section, i.e., narrow-base trapezoidal embankments. Graded bunds, however, are constructed across the land slope but not along contours. A significant difference between contour bunds and graded bunds is that the contour bunds are constructed for water conservation, whereas graded bunds are meant for the disposal of excess runoff. In graded bunds, the area between two consecutive bunds acts as a channel having a longitudinal gradient along the bund.

The longitudinal gradient may be uniform or variable. Uniformly graded bunds are preferred when the discharge is less, while variable-graded bunds are preferred

Fig. 5.1 Typical layout and cross-section of contour bunds

Fig. 5.2 Provision of the surplus weir in the contour bund

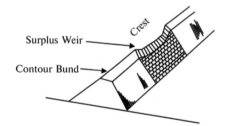

when the discharge is more. The flow velocity in the graded channels must be within the non-erosive limit. The flow velocity is considered as non-erosive if it is lower than 0.46 m s^{-1} for soils with high erodibility and 0.61 m s^{-1} for other soils.

Graded bunds are recommended in areas where land is susceptible to water erosion and waterlogging is an issue. These are usually preferred in medium to high rainfall areas having more than 700 mm of annual rainfall. The area may have 2–8% land slope with soils having low to moderate permeability. In addition to the contour and graded bunds, a few other types of bunds are also used in practice. These are described below:

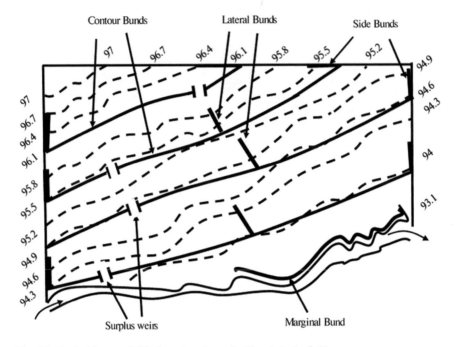

Fig. 5.3 Typical layout of side, lateral, and marginal bunds in the field

Side Bunds

Side bunds are constructed along the land slope at the extreme ends of contour bunds (Fig. 5.3). The primary function of these bunds is to check the outflow of the ponded water from the sides. The cross-section of these bunds may be kept the same as the contour bunds.

Lateral Bunds

Lateral bunds are constructed along the land slope in between two side bunds (Fig. 5.3). The primary function of these bunds is to break the length of contour bunds and prevent the concentration of water on one side. The cross-section of these bunds may also be kept the same as the contour bunds.

Marginal Bunds

Marginal bunds are constructed at the field and at road margins to demarcate the boundaries (Fig. 5.3).

Semi-circular Bunds

Semi-circular bunds involve the construction of soil bunds having a semi-circular shape, with the tip of the bunds kept at contours (Fig. 5.4).

These bunds are also referred to as crescent-shaped bunds or demi-lunes and used predominantly for planting tree seedlings in the semi-arid regions of Tanzania,

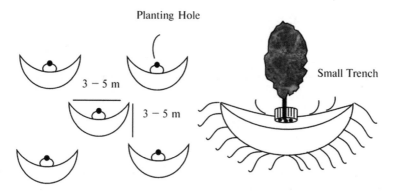

Fig. 5.4 Typical layout of the semi-circular bunds (After Hai 1998)

Kenya, and Ethiopia. The semi-circular bunds include a pit for planting the tree. The subsoil of the pit is excavated to construct the bund on the downstream side of the pit.

Typically, the radius and the depth of the pit are kept as 0.60 m. The radius of the bund ranges from 3.0 to 5.0 m, and its height usually is 0.25 m.

Contour Stone Bunds

Contour stone bunds involve the construction of porous stone barriers along the contours (Fig. 5.5). These bunds are preferred in the areas having an annual rainfall of 200–750 mm. The stone bunds are typically spaced 15–30 m apart, with the height of the bund ranging in between 0.20 and 0.30 m from the ground surface. As these bunds are porous, they reduce the water flow and increase the water spread area. Consequently, soil infiltration is enhanced, and soil erosion is reduced. Contour stone bunds are widely practised in Ethiopia, Eritrea, Kenya, Tanzania, Zimbabwe, and Botswana.

Fig. 5.5 Typical layout of the contour stone bund (After Hai 1998)

5.4 Common Design Considerations for Contour and Graded Bunds

Design of contour or graded bunds depends on different site-specific settings like land slope, rainfall and runoff magnitudes, soil texture and infiltration characteristics, and vegetative cover. Further, the design involves the determination of standard features like bund spacing in terms of vertical and horizontal intervals, bund cross-section in terms of the height, freeboard, side slope, bottom width and top width, and the volume of earthwork and cost of construction. The common considerations used while designing the contour or graded bunds are described below. The other specific considerations while designing the contour or graded bunds are presented later.

5.4.1 Bund Spacing

Like terraces, the bund spacing is expressed in terms of vertical interval (*VI*) and horizontal interval (*HI*). *VI* represents the vertical distance between two successive bunds, while *HI* represents the horizontal distance between two successive bunds (Fig. 5.6a). The equations used for terraces are also adopted for determining the *VI* and *HI* of bunds.

The vertical interval (*VI*) is first determined using any one of the relationships given by Ramser (Eq. (4.1)), Cox (Eq. (4.2)), or FAO (Eqs. (4.3) and (4.4)) (Chap. 4). The horizontal interval (*HI*) is subsequently determined using Eq. (4.5) (Chap. 4).

The computed values of *VI* and *HI* may be varied up to 25% to account for varying rainfall, soil, and crop cover conditions.

5.4.2 Bund Side Slopes

The side slopes of the bund depend on the type of soil used in the bund construction, i.e., the steepness of the side slope is governed by the stability of the soil material. Table 5.1 may be used to select the side slope.

5.4.3 Freeboard

A freeboard of 20% of the depth of water, d, is considered to get the bund height, H, i.e., $H = 1.2d$.

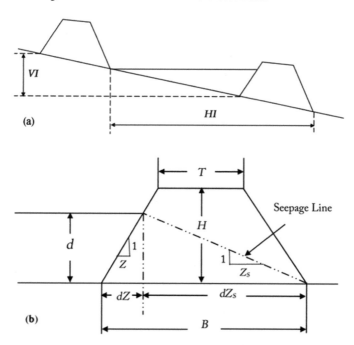

Fig. 5.6 Typical contour bund layout with trapezoidal bund cross-section and nomenclature. $d =$ depth of water; $VI =$ vertical interval; $HI =$ horizontal interval; $H =$ height of the bund; $T =$ top width; $B =$ bottom width; $Z =$ side slope; $Z_S =$ seepage line slope

Table 5.1 Side slope of the bund (Z)

Type of soil	Side slope, Z
Clay	1.0
Loam	1.5
Sand	2.0

5.4.4 Seepage Through Bund

Since the water accumulated behind the bund may seep through the body of the bund, the seepage through bunds must be given due consideration while designing the bund cross-section. The seepage through bund depends on the depth of impoundment, bund side slopes, and the permeability of the soil.

The *seepage line*, i.e., the top flow line below which seepage occurs (Fig. 5.6b), emerging from the downstream slope could cause slumping of the material and may lead to instability of the bund. Therefore, as a thumb rule, the seepage line must pass through the base of the bund, with the downstream toe of the bund representing the extreme limit. Table 5.2 presents the permissible slope of the seepage line for different types of soil.

Type of soil	Seepage line slope, Z_s
Clay	3.0
Loam	5.0
Sand	6.0

Table 5.2 Slope of the seepage line (Z_s)

5.5 Design of Contour Bunds

Besides vertical and horizontal intervals, the design of a contour bund system includes specifying the cross-section of the bund. The cross-section includes determining the bund height, including the freeboard, specifying the bund side slopes and the seepage line slope, and determining the bottom and the top widths. The vertical and horizontal intervals are determined using the empirical equations described in Sect. 4.4.1 (Chap. 4).

5.5.1 Height of Contour Bund

The height of the bund is decided based on the maximum amount of water to be stored behind the bund or the maximum permissible depth of impoundment, d (Fig. 5.6). In Fig. 5.7, W is the width of the water surface between the bunds (also referred to as the water spread length behind the bund), Z is the bund side slope, and S is the original land slope.

Considering similar triangles ABC and AMN (Fig. 5.7),

$$\frac{W - Zd}{HI} = \frac{d}{VI}$$

$$\text{Or, } W = \frac{HI \times d}{VI} + Zd \tag{5.1}$$

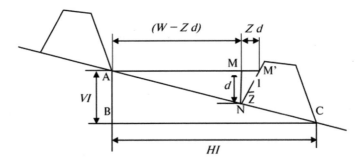

Fig. 5.7 Typical sketch showing neighbouring contour bunds with water stored between them

Thus, the maximum amount of water that can be stored.

$$= \text{Area of triangle AMN} + \text{Area of triangle NMM}'$$

$$= \frac{1}{2} \times \text{AM} \times \text{MN} + \frac{1}{2} \times \text{MN} \times \text{MM}'$$

$$= \frac{1}{2} \times (W - Zd) \times d + \frac{1}{2} \times d \times Zd$$

$$= \frac{1}{2} \times d \times [(W - Zd) + Zd] = \frac{1}{2} \times d \times W \qquad (5.2)$$

Substituting the value of W from Eq. (5.1) into (5.2), the maximum amount of water that can be stored

$$= \frac{1}{2} \times d \times \left[\frac{HI \times d}{VI} + Zd \right] = \frac{1}{2} \times d^2 \times \left[\frac{HI}{VI} + Z \right] \qquad (5.3)$$

As mentioned earlier, a storm having a 10-year return period is used to estimate the peak runoff volume for determining the depth of impoundment, d.

Let the depth of effective rainfall (peak rainfall minus losses) that needs to be stored be R_e (in m). Then, the maximum amount of water retained behind the bund per unit length of the bund would be

$$= R_e \times HI \qquad (5.4)$$

Equating Eqs. (5.3) and (5.4),

$$\frac{1}{2} \times d^2 \times \left[\frac{HI}{VI} + Z \right] = R_e \times HI$$

$$\text{Or, } d = \sqrt{\frac{2 \times R_e \times HI}{\left(\frac{HI}{VI} + Z \right)}} = \sqrt{\frac{2 \times R_e \times VI \times HI}{(HI + (Z \times VI))}} \qquad (5.5)$$

Equation (5.5) gives the general form of the equation for determining the depth of impoundment.

The height of the contour bund can be determined by adding a freeboard equal to 20% of d. Thus,

$$\text{Height of the contour bund} = 1.2d \qquad (5.6)$$

Example 5.1 Determine the vertical interval for the contour bunds to be constructed on a 4.5% land slope. The rainfall is moderate with an average infiltration rate of the soil, and cover conditions are good during the critical period.

Solution

Given: land slope $S = 4.5\%$; rainfall = moderate; soil infiltration rate = average; cover condition = good.

We need to estimate the vertical interval, VI, and based on the information available, we may use the Cox formula, i.e., Eq. (4.2) (Chap. 4),

$$VI = (xS + y) \times 0.3$$

From Tables 4.1 and 4.2 (Chap. 4), $x = 0.6$ and $y = 2.0$

Hence, $VI = (0.6 \times 4.5 + 2) \times 0.3 = 1.41$ m

Answer: The vertical interval of the bund = 1.41 m.

Example 5.2 Calculate the height of the contour bund in loam soil, having the original land slope of 2.5%. The region has moderate rainfall, with the maximum expected rainfall having 10-year return period being 180 mm. The soil absorbs 40% of the rainfall in the field. The cover condition is good during the critical periods. Also, determine the storage area required.

Solution

Given, land slope, $S = 2.5\%$; soil type = loam; rainfall = moderate; cover condition = good.

The maximum expected rainfall having 10-year recurrence interval = 180 mm = 0.18 m.

The depth of effective rainfall, $R_e = 60\%$ of the rainfall (as 40% of the rainfall is absorbed as infiltration).

Thus, $R_e = 60\%$ of rainfall = $0.6 \times 0.18 = 0.108$ m.

The effective rainfall also equals the maximum runoff potential which the bund system needs to handle.

Storage area required, A = Maximum runoff to be stored $\times HI$

The horizontal interval HI is given by Eq. (4.5) (Chap. 4), $HI = \frac{VI}{S} \times 100$

We need the vertical interval, VI, and based on the information available, we may use the Cox formula, i.e., Eq. (4.2) (Chap. 4),

$$VI = (xS + y) \times 0.3$$

From Tables 4.1 and 4.2 (Chap. 4), $x = 0.6$ and $y = 2.0$ (for loam soil, infiltration may be considered as average).

Hence, $VI = (0.6 \times 2.5 + 2) \times 0.3 = 1.05$ m

The horizontal interval, $HI = \frac{VI}{S} \times 100 = \frac{1.05}{2.5} \times 100 = 42$ m

Hence, storage area required,

$$A = \text{Maximum runoff to be stored} \times HI = 0.108 \times 42 = 4.54 \text{ m}^2$$

Using Eq. (5.5), the depth of impoundment, $d = \sqrt{\frac{2 \times R_e \times VI \times HI}{(HI+(Z \times VI))}}$.

Substituting the known values of R_e, VI, HI, and taking $Z = 1.5$ (for loam soil from Table 5.1), $d = \sqrt{\frac{2 \times R_e \times VI \times HI}{(HI+(Z \times VI))}} = \sqrt{\frac{2 \times 0.108 \times 1.05 \times 42}{42+(1.5 \times 1.05)}} = 0.47$ m

Hence, the height of the bund, $H = 1.2\ d = 1.2 \times 0.47 = 0.56$ m

Answers: Storage area required $= 4.54$ m^2; height of the bund $= 0.56$ m.

5.5.2 Bund Cross-Section

Besides determining the bund height, H, the cross-section includes specifying the bund side slopes and the seepage line slope and determining the bottom and the top widths.

As discussed earlier, the bund side slopes (Z) and the seepage line slope (Z_S) depend on the type of soil used in the bund construction. Tables 5.1 and 5.2 may be used to specify Z and Z_S, respectively. Once Z and Z_S are known, the bottom width, B (Fig. 6b), is calculated using the following equation:

$$B = Zd + Z_Sd \tag{5.7}$$

Knowing B, Z, and H, the top width, T, is calculated using the following equation:

$$T = B - 2ZH \tag{5.8}$$

The cross-section area of the bund is expressed as,

$$A_B = \frac{(B+T)}{2} \times H \tag{5.9}$$

Example 5.3 Calculate the dimensions of the contour bund for the data given in Example 5.2.

Solution

From the solution of Example 5.2, we know that the depth of impoundment, $d = 0.47$ m, and the bund height, $H = 0.56$ m.

Also, the soil is loam, hence, $Z = 1.5$ (Example 5.2) and $Z_S = 5.0$ (from Table 5.2).

Using Eq. (5.7), the bottom width,

$$B = Zd + Z_Sd = 1.5 \times 0.47 + 5.0 \times 0.47 = 3.06 \text{ m}$$

Using Eq. (5.8), the top width,

$$T = B - 2ZH = 3.06 - 2 \times 1.5 \times 0.56 = 1.38 \text{ m}$$

Answers: Bottom width of the bund $= 3.06$ m; top width of the bund $= 1.38$ m; bund side slopes $= 1.5$

Example 5.4 Design the dimensions of a contour bund to handle the peak rainfall intensity of 114 mm h^{-1} for a 10-year return period. The peak rainfall occurs for 54 min. A loam soil having a constant infiltration rate of 25 mm h^{-1} is used as the construction material. The land slope is 2.5%. The vertical interval and the length of bunds are 750 mm and 400 m, respectively.

Solution

Given: land slope, $S = 2.5\%$; length of bund, $l = 400$ m; vertical interval, $VI = 0.75$ m; peak rainfall intensity $= 114$ mm h^{-1}; duration of the peak rainfall $= 54$ min; soil type $=$ loam; constant soil infiltration rate $= 25$ mm h^{-1}.

Putting known values of VI and S in Eq. (4.5) (Chap. 4),

The horizontal interval, $HI = \frac{VI}{S} \times 100 = \frac{0.75}{2.5} \times 100 = 30$ m

The intensity of the effective rainfall $=$ maximum expected rainfall$-$constant infiltration rate $= 114 - 25 = 89$ mm h^{-1}

Hence, the depth of effective rainfall, $R_e =$ Effective rainfall intensity $\times$ rainfall duration $= 89 \times (54/60) = 0.08$ m

The effective rainfall also equals the maximum runoff potential which the bund system needs to handle.

Hence, storage area required,

$A =$ Maximum runoff to be stored $\times HI = 0.08 \times 30 = 2.4$ m^2

Knowing the storage area, the depth of impoundment, d, can be calculated in two different ways.

Option 1: Directly using Eq. (5.5)

Substituting the known values of R_e, VI, HI, and taking $Z = 1.5$ (for loam soil from Table 5.1) in Eq. (5.5),

$$\text{Depth of impoundment, } d = \sqrt{\frac{2 \times R_e \times VI \times HI}{(HI + (Z \times VI))}}$$

$$= \sqrt{\frac{2 \times 0.08 \times 0.75 \times 30}{30 + (1.5 \times 0.75)}}$$

$$= 0.34 \text{ m}$$

Option 2: Using the estimated storage area

From Fig. 5.8, the storage area behind the bund $=$ Area of triangle AMN, $a_1 +$ Area of triangle NMM', a_2

$$= \frac{1}{2} \times AM \times MN + \frac{1}{2} \times MM' \times MN$$

$$= \frac{1}{2} \times 40d \times d + \frac{1}{2} \times 1.5d \times d = 20.75d^2$$

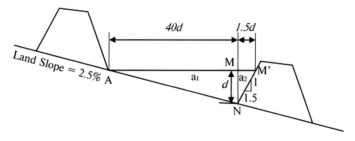

Fig. 5.8 Water storage area behind the contour bund (Example 5.4)

Since the storage area behind the bund should be equal to the storage area required, hence, $20.75d^2 = 2.4$ or $d = 0.34$ m.

As can be seen, the depth of impoundment, d, is the same by both options.

Hence, assuming a freeboard of 20% of d, the height of the bund,

$H = 1.2d = 1.2 \times 0.34 = 0.41$ m

Substituting the known values of d and Z, and taking $Z_S = 5.0$ (for loam soil from Table 5.2) in Eq. (5.7), the bottom width,

$$B = Zd + Z_S d = 1.5 \times 0.34 + 5.0 \times 0.34 = 2.21 \text{ m}$$

Using Eq. (5.8), the top width,

$$T = B - 2ZH = 2.21 - 2 \times 1.5 \times 0.41 = 0.98 \text{ m}$$

Answers: Depth of impoundment $= 0.34$ m; height of the bund $= 0.41$ m; bottom width of the bund $= 2.21$ m; top width of the bund $= 0.98$ m; bund side slopes $= 1.5$.

Example 5.5 Find the depth of impoundment immediately behind a contour bund built on 5% land slope. The vertical spacing is 1.25 m, and the bund needs to store a 3 h effective rainfall of 0.12 m. Also, find the water spread length behind the bund. The effect of the side slopes on the storage area may be neglected.

Solution

Given: land slope, $S = 5\%$, vertical interval, $VI = 1.25$ m, depth of rainfall excess, $R_e = 0.12$ m.

Substituting known values of S and VI in Eq. (4.5) (Chap. 4),

The horizontal interval, $HI = \frac{VI}{S} \times 100 = \frac{1.25}{5} \times 100 = 25$ m

Storage area required $= R_e \times HI = 0.12 \times 25 = 3.0$ m^2

Referring to Fig. 5.8, as per the problem statement, we may neglect a_2. Please note that the original land slope here is 5%. Hence, $AM = 20d$.

Thus, the storage area behind the bund $= \frac{1}{2} \times d \times (20d) = 10d^2$, with d being the depth of impoundment and $(20d)$ being the water spread length behind the bund.

Since the storage area behind the bund should be equal to the storage area required, hence,

$$10d^2 = 3 \text{ or } d = 0.55 \text{ m}$$

The water spread length behind the bund $= 20d = 11.0$ m

Answers: Depth of impoundment $= 0.55$ m; water spread length behind the bund $= 11.0$ m.

Example 5.6 For a contour bund, determine the weir dimensions for discharging excess runoff. The peak intensity of rainfall for a 10-year return period is 147 mm h^{-1}, and the peak rainfall occurs for an hour. The soil is sandy with a constant infiltration rate of 35 mm h^{-1}. The original land slope is 3%, and the bund dimensions are: height $= 650$ mm; top width $= 500$ mm; side slopes $= 2:1$; vertical interval $= 750$ mm; length $= 400$ m; height of weir crest from the ground surface $= 300$ mm.

Solution

Given: land slope, $S = 3.0\%$; top width of bund, $T = 0.5$ m; height of the bund, $H = 0.65$ m; depth of impoundment $=$ height of crest from the ground surface, $d = 0.3$ m; side slopes of bund, $Z = 2:1$; length of bund, $l = 400$ m; vertical interval, $VI = 0.75$ m; peak intensity of rainfall occurring for an hour $= 0.147$ m h^{-1}, soil $=$ sandy; constant infiltration rate of the soil $= 0.035$ m h^{-1}.

Substituting known values of S and VI in Eq. (4.5) (Chap. 4),

The horizontal interval, $HI = \frac{VI}{S} \times 100 = \frac{0.75}{3} \times 100 = 25$ m

The intensity of the effective rainfall $=$ peak intensity of rainfall$-$constant soil infiltration rate

$$= 0.147 - 0.035 = 0.112 \text{ m h}^{-1}$$

Since the duration of effective rainfall is one hour, the depth of effective rainfall, $R_e =$ Effective rainfall intensity $\times$ rainfall duration $= 0.112 \times 1 = 0.112$ m

The effective rainfall also equals the maximum runoff potential which the bund system needs to handle.

Hence, the maximum runoff volume generated $=$ Maximum runoff potential $\times HI \times l$

$$= 0.112 \times 25 \times 400 = 1120 \text{ m}^3$$

Referring to Fig. 5.9, storage area behind the bund,

$$A = a_1 + a_2 = \frac{1}{2} \times d \times 33.33d + \frac{1}{2} \times d \times 2d$$

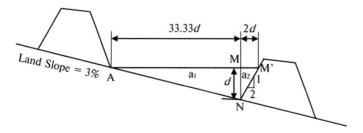

Fig. 5.9 Water storage area behind the contour bund (Example 5.6)

$$= \frac{1}{2} \times 0.3 \times 33.33 \times 0.3 + \frac{1}{2} \times 0.3 \times 2 \times 0.3 = 1.59\,\text{m}^2$$

Hence, total volume of water that could be stored $= A \times l = 1.59 \times 400 = 636\,\text{m}^3$

Hence, excess runoff to be disposed of by the weir $=$ Maximum runoff volume generated $-$ the total volume of water that could be stored

$$= 1120 - 636 = 484\,\text{m}^3$$

Since the excess runoff needs to be disposed of in one hour, the flow rate over the weir $= \frac{484}{3600} = 0.134\,\text{m}^3\text{s}^{-1}$.

Referring to Fig. 5.10, the weir formula, $Q = CL_w H_w^{1.5}$.

Where $Q =$ discharge over the weir crest; $L_W =$ crest length; $H_W =$ head over the crest; and $C =$ discharge coefficient $= (2/3)^{1.5} \sqrt{g} = 1.71$ (in SI units).

Assuming, $H = 0.30$ m, $0.134 = 1.71 \times L_W \times (0.30)^{1.5}$; or $L_W = 0.48$ m

Assuming a freeboard of 0.05 m, the total depth of weir $= 0.35$ m

Answers: Length of the weir crest $= 0.48$ m; depth of the weir $= 0.35$ m.

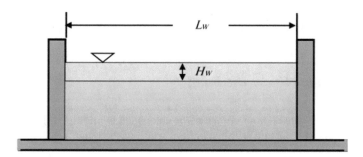

Fig. 5.10 Cross-sectional view of a suppressed rectangular weir

5.5.3   *Length of the Bund*

The linear length of the bund per hectare is given as follows:

$$l = \frac{10,000}{HI} = \frac{100S}{VI} \tag{5.10}$$

where l = linear length of the bund per hectare, m.

5.5.4   *Earthwork*

The earthwork for the contour bund may be calculated by multiplying the length of contour bund per hectare with the cross-section area of the bund, i.e.,

Earthwork = Length of the bund per hectare × cross-section area = $l \times A_B$

Substituting l and A_B from Eqs. (5.10) and (5.9),

$$Earthwork = l \times A_B = \left[\left(\frac{10000}{HI} \right) \left\{ \left(\frac{(B+T)}{2} \right) \times H \right\} \right] m^3 \tag{5.11}$$

The earthwork for contour bunds, however, also includes the earthwork involved in constructing the side bunds and the lateral bunds. For ease of calculation, the cross-sections of the side bunds and the lateral bunds are assumed to be the same as the contour bund, with their combined length being equal to 30% of the total length of the contour bunds.

Hence, if the side bunds and the lateral bunds are present, then for the earthwork calculation, the total length of the bund will be 1.3 l, and in that case, l in Eq. (5.11) would be replaced by 1.3 l.

5.5.5   *Area Lost Under the Bund*

The area lost under the contour bund may be calculated by multiplying the length of the contour bund per hectare with the bottom width of the bund, i.e.,

Area lost = Length of the bund per hectare × Bottom width = $l \times B$

$$Area\ lost\ per\ hectare = l \times B = \left[\left(\frac{10000}{HI} \right) \times B \right] = \left[\left(\frac{100S}{VI} \right) \times B \right] m^2 \tag{5.12}$$

Like in the case of the earthwork calculation, if the side bunds and the lateral bunds are present, then l in Eq. (5.12) would be replaced by 1.3 l.

Example 5.7 Determine the volume of earthwork involved in the construction of contour bunds in a 100-hectare watershed having a land slope of 2.5%. The bund dimensions are: height $= 700$ mm; top width $= 500$ mm; bottom width $= 2.25$ m; and vertical interval $= 1.3$ m. Also, determine the total area lost due to bund construction.

Solution

Given: land slope, $S = 2.5\%$; height of the bund, $H = 0.7$ m; top width of bund, $T = 0.5$ m; bottom width of the bund, $B = 2.25$ m; vertical interval, $VI = 1.3$ m; area of the watershed $= 100$ ha.

Using Eq. (5.9), the cross-section area of the bund,

$$A_B = \frac{(B + T)}{2} \times H = \frac{(2.25 + 0.5)}{2} \times 0.7 = 0.96 \text{ m}^2$$

Using Eq. (5.10), the length of the contour bund per hectare,

$$l = \frac{100S}{VI} = \frac{100 \times 2.5}{1.3} = 192.3 \text{ m}$$

The total length of the bunds per hectare (including the side and the lateral bunds).

$$= 1.3\, l = 1.3 \times 192.3 = 250 \text{ m}$$

The total length of the bund in the 100-hectare watershed $= 250 \times 100$

$$= 25000 \text{ m}$$

Hence, the total volume of earthwork in the watershed $=$ Total length of the bund $\times A_B = 25000 \times 0.96 = 24000$ m^3
Using Eq. (5.12),

$$\text{Area lost per hectare} = \left[1.3 \times \left(\frac{100\, S}{VI} \right) \times B \right]$$

$$= 1.3 \times \left(\frac{100 \times 2.5}{1.3} \right) \times 2.25$$

$$= 562.5 \text{ m}^2$$

Hence, total area lost in the watershed $= 562.5 \times 100 = 56250$ m$^2 = 5.625$ ha

Answers: Total volume of the earthwork $= 24,000$ m^3; total area lost $= 5.625$ hectares.

5.6 Design of Graded Bunds

As mentioned earlier, graded bunds are similar to contour bunds in cross-section. There are two significant differences between graded bunds and contour bunds: (i) graded bunds are constructed across the land slope but not along contours; (ii) graded bunds are constructed with a longitudinal gradient along the bund (Singh et al. 2010). The longitudinal gradient may be uniform or variable.

The graded bunds are designed following the same principles and procedures that are adopted while designing contour bunds. The vertical and horizontal intervals and the cross-section of the bund are determined using the procedure adopted while designing contour bunds. The bund side slopes and the seepage line slope are referred from Tables 5.1 and 5.2, respectively. The length of the bund per hectare, the volume of the earthwork, and the area lost under the bund are also determined using the procedure adopted while designing contour bunds. It may, however, be noted that the side bunds and the lateral bunds are absent in the graded bund systems. Thus, for calculating the volume of the earthwork or the area lost, Eqs. (5.11) or (5.12) are used.

Moreover, there are two notable differences in the graded bund design:

(i)   The design is based on the value of estimated peak runoff rate (peak runoff volume in case of contour bunds) due to a storm of 10-year recurrence interval. The peak runoff rate is used to determine the depth of flow, d (depth of impoundment in case of contour bunds).

(ii) The graded bunds are provided with a uniform or variable grade along the length of the bund to transport the runoff at non-erosive velocities to a safe outlet. The grade should be sufficient to avoid prolonged ponding and, thus, preventing the damages to crops or delay in the farming operations. The following guidelines are used while deciding the grade:

   (a)   A uniform grade is used for short bunds of 100–150 m length, while a variable grade is recommended for longer bunds.
   (b)   In general, for most of the soils, the grade ranges over 0.2–0.4%.
   (c)   In case of long bunds with impermeable soils, the grades may be zero in the beginning and 0.5% at the tail end.

Like terrace channels, the flow velocities are kept lower than 0.46 m s^{-1} for soils with high erodibility and 0.61 m s^{-1} for other soils. The Manning equation is usually adopted for the channel design with the roughness coefficient value of 0.035 or less.

Example 5.8 Determine the dimensions of a graded bund proposed to be constructed on land having 3% slope. The length of the bund is 350 m. The soil is sandy. The rainfall intensity of the storm having 10-year recurrence interval is 200 mm h^{-1}. The rainfall duration is the same as the time of concentration. A uniform grade of 0.18% is to be maintained along the bund. Assume, the vertical interval $= 1.25$ m and the runoff coefficient $c = 0.35$.

Solution

Given: length of the bund, $l = 350$ m; land slope, $S = 3\%$; 10-year storm rainfall intensity having same duration as the time of concentration $= 200$ mm h^{-1}; grade of the bund, $s = 0.18\%$; soil $=$ sandy; vertical interval, $VI = 1.25$ m; runoff coefficient, $C = 0.35$.

Putting known values of VI and S in Eq. (4.5) (Chap. 4),

The horizontal interval, $HI = \frac{VI}{S} \times 100 = \frac{1.25}{3} \times 100 = 42$ m.

Catchment area formed between two consecutive bunds: $A = HI \times l = 42 \times 350 = 14{,}700$ m^2

Peak runoff rate from the catchment area of 14,700 m^2 can be estimated using the rational formula (Eq. (3.19); Chap. 3): $Q = ci\,A$.

Substituting, runoff coefficient, $c = 0.35$, rainfall intensity,

$$i = 200 \text{ mm h}^{-1} = \left(\frac{200 \times 10^{-3}}{3600}\right) \text{m s}^{-1}, \text{ and } A = 14{,}700 \text{ m}^2$$

$$Q = ci\,A = 0.35 \times \left(\frac{200 \times 10^{-3}}{3600}\right) \times 14700 = 0.29 \text{ m}^3\text{s}^{-1}$$

Thus, the graded bund system needs to dispose of a flow of 0.29 m^3 s^{-1}

Referring to Fig. 5.11, the drainage area behind the bund

$$= \text{Area of triangle AMN, } a_1 + \text{Area of triangle M'MN, } a_2$$
$$= (\tfrac{1}{2} \times d \times 33.33d) + (\tfrac{1}{2} \times d \times 2d) = 17.67d^2$$

Note that the drainage area behind the bund represents the cross-section of the channel that would dispose of the estimated flow of 0.29 m^3 s^{-1},

Assuming $d = 0.25$ m, drainage area behind the bund $=$ cross-section area of flow

$$= 17.67 \times 0.25 \times 0.25 = 1.1 \text{ m}^2$$

The wetted perimeter $= AN + NM' = \sqrt{(AM^2 + MN^2)} + \sqrt{(MM'^2 + MN^2)}$

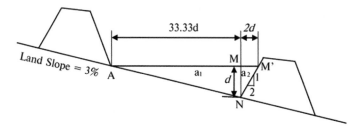

Fig. 5.11 Drainage area behind the graded bund (Example 5.8)

$$= \sqrt{\left((33.33 \times 0.25)^2 + (0.25)^2\right)} + \sqrt{\left((2 \times 0.25)^2 + (0.25)^2\right)} = 8.9 \text{ m}$$

$$\text{Hydraulic radius, } R = \frac{\text{Cross-section area of flow}}{\text{Wetted perimeter}} = \frac{1.1}{8.9} = 0.12 \text{ m}$$

Flow velocity from Manning equation (assuming $n = 0.035$),

$$V = \frac{1}{n} R^{2/3} s^{1/2} = \frac{1}{0.035} (0.12)^{2/3} (0.0018)^{1/2} = 0.29 \text{ m s}^{-1}$$

Using the continuity equation, the discharge that could be drained by the channel formed behind the bund

$$= \text{cross-section area of flows} \times \text{velocity of flow} = 1.1 \times 0.29 = 0.32 \text{ m}^3 \text{ s}^{-1}$$

Since the velocity of flow is non-erosive (<0.46 m s^{-1} for the sandy soil) and the capacity of the channel formed behind the bund is greater than the discharge to be disposed of, i.e., 0.29 m^3 s^{-1}, the assumed depth of flow, i.e., 0.25 m, is satisfactory.

Thus, assuming a freeboard of 20% of d,

The height of the bund, $H = 1.2d = 1.2 \times 0.25 = 0.3$ m

Using Eq. (5.7), the bottom width of the bund

$$B = Zd + Z_S d$$

Since soil type is sandy, from Tables 5.1 and 5.2, the bund side slopes and the seepage line slope will be 2:1 and 6:1, respectively. Thus, $Z = 2$ and $Z_S = 6$.

Substituting d, Z, and Z_S in Eq. (5.7),

$$B = Zd + Z_S d = 2 \times 0.25 + 6 \times 0.25 = 2.0 \text{ m}$$

Knowing B, Z, and H, the top width, T, can be calculated using Eq. (5.8):

$$T = B - 2ZH = 2 - (2 \times 2 \times 0.3) = 0.8 \text{ m}$$

Answers: Height of the bund $= 0.3$ m; top width of the bund $= 0.8$ m; bottom width of the bund $= 2.0$ m; side slopes of the bund $= 2$:1.

5.7 Planning Considerations for Bunds

The successful layout and construction of bunds require a well-organised plan of the area. A topographic map or digital elevation model may be used for developing the preliminary plans. The map or the digital elevation model should include various natural features like watercourses, gullies, cultivated lands with their field boundaries, and roads. Two different approaches may be adopted for positioning the bunds in the field. In the first approach, the bund positions may be marked on the map or the digital elevation model and then implemented in the field. Alternatively, the bund positions may be marked directly on the field. The latter, however, requires considerable expertise and experience. The vertical and horizontal interval values help in marking the bund positions.

In the case of graded bunds, the planning begins by selecting an appropriate outlet for disposing of the excess runoff from the area. Like terraces, natural drainage channels, especially the grassed waterways, are usually preferred as the outlet. In cases where natural drainage outlets are not available, vegetated waterways may be constructed along field boundaries. The capacity of the outlet must be sufficient to handle the excess runoff being disposed of from the area.

5.8 Alignment of Bunds

For aligning contour bunds, the topographic map or the digital elevation model is divided into several blocks of an appropriate size. The recommended block size in India is 20–25 hectares. The bund alignment starts at the ridgeline of the area, and the first bund is located directly in the field. The position of subsequent bunds is marked considering the vertical interval. The bunds should follow the contour line as far as possible. In case an undulation in the form of a gully or depression falls in the path, the bunds should usually cross these at right angles. A deviation of 0.10 and 0.20 m from the contour elevation is permitted on the upper and lower sides, respectively, for contour bunds to tackle the topographic undulations.

Side bunds are aligned along the slope on either end of the contour bunds. The side bunds are typically planned up to 0.30 m above the contour elevation to facilitate impoundment of 0.30 m depth of water behind the bunds. Lateral bunds are placed between side bunds when the length of contour bunds exceeds 300 m. These bunds prevent the concentration of water on one side by breaking the length of contour bunds.

The graded bunds are also aligned following the procedure similar to the one adopted for the contour bunds. A notable difference, however, is that the alignment of the graded bunds begins at the outlet end and moves upstream to the edge of the field. The elevation of outlet is fixed based on vertical interval and the highest elevation of the field.

5.9 Construction of Bunds

Like the bund alignment, the bund construction also starts from the ridgeline, and the bund nearest to the ridge is constructed first. The construction from the top helps in avoiding excessive runoff during heavy rains while the construction is in progress. First, the upstream toe of the bund is located on the ground as per the plan. Subsequently, the base width is marked, and the ground is cleared of the vegetation. The soil needed for the construction of the bunds is excavated from the burrow pits located on the upstream side. The stones and clods are removed from the excavated soil, and then the soil is placed in layers of 0.15 m and compacted to develop the bund cross-section of the designed dimensions. The bund dimensions are finally checked using templates, and the construction completed. The field is subsequently ploughed, and the burrow pits are filled.

Questions

1. Classify bunds, and describe specific features of different types of bunds.
2. Based on the information from the treated watersheds near your home town, enlist the types of bunds used, and the conditions under which a particular type of bund is preferred.
3. Describe the common design considerations briefly, in terms of storm frequency, bund spacing, bund side slopes, and seepage through the bunds, which play a vital role while designing bunds.
4. Determine the vertical and horizontal intervals of bunds for land having a 5% slope. The rainfall is heavy, and soils have an average infiltration rate. The cover conditions are good during the critical period.
5. Contour bunds, having a vertical interval of 1.35 m, are constructed on land having 3% slope. The effective rainfall during 24-h is 0.1 m. Determine the depth of impounding immediately behind the bund. Assume that the bund side slopes have a negligible effect on the storage area behind the bund.
6. Calculate the dimensions of a contour bund constructed on land having 4% slope. The rainfall excess for the 24-h duration is 100 mm. The horizontal interval of the bunds is 10 m, and the depth of water flowing over the weir is 300 mm.
7. Determine the excess runoff to be disposed of through the waste weir of a contour bund, constructed on land having 2.5% slope. The height, top width, and the side slopes of the bund are 0.7, 0.5 m, and 1.5:1, respectively, and the height of the weir crest is 0.3 m from the ground level. The vertical interval is 0.80 m, and the length of the bund is 300 m. The peak intensity of rainfall for a 10-year return period is 75 mm h^{-1}, and the peak rainfall occurs for 49 min. The constant soil infiltration rate during the peak rainfall is 20 mm h^{-1}. Assume no storage behind the bund before the peak rainfall.
8. Determine the dimensions of a rectangular weir to dispose of excess runoff through a contour bund. The horizontal interval and the length of the bund are 60 and 250 m, respectively, and the height of the weir crest is 0.2 m from the ground level. The peak intensity of rainfall for a 10-year return period is 90 mm h^{-1}, and the peak rainfall occurs for a duration equal to the time of concentration. The constant soil infiltration rate during the peak rainfall is 20 mm h^{-1}. The volume of water stored behind the bund during the storm is 750 m^3. Assume no storage behind the bund before the storm.

9. Determine the area occupied by a contour bund system on land with 5% slope. The horizontal spacing of the bund is 30 m. The base width of the bund is 1.0 m. Determine the area lost due to the bund system if the side and lateral bunds occupy 30% of the area.

10. Determine the dimensions of the graded bunds constructed on land having a slope of 5%. The vertical interval of the bund is 1 m. The soil in the area is sandy, and bund side slopes of 2:1 and the seepage line slope of 6:1 are recommended. The channel formed between the bunds is under vegetation with Manning roughness coefficient of 0.04. The bunds are expected to drain flow of 3.5 m^3 s^{-1} from the area. Assume a freeboard of 0.15 m. Also, calculate the volume of earthwork per hectare.

Multiple Choice Questions

Answer the questions by choosing the correct option.

1. Among the following, the most effective measure for soil erosion control is

    (A) contour farming   (B) fertilising
    (C) strip cropping   (D) terracing and bunding

2. Contour bunds are recommended for areas having

    (A) low rainfall   (B) medium rainfall
    (C) medium to high rainfall   (D) high rainfall

3. Graded bunds are useful in areas having

    (A) low rainfall   (B) medium rainfall
    (C) medium to high rainfall   (D) high rainfall

4. Contour bunds having a linear length L are being planned on a farm. The average land slope is S %, and the vertical interval between the bunds is D. The number of bunds per hectare of land would be

    (A) S/LD (B) LD/S
    (C) $100S/LD$ (D) $10000S/LD$

5. The line within a bund section, below which there is positive hydrostatic pressure, is known as

    (A) equipotential line   (B) flow line
    (C) isopluvial line   (D) seepage line

6. Contour bunds are constructed on land having a 5% slope. The top width, bottom width, and height of the bunds are 0.5 m, 2.5 m, and 1.9 m, respectively. The vertical interval between the bunds is 1.2 m. The side and lateral bunds have the same cross-section as the contour bund, and their combined length is equal to 30% of the length of the contour bunds. The total area lost (%) due to the bund system would be

(A) 4.2   (B) 5.2
(C) 10.8   (D) 13.5

7. Contour bunds are constructed to store water on land having a 6% slope. The horizontal interval between the bunds is 30 m, and the side slopes of the bund are 1.5:1. The rainfall having 10-year recurrence interval is 180 mm. If the infiltration loss is neglected, the required height of the bund (in mm) would be

(A) 540   (B) 770
(C) 780   (D) 805

8. Contour bunds are constructed on land having a 6% slope. The top width, bottom width, and the height of the bunds are 0.5 m, 1.25 m, and 1.0 m, respectively. The horizontal interval between the bunds is 25 m. The volume of the earthwork required for the construction of the bunds per hectare of land, in m^3, would be

(A) 225   (B) 350
(C) 375   (D) 650

9. Identify the incorrect statement:

(A) Side bunds are constructed along the land slope at the extreme ends of contour bunds.
(B) Lateral bunds are constructed along the land slope in between two side bunds.
(C) Semi-circular bunds are also referred to as crescent-shaped bunds.
(D) Contour stone bunds are also known as marginal bunds.

10. Match Column I and Column II and identify the most appropriate combinations:

Column I		Column II	
P	Contour bunds	1	Demi-lunes
Q	Graded bunds	2	Constructed along the land slope in between two side bunds
R	Side bunds	3	Narrow-base trapezoidal embankments constructed along contours

(continued)

(continued)

Column I		Column II	
S	Lateral bunds	4	Porous barriers constructed along the contours
T	Semi-circular bunds	5	Constructed at the extreme ends of contour bunds
U	Contour stone bunds	6	Constructed across the land slope but not along contours

(A) P-3, Q-6, R-1, S-2, T-5, U-4
(B) P-3, Q-6, R-5, S-2, T-1, U-4
(C) P-4, Q-3, R-5, S-1, T-6, U-2
(D) P-3, Q-6, R-5, S-2, T-1, U-4

References

Hai MT (1998) Water harvesting: an illustrative manual for development of microcatchment techniques for crop production in dry areas. Regional Land Management Unit (RELMA) Handbook Series 16, Swedish International Development Authority, Nairobi

Mati BM (2007) 100 Ways to manage water for smallholder agriculture in Eastern and Southern Africa. SWMnet Working Paper 13, Improved Management in Eastern & Southern Africa (IMAWESA), Nairobi

Pathak P, Laryea KB, Singh S (1989) A modified contour bunding system for Alfisols of the semi-arid tropics. Agric Water Manag 16:187–199

Singh G, Venkataramanan C, Sastry G, Joshi BP (1990) Manual of soil and water conservation practices. Oxford and IBH Publishing, New Delhi

Singh PV, Kumar A, Beg MKA, Gupta A (2010) Nomographs for graded bund design in heavy soils. Indian J Soil Conserv 38:155–160

Chapter 6
Vegetated Waterways

Abstract Vegetated waterways are natural or constructed channels having vegetative cover to dispose of runoff safely without causing erosion. These waterways are designed using the *permissible velocity approach* and constructed along the natural slope. This chapter presents the preliminary design considerations for vegetated waterways and elaborates on the design processes to decide the size, shape, vegetation, permissible velocity, and roughness coefficient. Solved examples are included to demonstrate the design procedure. The chapter also contains the layout, construction, and maintenance of the waterways.

6.1 Vegetated Waterways

Vegetated waterways are natural or constructed channels lined with suitable vegetation to dispose of runoff safely without causing erosion. Besides conveying runoff, vegetated waterways prevent gully formation, act as outlets for terraces or graded bunds, and protect and improve water quality. Vegetated waterways are constructed along the natural slope and provided with a carrying capacity to handle the peak runoff rate generated from the area. These waterways are usually designed using the *permissible velocity approach* to carry the peak flow without damaging the vegetal lining.

The usage of vegetation, mainly grass, as a lining material for earthen channels was propagated by Soil Conservation Service (SCS) (now Natural resources Conservation Service (NRCS)) in the 1940s. SCS proposed the permissible velocity approach for designing waterways, which is widely adopted (SCS 1966). Researchers found that in a vegetated waterway, the vegetation length and density influenced the Manning roughness n. The Manning roughness n was found to vary with a flow parameter represented by the product of the velocity and hydraulic radius, VR (Cox 1942; Ree 1949; Ree and Palmer 1949; SCS 1966; Yong and Stone 1967; Eastgate 1969; HR Wallingford 1992).

© The Author(s), under exclusive license to Springer Nature Singapore Pte Ltd. 2023   121
R. Singh, *Soil and Water Conservation Structures Design*, Water Science and Technology
Library 123, https://doi.org/10.1007/978-981-19-8665-9_6

6.2 Vegetated Waterway Design

6.2.1 Preliminary Design Considerations

While designing the vegetated waterways, the site conditions must be first assessed in terms of location, climate, topography, soil characteristics, and vegetation, and then the channel shape is chosen.

1. Location: Typically, a natural drainage channel that receives flow from all sides may be shaped and used as a vegetated waterway. Alternatively, vegetated waterways could be located along the farm boundary or an existing road.
2. Climate: It governs the hydrology of the site and the design discharge. It also influences vegetation selection. The capacity of the channel is selected to handle the expected peak flow generated from the contributing drainage area. Typically, a storm having a 10-year frequency is used to estimate the peak flow rate.
3. Topography: It governs the alignment and bed slope of the waterway. If the slope at the site changes considerably, the bed slope also changes. If the changes in the channel bed slope are significant, the channel may be broken into reaches. Also, if the length of the waterway is significantly long, it may be necessary to consider multiple reaches to account for the flow rate variations.
4. Soil Properties: These govern the permissible flow velocity in the waterway. Besides, these also influence the expected peak flow generated from the contributing drainage area.
5. Vegetation: It protects the waterway and governs the permissible flow velocity in the waterway. Also, the height of vegetation governs the resistance to flow and affects the design capacity of the waterway.

6.2.2 Design Process

The design of a vegetated waterway includes determining the waterway size and its cross-sectional shape and dimensions. It also includes selecting the appropriate vegetation and subsequently determining the roughness coefficient and the design velocity.

6.2.2.1 Size of Waterway

As mentioned earlier, the vegetated water is typically designed to handle the peak flow rate due to a 10-year frequency storm. The peak flow rate can be determined using the rational method, described in Sect. 3.2.4 (Chap. 3). Once the peak flow rate is determined, the cross-sectional area of the waterway can be estimated using the continuity equation:

$$A = \frac{Q}{V} \tag{6.1}$$

where A = cross-sectional area of the waterway, m^2; Q = peak flow rate, m^3 s^{-1}; and V = velocity of flow, m s^{-1}.

6.2.2.2 Shape of Waterway

The three typical cross-sectional shapes adopted for vegetated waterways are trapezoidal, triangular, or parabolic (Fig. 6.1). The parabolic shape is the most common as it represents a natural channel. Even the trapezoidal or triangular waterways assume a near parabolic shape after some time due to bed and bank erosion and sedimentation. Consequently, parabolic waterways are more stable and economical. Parabolic or trapezoidal shapes with flat side slopes are preferred where farm equipment has to cross the waterways.

The cross-sectional shapes are chosen based on the site conditions and the construction equipment available. Small waterways can easily be constructed using standard agricultural equipment like ploughs or blades. However, large waterways may require graders, scrapers, or excavators. Trapezoidal cross-sections are preferred with blade-type machines if the blade width is smaller than the desired bottom width of the waterway. Standard equipment may also construct a triangular or parabolic-shaped waterway, provided the side slopes are flatter than 4:1.

Table 6.1 presents the geometrical characteristics of the typical cross-sections shown in Fig. 6.1.

The bed slope of the vegetated waterways is chosen to satisfy the velocity, capacity, and lining requirements. The bed slope should be selected to minimise the sediment

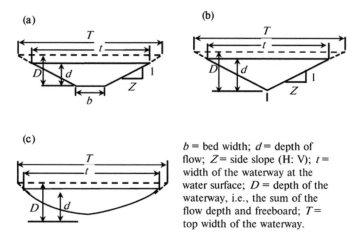

b = bed width; d = depth of flow; Z = side slope (H: V); t = width of the waterway at the water surface; D = depth of the waterway, i.e., the sum of the flow depth and freeboard; T = top width of the waterway.

Fig. 6.1 Typical cross-sections of the vegetated waterways, **a** trapezoidal, **b** triangular, **c** parabolic

Table 6.1 Geometric characteristics of typical cross-sections of vegetated waterways

Cross-section	Cross-sectional area (A)	Wetted perimeter (P)	Top width at the water surface (t)	Top width (T)
Trapezoidal	$bd + zd^2$	$b + 2d\sqrt{z^2 + 1}$	$b + 2zd$	$b + 2zD$
Triangular	zd^2	$2d\sqrt{z^2 + 1}$	$2zd$	$\frac{tD}{d}$
Parabolic	$\frac{2}{3}td$	$t + \frac{8d^2}{3t}$	$\frac{A}{0.67d}$	$t\sqrt{\frac{D}{d}}$

deposition during usual flows and erosion during extreme flow events. Typically, the waterway bed slope follows the natural slope of the site. However, adjustments may be needed to meet the velocity or the capacity criteria.

6.2.2.3 Vegetation Selection for Waterway

The selection of the most appropriate vegetation for the waterway depends on the climate and soil characteristics of the site. Besides, the ease of vegetation establishment and time required to grow a protective cover may also govern the vegetation selection.

Usually, the rhizomatous grasses having the characteristic of producing many rhizomes, e.g., Bermuda grass (*Cynodon dactylon* (L) Pers), a native to India with the popular name doob, are preferred for the waterway because they establish easily and grow fast to provide a protective cover to the waterways. In India, besides Bermuda grass or doob, a combination of two species of vetiver grass, i.e., *Vetiveria zizanioides* L Nash (now reclassified as *Chrysopogon zizanioides* L Roberty) and *Chrysopogon lawsonii* (Hook f) Veldkamp, and Cannas (*Canna indica* L) are popularly used in vegetated waterways. *Vetiveria zizanioides* L, commonly known as Khas, and *Chrysopogon lawsonii* (Hook f) are inexpensive, low-maintenance grasses that are highly effective in slowing down the flow in the waterway and preventing soil erosion. *Canna indica* L, perennial herbs with a rhizomatous rootstock, also act like vetiver grasses, reducing the flow velocity and preventing soil erosion. Cannas, commonly known as Indian shot or Canna lily, is one of the most popular garden plants worldwide (Trivedi 2002; Omkar et al. 2018).

6.2.2.4 Permissible Velocity in Waterway

The permissible flow velocity in the grassed waterway depends on the type and condition of vegetation and the erosion characteristics of the soil. Table 6.2 presents the permissible flow velocities based on vegetation covers and soil types (SCS 1979). The permissible velocity values apply to average, uniform stands of each type of cover.

Table 6.2 Permissible flow velocities in vegetated waterways (after SCS 1979)

Cover	Slope range (%)[b]	Permissible velocity (m s^{-1})[a]	
		Erosion-resistant soils	Easily eroded soils
Bermuda grass	0–5	2.4	1.8
	5–10	2.1	1.5
	> 10	1.8	1.2
Bahia grass	0–5	2.1	1.5
Buffalo grass	5–10	1.8	1.2
Kentucky bluegrass	> 10	1.5	0.9
Blue grama			
Tall fescue			
Grass mixtures	0–5	1.5	1.2
Reed canary grass	5–10	1.2	0.9
Sericea lespedeza	< 5[c]	1.0	0.8
Weeping love grass			
Yellow bluestem			
Redtop			
Alfalfa			
Red fescue			
Common lespedeza[d]	< 5[e]	1.0	0.8
Sudan grass[d]			

[a]Use velocities exceeding 1.5 m s^{-1} only where good covers and proper maintenance can be obtained
[b]Do not use on slopes steeper than 10% except for vegetated side slopes in combination with a stone, concrete, or highly resistant vegetative centre section
[c]Do not use on slopes steeper than 5% except for vegetated side slopes in combination with a stone, concrete, or highly resistant vegetative centre section
[d]Annuals use on mild slope or as temporary protection until permanent covers are established
[e]Use on slopes steeper than 5% is not recommended

Erosion-resistant soils are cohesive (clayey) fine-grained and coarse-grained soils that have cohesive fines and a plasticity index of 10–40. On the contrary, easily eroded soils do not have the characteristics specified for erosion-resistant soils. Soil density also affects the erosion resistance.

The flow velocity in a grassed waterway is typically determined using the Manning equation, given as,

$$V = \frac{1}{n} R^{2/3} S^{1/2} \tag{6.2}$$

where R = hydraulic radius of the channel, m; and S = channel bed slope, fraction.

6.2.2.5 Roughness Coefficient of Waterway

The roughness coefficient of a vegetated waterway is significantly influenced by the type and height of the vegetation, besides the flow depth and the bed slope of the waterway. Short vegetation will offer lesser resistance than tall vegetation. Also, a waterway will have higher resistance with low flows than with high flows. Similarly, a waterway with a steeper slope will have less resistance as higher velocity will flatten the vegetation. A series of experimental studies conducted by various researchers at the US National Resources Conservation Service (earlier SCS) concluded that the vegetation length and density influenced the Manning roughness n. Also, the Manning roughness n varied with a flow parameter represented by the product of the velocity, V, and flow depth, D. However, to accommodate various cross-sectional shapes of the waterway, D was replaced by hydraulic radius, R (Cox 1942; Ree 1949; Ree and Palmer 1949; SCS 1966).

The use of product VR as the flow parameter to represent the channel retardance was also propagated by other researchers (Yong and Stone 1967; Eastgate 1969; HR Wallingford 1992). Usually, product VR is inversely proportional to the resistance offered by vegetation.

NRCS (2021) categorised vegetation into five retardance groups, A to E (Table 6.3). Figure 6.2 presents the n-VR curves for these five retardance groups (SCS 1979). While using the Manning equation for determining the flow velocity, the Manning n, representing the retardance, has to be obtained for the vegetation from Table 6.3 and Fig. 6.2.

6.2.3 Design Procedure

Considering a typical case where the cross-sectional shape, the carrying capacity, Q, the bed slope, S, and vegetation type are known, the design procedure comprises the following steps.

1. Determine the permissible velocity of flow, $V_{permissible}$, based on the vegetation condition and soil type (Table 6.2).
2. Compare the calculated V and Q with the permissible velocity, $V_{permissible}$, and design Q. If the values are acceptable, go to the next step; otherwise, adjust the flow depth to recalculate V and Q.
3. Determine VR and obtain Manning n using the NRCS procedure (Table 6.3 and Fig. 6.2).
4. Compare the calculated n with the assumed n. If the values match, okay; else, repeat steps 4 and 5.
5. Please note that steps 3–5 are iterative.

Add a freeboard of 0.10–0.15 m to the flow depth, d, to obtain the depth of the waterway, D. Determine the top width, T, of the waterway to complete the design.

Example 6.1 Design a trapezoidal-shaped grassed waterway to carry the peak flow of 8 m^3 s^{-1}. The bed slope of the waterway is 5%. The waterway has Bermuda grass cover and the easily eroded soil. Assume a side slope of 4:1.

Table 6.3 Vegetal retardance groups (after NRCS 2021)

Retardance	Cover	Condition
A	Weeping love grass	Excellent stand, tall (average 762 mm)
	Reed canary grass or Yellow bluestem ischaemum	Excellent stand, tall (average 914 mm)
	Smooth bromegrass	Good stand, mowed (average 305–381 mm)
	Bermuda grass	Good stand, tall (average 305 mm)
	Native grass mixture (little bluestem, blue grama, and other long and short Midwest grasses)	Good stand, unmowed
B	Tall fescue	Good stand, unmowed (average 457 mm)
	Sericea lespedeza	Good stand, not woody, tall (average 483 mm)
	Grass-legume mixture—Timothy, smooth bromegrass, or orchard grass	Good stand, uncut (average 508 mm)
	Reed canary grass	Good stand, uncut (average 305–381 mm)
	Tall fescue, with birdsfoot trefoil or ladino clover	Good stand, uncut (average 457 mm)
	Blue grama	Good stand, uncut (average 330 mm)
	Bahiagrass	Good stand, uncut (152–203 mm)
	Bermuda grass	Good stand, mowed (average 152 mm)
C	Redtop	Good stand, headed (381–508 mm)
	Grass-legume mixture—summer (orchard grass, redtop, Italian ryegrass, and common lespedeza)	Good stand, uncut (152–203 mm)
	Centipede grass	Very dense cover (average 152 mm)
	Kentucky bluegrass	Good stand, headed (152–305 mm)
	Bermuda grass	Good stand, cut to 64 mm height
	Red fescue	Good stand, headed (305–457 mm)
	Buffalo grass	Good stand, uncut (76–154 mm)
D	Grass-legume mixture—fall, spring (orchard grass, redtop, Italian ryegrass, and common lespedeza)	Good stand, uncut (102–127 mm)
	Sericea lespedeza or Kentucky bluegrass	Good stand cut to 51 mm height. Very good stand before cutting
E	Bermuda grass	Good stand, cut to 38 mm height
	Bermuda grass	Burned Stubble

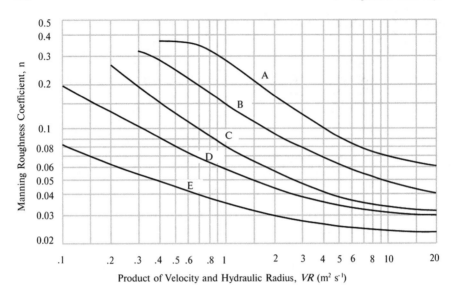

Fig. 6.2 Manning n as a function of VR for different vegetation retardance groups (after SCS 1979)

Solution Given: Waterway cross-section = trapezoidal; design capacity, $Q = 8$ m³ s⁻¹; bed slope, $S = 5\%$; side slope, $Z = 4$; vegetation = Bermuda grass cover; soil type = easily eroded.

We may follow the steps given in the design procedure.

1. From Table 6.2, the permissible velocity for Bermuda grass on easily eroded soils, having 5–10% slope, is 1.5 m s⁻¹. Therefore, assuming $V_{permissible} = 1.5$ m s⁻¹.
2. Using Eq. (6.1), the minimum cross-sectional area,

$$A = \frac{Q}{V} = \frac{8}{1.5} = 5.33\ \text{m}^2$$

3. Assume $d = 0.1$ m, and determine b, P, and R (Table 6.4; iteration 1).
4. Assume $n = 0.036$ and calculate flow velocity using Eq. (6.2):

$$V = \frac{1}{n}R^{2/3}S^{1/2} = \frac{1}{0.036} \times (0.10)^{2/3} \times (0.05)^{1/2} = 1.33\ \text{m s}^{-1}$$

Using Eq. (6.1), $Q = AV = 5.33 \times 1.33 = 7.10$ m³ s⁻¹

Since both Q and V are lower than the desired values, we go for the second iteration, as shown in Table 6.4.

In iteration 2, let us assume $d = 0.12$ m. As evident from the Table, both Q and V match their desired values.

5. We now calculate VR (column 9) and find the corresponding n (column 10) from Fig. 6.2 for retardance class E. The value of n (column 10) obtained from Fig. 6.2

Table 6.4 Iterative calculations following steps 3–5 (Example 6.1)

Iteration	d (m)	b (m)	P (m)	R (m)	n	V (m s^{-1})	Q (m^3 s^{-1})	VR (m^2 s^{-1})	n	Remarks
(1)	(2)	(3)	(4)	(5)	(6)	(7)	(8)	(9)	(10)	(11)
1	0.10	52.93	53.76	0.10	0.036	1.33	7.10			V and Q are lower than design values
2	0.12	43.96	44.95	0.12	0.036	1.50	8.00	0.18	0.062	n in Col. (10) higher than in Col. (6)
3	0.12	43.96	44.95	0.12	0.062	0.87	4.64	0.10	0.090	V and Q lower than design values
4	0.22	23.36	25.18	0.21	0.062	1.28	6.84	0.27	0.056	V and Q lower than design values
5	0.22	23.36	25.18	0.21	0.052	1.53	8.15	0.32	0.052	V and Q are close to their design values, and n in Columns (6) and (10) matches

is higher than the value of n assumed in step 4 (column 6); we go for iteration (3) with $n = 0.062$.

As evident, both Q and V fall below the desired values, and also n in column (10) differs from its value in column (6). Therefore, in iteration (4), we assume $d = 0.56$ and find that both Q and V are below their desired values, and also n in column (10) differs from its value in column (6). In the next iteration, we change d and find that Q and V have increased values, and also n in column (10) is closer to the assumed value in column (6). In the next iteration, we change n value in column (6) to 0.052 and find that Q and V are close to their desired values, and also n in column (10) matches the assumed value in column (6). Now, the iterative process involved in steps 3–5 is complete.

6.   Add a freeboard, f, of 0.13 m. Hence,

Depth of the waterway, $D = d + f = 0.22 + 0.13 = 0.35$ m.
The top width of the waterway,

$$T = b + 2ZD = 23.26 + 2 \times 4 \times 0.35 = 25.12\,\text{m}$$

Answer: The trapezoidal waterway has the following dimensions: Depth, $D = 0.35$ m; base width, $b = 23.26$ m; side slope, $Z = 4$; top width, $T = 25.12$ m; bed slope, $S = 5\%$.

Example 6.2

a.   Determine the peak runoff rate expected to occur once in 10 years in a 35 ha watershed. The time of concentration of the watershed is 30 min, and the maximum rainfall for 10-year, 30-min storm is 50 mm. The runoff coefficient of the watershed is 0.40.
b.   Design a trapezoidal grassed waterway to carry the peak runoff rate.

The bed width, bed slope, and the side slopes are 5 m, 2%, and 4:1, respectively. Assume the Manning roughness coefficient as 0.04.

Solution   Given: Watershed area, $A = 35$ ha; time of concentration, $T_c = 30$ min; maximum rainfall for 10-year, 30-min storm $= 50$ mm; runoff coefficient $c = 0.40$; waterway cross-section $=$ trapezoidal; bed width, $b = 5$ m; bed slope, $S = 2\%$; side slope, $Z = 4$; Manning roughness coefficient, $n = 0.04$.

(a)   The peak runoff rate can be determined using the rational formula (Eq. (3.19), Chap. 3) as $Q = c\,i\,A$

From the given rainfall data, the rainfall intensity for use in the rational formula,

$$i = \frac{50}{30} \times 60 = 100\,\text{mm h}^{-1} = \frac{100 \times 10^{-3}}{3600} = 2.78 \times 10^{-5}\,\text{m s}^{-1}$$

Putting the values of c, I, and A in Eq. (3.19)

$$Q = ci\,A = 0.40 \times (2.78 \times 10^{-5}) \times (35 \times 10^4) = 3.89\,\text{m}^3\,\text{s}^{-1}$$

(b)   From (a), the trapezoidal waterway has a discharge capacity, $Q = 3.89\,\text{m}^3\,\text{s}^{-1}$

Assuming the depth of flow, $d = 0.50$ m, A, P, and R can be determined using the geometric characteristics relationships for the trapezoidal waterway (Table 6.1):

$$A = bd + Zd^2 = 5 \times 0.5 + 4 \times (0.5)^2 = 3.5\,\text{m}^2$$

$$P = b + 2d\sqrt{Z^2 + 1} = 5 + 2 \times 0.5 \times \sqrt{4^2 + 1} = 9.12\,\text{m}$$

$$R = \frac{A}{P} = \frac{3.5}{9.12} = 0.38\,\text{m}$$

Putting n, R, and S in Eq. (6.2) to calculate the flow velocity,

$$V = \frac{1}{n} R^{2/3} S^{1/2} = \frac{1}{0.04} \times (0.38)^{2/3} \times (0.02)^{1/2} = 0.84\,\text{m s}^{-1}$$

Using Eq. (6.1), $Q = AV = 3.5 \times 0.84 = 2.94\,\text{m}^3\,\text{s}^{-1}$
Since the calculated Q is smaller than the peak flow estimated in (a), we need to iterate.
Now, assuming $d = 0.56$ m,

$$A = bd + Zd^2 = 5 \times 0.56 + 4 \times (0.56)^2 = 4.05\,\text{m}^2$$

$$P = b + 2d\sqrt{Z^2 + 1} = 5 + 2 \times 0.56 \times \sqrt{4^2 + 1} = 9.62\,\text{m}$$

$$R = \frac{A}{P} = \frac{4.05}{9.62} = 0.42\,\text{m}$$

Putting n, R, and S in Eq. (6.2) to calculate the flow velocity,

$$V = \frac{1}{n} R^{2/3} S^{1/2} = \frac{1}{0.04} \times (0.42)^{2/3} \times (0.02)^{1/2} = 0.97\,\text{m s}^{-1}$$

Using Eq. (6.1), $Q = AV = 4.05 \times 0.97 = 3.92\,\text{m}^3\,\text{s}^{-1}$
Since the calculated Q is almost the same as the desired $Q = 3.89\,\text{m}^3\,\text{s}^{-1}$, the assumed flow depth is acceptable.
Adding a freeboard, f, of 0.14 m,
Depth of the waterway, $D = d + f = 0.56 + 0.14 = 0.70$ m.
The top width of the waterway, $T = b + 2ZD = 5 + 2 \times 4 \times 0.7 = 10.60$ m.

Answer:

(a)   The peak rate of the runoff from the watershed is $3.89\,\text{m}^3\,\text{s}^{-1}$.
(b)   The designed trapezoidal waterway has the following dimensions:

Depth of flow, $d = 0.56$ m; freeboard, $f = 0.14$ m; depth, $D = 0.70$ m; base width, $b = 5.0$ m; side slope, $Z = 4$; top width, $T = 10.60$ m; bed slope, $S = 1\%$.

6.3   Waterway Layout and Construction

The waterway layout usually begins at the outlet. However, any other prominent landscape feature may also be used as the starting point. Subsequently, the centreline and the construction area are staked. The best time to construct a waterway is the dry season when the flow will be minimal, and vegetation will have enough time to establish before the storms occur.

As mentioned earlier, for small waterways, standard agricultural equipment like ploughs or blades may be sufficient. However, large waterways may require scrapers, graders, bulldozers, or excavators. Waterways must be shaped correctly to ensure design features and reduce maintenance. The establishment of the vegetation follows the waterway construction. The seeds of the mixture of grasses and legumes, chosen based on the soil and climate conditions, are either broadcasted or drilled perpendicular to the flow direction. Sodding may be done in cases where an immediate cover is needed.

6.4   Waterway Maintenance

It is essential to maintain and keep the waterways in good working condition. Regular maintenance may include proper cutting and removing the vegetation to avoid obstruction to the flow and excessive sedimentation. The movement of humans or animals should not be allowed in the vegetated waterways. The crossing of the waterway by vehicular traffic should not be permitted, except at designated crossings. If tillage implements have to cross the waterways, proper precautions should be taken to avoid harming the waterway's surface.

The waterways must be checked frequently during the rainy season, immediately after construction. If there is any erosion or damage, it should be repaired promptly. Besides routine maintenance checks, the additional inspection must be carried out following a heavy storm to ensure proper functions of the waterways. The outlet of the waterway must be given special attention to avoid erosion in the downstream channel.

Questions

1.   Define a vegetated waterway. Discuss various factors that must be considered while designing a vegetated waterway.
2.   Based on the information from the watersheds closer to your home town, managed under the soil management programmes, describe various vegetation preferred as a lining material for vegetated waterways.
3.   Two triangular waterways A and B have the same depth of flow and the bed slope. The Manning roughness coefficient of A and B is also the same. The side slopes of waterways A and B are 1:1 (horizontal: vertical) and 2:1, respectively. Determine the ratio of discharges in waterways A and B.

4. A parabolic vegetated waterway has a top width of 5.0 m, depth of flow 0.4 m, and bed slope of 4%. The roughness coefficient of the vegetation is 0.04. What is the discharge capacity of the waterway? What would be the per cent change in the discharge capacity of the waterway if the roughness coefficient of the vegetation changes to 0.05?

5. Determine the dimensions of a trapezoidal-shaped vegetated waterway to carry the peak flow of 4 m^3 s^{-1}. The bed slope of the waterway is 0.2%. Assume a permissible flow velocity of 0.9 m s^{-1} and the Manning roughness coefficient of 0.45.

6. Design a trapezoidal-shaped grassed waterway to carry the peak flow of 5 m^3 s^{-1}. The bed slope of the waterway is 2%. The waterway has a Bermuda grass cover and easily eroded soil. Assume a side slope of 6:1.

7. Design a parabolic-shaped vegetated waterway having a Bermuda grass cover to convey the peak flow of 3.0 m^3 s^{-1}. The soil is erosion-resistant, and the land slope is 3%.

8. A triangular vegetated waterway is designed to carry 2.0 m^3 s^{-1} with a permissible velocity of 1.0 m s^{-1}. The bed slope of the waterway is 2%. The side slopes of the channel are 2:1 (horizontal: vertical). Determine the top width of the waterway without considering the freeboard.

9. Design the most economical trapezoidal section of a grassed waterway to carry 5 m^3 s^{-1} in a bed slope of 1%. For the waterway, the side slopes are 1.5:1, and the Manning n is 0.04.

10. Design the most economical triangular section of a grassed waterway to carry 3 m^3 s^{-1} in a bed slope of 1%. For the waterway, the side slopes are 2:1, and the Manning n is 0.03.

Multiple Choice Questions

Answer the questions by choosing the correct option.

1. The most stable and economic cross-sectional shape of a vegetated waterway is

    (A) parabolic   (B) triangular
    (C) trapezoidal   (D) any regular shape

2. For a hydraulically efficient trapezoidal waterway with a depth of flow (d), the bed width (B) is

    (A) $2\sqrt{3}d$ (B) $d/2$ (C) $(2/\sqrt{3})d$ (D) $d/2\sqrt{3}$

3. In vegetated waterways, the Manning roughness n usually varies with the vegetation length, density, and flow parameter. The flow parameter is represented by the product (V, D, R, and S represent flow velocity, flow depth, hydraulic radius, and the bed slope)

    (A) VD (B) VR
    (C) $R^{2/3}S^{1/2}$ (D) $\frac{1}{V}R^{2/3}S^{1/2}$

4. While estimating the peak flow for designing vegetated waterways, the

recurrence interval (in years) used is

(A) 10   (B) 20
(C) 25   (D) 50

5. For Bermuda grass cover in a vegetated waterway, having < 5% bed slope, the permissible flow velocity, in m s^{-1}, is taken as

(A) 0.6–1.2   (B) 1.2–1.8
(C) 1.8–2.4   (D) 2.4–3.0

6. A triangular-shaped vegetated waterway is designed to carry 2.0 m^3 s^{-1} at a permissible velocity of 1.5 m s^{-1}. The bed slope and the side slopes of the waterway are 2% and 1.5:1 (horizontal: vertical), respectively. Without considering freeboard, the top width of the waterway, in metres, is

(A) 0.89   (B) 1.33
(C) 2.67   (D) 2.83

7. A trapezoidal-shaped grassed waterway has a bed slope of 2%. In a typical flow condition in the waterway, the wetted perimeter is 7.2 m, and the cross-sectional flow area is 2.64 m^2. If the Manning roughness coefficient for the waterway is 0.03, the flow through the waterway in m^3 s^{-1} is

(A) 3.93   (B) 6.38
(C) 10.71   (D) 17.39

8. A parabolic-shaped grassed waterway has a top width of 5 m, a maximum depth of 0.50 m, and a bed slope of 2.0%. The Manning roughness coefficient is 0.04, and there is no provision of freeboard. The discharge carrying capacity of the waterway in m^3 s^{-1} is

(A) 1.38   (B) 2.69
(C) 2.78   (D) 7.87

9. A trapezoidal-shaped vegetated waterway carries 6 m^3 s^{-1}. The waterway has a bottom width of 5 m, side slopes of 2:1, and a bed slope of 1%. If the Manning roughness coefficient is 0.025, the section factor of the waterway is

(A) 1.5   (B) 6
(C) 15   (D) 60

10. A hydraulically efficient trapezoidal waterway has a depth of 1.0 m and side slopes of 2:1. If the velocity of flow in the waterway is 0.5 m s^{-1}, the discharge in m^3 s^{-1} is

(A) 0.58   (B) 1.25   (C) 1.58   (D) 2.81

References

Cox MB (1942) Tests on vegetated waterways. Technical Bulletin T-15, Oklahoma Agricultural Ex-perimental Station, Oklahoma

Eastgate W (1969) Vegetated stabilisation of grassed waterways and dam by washes. Bulletin 16, Water Research Foundation of Australia, Den-ver, USA

HR Wallingford (1992) The hydraulic roughness of vegetated channels. Report SR 305, Hydraulics Research Ltd., Wallingford, Oxfordshire, UK

Natural Resources Conservation Service (NRCS) (2021) Grassed Waterways. Chapter 7: In: National Engineering Handbook, Part 650 Engineering Field Handbook, 210–650-H, 2nd ed. USDA Natural Re-sources Conservation Service, Washington DC

Omkar D, Ravikumar V, Joyce M (2018) Vegetative waterways. Int J Adv Sci Eng Tech 6:23–26

Ree WO (1949) Hydraulic characteristics of vegeta-tion for vegetated waterways. Agric Eng 30:184–189

Ree WO, Palmer VJ (1949) Flow of water in chan-nels protected by vegetative linings. Technical Bul-letin 967, USDA Natural Resources Conservation Service, Washington DC

Soil Conservation Service (SCS) (1966) Handbook of channel design for soil and water conservation structures. TP61, USDA Soil Conservation Service, Washington DC

Soil Conservation Service (SCS) (1979) Engineer-ing Field Manual for Conservation Practices. USDA Soil Conservation Service, Washington DC

Trivedi BK (2002) Grasses and legumes for tropical pastures. Indian Grassland and Fodder Research Institute, Jhansi, India

Yong KC, Stone DM (1967) Resistance of low-cost surfaces for farm dam spill-ways. Wales Research Laboratory Report 95, University of New South Wales, Australia

Chapter 7
Gully Control Structures

Abstract Gully control structures, i.e., the check dams, have been used since the twelfth century for soil and water conservation and more frequently over the past 150 years. These are employed in severely eroded gullies that cannot be managed with biological or vegetative erosion control measures. The temporary or permanent structures are constructed across the gully to reduce the channel gradient and stabilise the gully to prevent further erosion. This chapter presents the design principles used in designing temporary gully control structures, i.e., different check dams, preferred in areas where labour is inexpensive and the appropriate construction materials are readily available. The design includes the number of structures, spacing between structures and a spillway to handle the peak runoff due to a 10-year return period storm. Subsequently, the chapter introduces three established permanent gully control structures, i.e., the drop spillway, drop-inlet spillway and chute spillway, preferred in medium to large gullies with significantly high flows that the temporary structures cannot handle. The hydrologic, hydraulic and structural design principles of the permanent structures are introduced. The chapter also includes the prerequisites, viz., the specific energy considerations, critical flow characteristics and hydraulic jump, for designing permanent structures.

7.1 Background

Gully control structures are essentially used in gullies facing severe erosion to such an extent that these cannot be tackled with biological or vegetative erosion control measures. These structures are built across the gully to reduce the channel gradient and stabilise the gully to prevent further erosion. The structures could be either temporary or permanent.

Temporary structures are preferred in areas where labour is inexpensive, and the construction materials like planks, logs, stones, brush, woven wire or earth are readily available. On the other hand, permanent structures need technical skills for their design and construction and are built using concrete or masonry.

The use of the gully control structures may be traced back to the twelfth century when the aborigines of Mexico used trincheras, check dams built from loose rock, to

© The Author(s), under exclusive license to Springer Nature Singapore Pte Ltd. 2023 137
R. Singh, *Soil and Water Conservation Structures Design*, Water Science and Technology Library 123, https://doi.org/10.1007/978-981-19-8665-9_7

develop an intensive agricultural system. Similarly, the first torrent control structures may be traced back to the sixteenth century when a retention dam was constructed in Trent, Italy using wood and masonry. The check dams, however, have been used popularly for soil and water conservation over the past 150 years in China, India, Iran, Japan, and Thailand. In the USA, the gully management practice on agricultural lowlands started in the 1930s. The first textbook on gully control was published in France in the 1860s. In India, ancient techniques have been employed to construct check dams. The first check dam of India dates back to the second century A.D. when the Grand Anicut was built in Tamil Nadu. Since 1960, around 200,000 check dams have been constructed in India for water harvesting and erosion control.

7.2 Temporary Gully Control Structures

The temporary gully control structures are constructed to retard the flow of water, reduce the channel erosion, and stabilise gullies by encouraging the vegetative growth in the gully. These are built using locally available materials and are simple to construct and maintain. These usually have a life span of about 3–8 years.

The temporary gully control structures include various types of check dams listed below.

1. Woven-wire check dams
2. Brushwood check dams
3. Loose rock dams
4. Log check dams
5. Gabion check dams.

7.2.1 Design of Temporary Gully Control Structures

The followings points must be taken into considerations while designing the temporary gully control structures or check dams.

1. The life span of the structures being 3–8 years, a storm with a 10-year return period should be considered while deciding the spillway capacity.
2. The effective height of temporary structures, defined as the elevation difference between the spillway crest and the gully bottom, is usually limited to 1.0 m.
3. The crest of the structure should be lower in the centre.
4. The structures should be *keyed* properly into the gully bottom and sides. The *keying* is essential for preventing the tunnelling and scouring under and around the structures. The extent of keying depends on the soil characteristics and could range over 0.3–1.0 m.
5. An apron or platform of rocks, sod or brush should be provided on the downstream of these structures to prevent the scouring caused by the flow passing over the

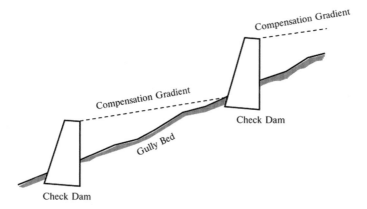

Fig. 7.1 Compensation gradient between temporary gully control structures (After Geyik 1986; Reproduced with permission from FAO)

structure. The length of the apron or the platform maybe 1.5–2 times of the structure height.

6. The structures should be spaced so that the upstream structure's lowest elevation should be the same as the crest elevation of the adjacent downstream structure.

The *compensation gradient*, defined as the gradient between the lowest elevation of the upstream structure and the crest of the adjacent downstream structure (Fig. 7.1) is the target gradient of the gully channel. The target gradient would be achieved when the structure would be filled up to the spillway level by sediments. It is useful in deciding the spacing between the structures or the number of structures for a gully.

The value of the compensation gradient depends on the soil texture, and may be chosen as 2% for coarse sand with small gravel, 1% for fine sand and silt loam, and 0.5% for silt and clay (Ayres and Scoates 1939).

7.2.2 Number of Temporary Structures

A gully erosion control programme should have provisions for multiple temporary structures as it is observed that isolated structures are less effective in controlling the erosion in the long run. The number of temporary structures to be constructed in a gully depends on the gully channel gradient, the compensation gradient to be maintained and the average effective height of the structures proposed.

The following relationship may be used to determine the number of temporary structures for a section of the gully (Geyik 1986):

$$N_{TS} = \frac{a - b}{H_E} \qquad (7.1)$$

where N_{TS} = number of temporary structures; a = the vertical distance between the first and last check dam in a section of the gully based on the average channel gradient (= the average channel gradient × the distance between the structures); b = the vertical distance between the first and last check dam in a section of the gully based on the compensation gradient (= compensation gradient × the distance between the structures); and H_E = the effective height of the structures (= elevation difference between the spillway crest and the gully bottom).

As evident from Eq. (7.1), the number of structures would increase with increasing gully gradient and decrease with increasing structure height.

7.2.3 Spacing Between Structures

As mentioned under the design considerations, the temporary structures should be spaced so that the upstream structure's lowest elevation should be the same as the crest elevation of the adjacent downstream structure. The spacing is, therefore, governed by the compensation gradient and the effective height of the structure. It may, however, be noted that a narrow gully width is preferred for constructing a temporary structure to minimise the construction cost. Therefore, the spacing between structures may be adjusted based on the site conditions.

7.2.4 Design of Spillway

As mentioned under the design considerations, the temporary structures are typically designed to handle the peak runoff resulting due to a storm having a 10-year return period. Thus, we first need to determine the 10-year return period peak runoff for deciding the spillway capacity. Subsequently, we estimate the spillway dimensions. A rectangular spillway with the length and depth of the spillway as the design variables is shown in Fig. 7.2.

The following weir formula is used to design the rectangular spillway (Geyik 1986):

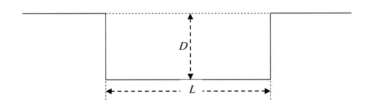

Fig. 7.2 A typical rectangular spillway

$$Q = CLD^{3/2} \tag{7.2}$$

where $Q =$ discharge, m^3 s^{-1}; $C =$ a coefficient; $L =$ length of the spillway crest, m; and $D =$ depth of the spillway crest, m.

The value of C is taken as 3.0 for brushwood, log and loose rock check dams, and 1.8 for cement masonry and gabion check dams.

7.2.5 Types of Temporary Gully Control Structures

7.2.5.1 Woven-Wire Check Dams

Woven-wire check dams are made across the gully bed to trap the fine soil particles and facilitate the vegetative growth in the gully (Fig. 7.3). These dams are preferred in gullies having moderate slope (limited to 10%) and a small drainage area. They usually have a crescent shape with the opening towards the upstream.

The crescent shape makes it possible to have a longer spillway. The curvature of the structure is limited to an offset of about one-sixth of the gully width. For example, an offset of 1 m would be provided if the gully width is 6.0 m at the proposed dam site.

For constructing the woven wire dam, a 0.60–0.90 m deep row of wooden posts is set at an interval of about 1.2 m along the proposed curvature of the dam. A trench having 0.20 m depth and 0.15 m width is then excavated to place a heavy gauge woven wire against the upstream side of the posts. The top of the woven wire is maintained up to a height of about 0.25–0.30 m above the ground surface. The wire is stapled tightly to the posts, and the structure is sealed up to the spillway crest by placing straw, fine brush or similar materials on the upstream side of the wire. The sealing is further strengthened by providing a well-packed earth fill having 2:1 slope. The spillway crest is also protected by placing sods along the crest.

7.2.5.2 Brushwood Check Dams

Brushwood check dams are made of brushwood and posts. These are constructed across the gully with the focal objectives of holding the fine soil particles and facilitating the vegetative growth in the gully. The brushwood dams are the least stable among all types of check dams and are best suited for gullies having deep soils, moderate slope and a smaller drainage area. The effective height of the dam should be limited to 1.0 m. The dam height is kept lower at the centre than at sides to confine the flow over the dam. The spillway is usually rectangular or concave. The brushwood check dams are of two types: single row and double row. The brushwood availability and the flow rate and volume govern the type of brushwood dam for a particular site.

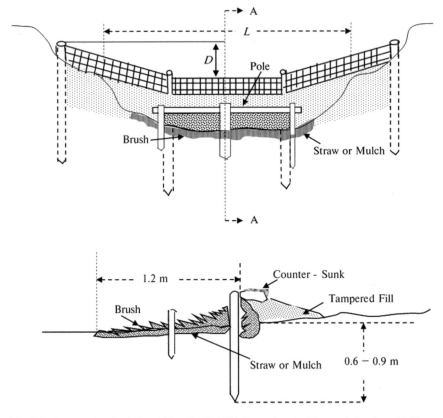

Fig. 7.3 Woven-wire check dam (After Geyik 1986; Reproduced with permission from FAO)

Figure 7.4 shows a typical single-row brushwood check dam. These are preferred where the flow rate is limited to 0.5 m³ s⁻¹. The gully bed is excavated along its width to a depth of 0.15–0.30 m for placing the brushwood for constructing these dams. Subsequently, thicker brushwoods of 100 mm diameter are used as posts and set vertically in a line across the gully at an interval of 0.30–0.50 m. The bottom of the posts should be buried up to a depth of 0.50–1.0 m in the gully bed, and the top should be kept at sufficient height to form a spillway of designed dimensions. Flexible wooden branches are woven around the vertical posts to provide stability. The branches should extend into the gully banks to protect them against erosion. The brushwoods are then tied to the upstream side of the posts using galvanised wire for keeping them intact.

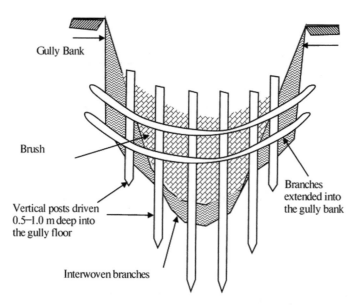

Gully Bank

Brush

Vertical posts driven
0.5–1.0 m deep into
the gully floor

Branches
extended into
the gully bank

Interwoven branches

Fig. 7.4 Single-row brushwood check dam (After Geyik 1986; Reproduced with permission from FAO)

7.2.5.3 Log Check Dams

Log check dams (Fig. 7.5) are made by placing logs and posts across the gully. These are preferred in areas where large size timbers are readily available to be used as logs. About 1.5–2.0 m long posts of 100 mm diameter are fixed vertically in two rows, with rows spaced 0.50 m apart and posts in a row spaced 1.0 m apart. The posts are usually buried up to a depth of 0.75 m–1.0 m in the gully bed. A rectangular spillway is created in the gully centre by placing logs between the two rows of posts. The logs should extend at least 0.50 m into the gully banks to protect them against erosion. The effective height of the dam should be lower than 1.5 m.

7.2.5.4 Loose Rock Check Dams

Loose rock check dams (Fig. 7.6) are made of small stones placed across the gully bed. These dams aim to control the bed erosion and check the waterfall erosion by stabilising the gully head. These dams are preferred in areas where loose rocks are readily available. For constructing these dams, about 0.50 m deep foundation is provided by excavating the gully bed along its width. The foundation is also extended up to 0.5 m on either side of the gully. Loose rocks or stones are then placed to interlock easily to form a dense structure, with a trapezoidal or parabolic spillway of designed dimensions. The smaller stones are usually placed in the centre and covered with larger stones to strengthen the dams. The upstream wall and the crest

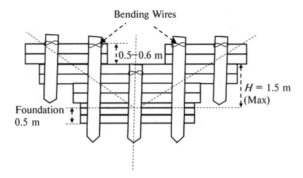

Fig. 7.5 Log check dam (After Geyik 1986; Reproduced with permission from FAO)

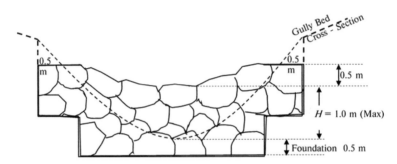

Fig. 7.6 Loose rock check dam (After Geyik 1986; Reproduced with permission from FAO)

of the structure are covered with bamboo, mat or soil excavated from the foundation for further strengthening. The effective height of the dam should be lower than 1.0 m, and the thickness of the dam at the spillway level should be 0.50–0.70 m.

7.2.5.5 Gabion Check Dams

Gabions are welded wire cage or boxes, usually rectangular, filled with stones, rocks or soil (Fig. 7.7). Gabions are used as check dams to stabilise gully sides and gully heads. Gabions are usually filled with larger stones along the sides and the smaller ones in the middle. The gaps between the stones should be minimum to avoid the concentrated flow through them undermining the gully bed. Though gabions offer the advantages of being tough and long-lasting, and useful even when the surface is uneven, they are expensive. They should be used only if inexpensive alternatives are not available.

For constructing these check dams, a layer of box gabions is first placed along the gully width in a trench having a depth equal to the half of the effective height of the dam. The foundation is also extended up to 0.5 m on either side of the gully. The

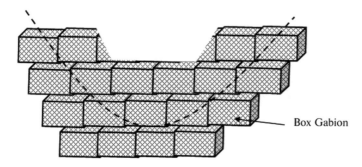

Box Gabion

Fig. 7.7 Gabion check dam

other layers are put on the top, and a rectangular spillway is formed in the centre. The gabion layers should be laced tightly for stability.

7.3 Permanent Gully Control Structures

Permanent gully control structures are proven measures adopted in any soil and water conservation programme for efficient erosion control. These are usually used in medium to large gullies having significantly high flows that could not be handled by the temporary structures. These are typically constructed using masonry or rein-forced concrete and used in the channel reach to handle steep slopes, at the head of large gullies to control the over fall, or as outlet structures of a vegetated waterway discharging into a drainage ditch.

The permanent gully control structures are mainly of the following three types:

1. Drop spillway
2. Drop-inlet spillway
3. Chute spillway.

All these permanent structures may be used in a gully control programme, though the choice of a particular structure is site-specific. The choice is mainly governed by the drop or fall. Drop spillways are used when the drop or fall is less than 3.0 m, while drop-inlet and chute spillways are chosen if the drop or fall exceeds 3.0 m. Drop-inlet spillways are preferred over chute spillways at locations in the gully where it is feasible to store water behind the structure.

The permanent structures have three essential components: the inlet, the conduit and the outlet. A box or a weir is typically used as an inlet for receiving water from the upstream channel. The conduit is essentially used in the drop-inlet and chute spillways to receive water from the inlet and carry it through the structure. A box or pipe-shaped closed conduit may be used in the drop-inlet spillway, while a rectangular or trapezoidal shaped open channel may be used as a conduit in the chute spillway. The outlet releases the water into the downstream channel at a non-erosive velocity. An apron or a stilling basin is usually provided in the downstream section for dissipating the kinetic energy.

7.3.1 Design of Permanent Gully Control Structures

While designing the permanent gully control structures, the following points must be considered:

1. The structure must have sufficient capacity to handle the design discharge.
2. The structure must have sufficient strength to withstand the pressure exerted by the flowing water.
3. The structure must have provisions for dissipation of the kinetic energy to protect both the structure and the downstream channel from damage.

The design of a permanent gully control structure involves the following three phases:

1. Hydrologic design
2. Hydraulic design
3. Structural design

Hydrologic Design

The hydrologic design involves determining the peak flow rate for drop and chute spillways and flow volume for the drop-inlet spillway. Though the peak flow rate or the flow volume depends on the expected life of the structure and the extent of damage that may be caused due to the failure of the structure, typically a return period of 25–50 years is considered while determining the flow rate or volume.

Hydraulic Design

The hydraulic design includes the determination of the dimensions of various components of the structure for handling the design flow rate or volume. It also includes studying the effect of flow in the channel reach, and the dissipation of the kinetic energy on the downstream side of the structure.

Structural Design

The structural design involves the analysis of various forces acting on the structure, and then to provide the required strength and stability to the various component of the structure. The forces may be due to overflow over the structure, flow underneath the structure, and static and dynamic water pressure acting on the structure. The structure must also be safe against the sliding forces resulting due to its weight.

7.3.2 Energy Considerations in Design of Permanent Structures

7.3.2.1 Energy Relationships in Open Channel Flow

The flow in a gully (open channel) is typically steady and non-uniform. Thus, in an open channel, if we consider two sections Δx distance apart (Fig. 7.8), the one-dimensional energy equation may be written as,

$$\frac{p_1}{\gamma} + \alpha_1 \frac{V_1^2}{2g} + z_1 = \frac{p_2}{\gamma} + \alpha_2 \frac{V_2^2}{2g} + z_2 + h_L \tag{7.3}$$

where p = pressure; V = velocity of flow; z = elevation above a fixed reference point; γ = specific weight of water; α = energy coefficient; g = acceleration due to gravity; and h_L = head loss due to friction. Suffixes 1 and 2 represent sections 1 and 2, respectively.

From Fig. 7.8,

$$\frac{p_1}{\gamma} + z_1 = y_1 + S_0 \Delta x, \text{ and } \frac{p_2}{\gamma} + z_2 = y_2$$

where S_0 = channel bed slope; and y = flow depth.

Assuming $\alpha_1 = \alpha_2 = 1.0$, Eq. (7.3) can be written as

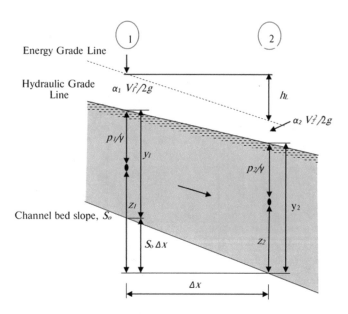

Fig. 7.8 Steady, non-uniform flow in a channel

$$y_1 + \frac{V_1^2}{2g} + S_0 \Delta x = y_2 + \frac{V_2^2}{2g} + h_L \tag{7.4}$$

Now, if we consider a case with horizontal channel bed ($S_0 = 0$), and zero head loss ($h_L = 0$), then Eq. (7.4) would reduce to

$$y_1 + \frac{V_1^2}{2g} = y_2 + \frac{V_2^2}{2g} \tag{7.5}$$

The sum of the flow depth and velocity head is termed as the *Specific Energy* (E), i.e.,

$$E = y + \frac{V^2}{2g} \tag{7.6}$$

Thus, from Eq. (7.5), we see that *specific energy at section 1 = specific energy at section 2*.

The continuity equation between sections 1 and 2 can be written as,

$$Q = A_1 V_1 = A_2 V_2 \tag{7.7}$$

where Q = discharge; and A = cross-section area of flow.

Substituting V_1 and V_2, in terms of Q and A from Eq. (7.7), in (7.5),

$$y_1 + \frac{Q^2}{2g A_1^2} = y_2 + \frac{Q^2}{2g A_2^2} \tag{7.8}$$

Since A depends on y, the specific energy depends only on depth at each section.

For a given Q in a channel, y *versus specific energy* plot is known as the *specific energy curve* (Fig. 7.9).

It is evident from Fig. 7.9 that for a given specific energy, there are two possible depths, y_1 and y_2. The depths y_1 and y_2 are known as *alternate depths*. The specific

Fig. 7.9 Typical specific energy curve

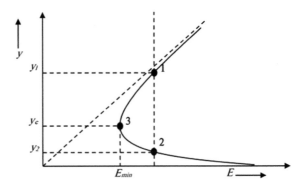

energy curve always lies below 1:1 (45°) line because as per Eq. (7.6), E would always be larger than y. Point (3) on the specific energy curve shows minimum specific energy with a single flow depth. The flow at this point is *critical*. Thus, *critical flow* is the flow that occurs with the minimum specific energy for a given discharge. The flow depth at this point is termed as the *critical depth* (y_c).

When the flow depth is less than the critical depth, or when the flow velocity is more than critical, the flow is known as the *supercritical flow*. On the contrary, when the flow depth is more than the critical depth, or when the flow velocity is less than critical, the flow is known as the *subcritical flow*.

7.3.2.2 Characteristics of Critical Flow

We know that for a given Q, the specific energy is minimum for the critical flow. Therefore, on differentiating E in Eq. (7.6) with respect to y [after substituting v with (Q/A)], we get,

$$\frac{dE}{dy} = 1 - \frac{Q^2}{gA^3}\frac{dA}{dy} \tag{7.9}$$

If T = channel width at the water surface, then $dA = T\,dy$. Putting this in (7.9), and setting $dE/dy = 0$ for critical condition, we get

$$\frac{Q^2 T_c}{gA_c^3} = 1 \tag{7.10}$$

where T_c and A_c represent the channel width at the water surface and cross-sectional area of flow corresponding to the critical flow, respectively.

Equation (7.10) may also be expressed as

$$\frac{A_c}{T_c} = \frac{Q^2}{gA_c^2} \tag{7.11}$$

$$\text{Or,} \quad \frac{A_c}{T_c} = \frac{V^2}{g}. \tag{7.12}$$

For a rectangular channel, (A_c/T_c) is equal to the critical depth, y_c. Also, if we define, q = discharge per unit channel width = Q/B, with $B = T_c$, then Eq. (7.11) may be written as

$$y_c = \frac{q^2}{gy_c^2} \tag{7.13}$$

$$\text{Or,} \quad y_c = \left(\frac{q^2}{g}\right)^{1/3} \tag{7.14}$$

Also, applying Eq. (7.12) to a rectangular channel, we get

$$y_c = \frac{V_c^2}{g} \text{ or } \frac{V_c^2}{g y_c} = 1$$

$$\text{OR,} \quad \frac{V_c}{\sqrt{g y_c}} = 1 \tag{7.15}$$

The term in Eq. (7.15), $\frac{V}{\sqrt{gy}}$, is the *Froude Number*, (F), defined as the ratio of the inertia forces to the gravity forces.

Thus, from Eq. (7.15), for the critical flow, Froude number, $F = 1$. Therefore, for subcritical flow, i.e., when $V < V_c$, $F < 1$, and for supercritical flow, i.e., when $V > V_c$, $F > 1$.

Example 7.1 A trapezoidal channel, having 6.0 m bottom width and 1:1 side slopes, is carrying 14 m^3 s^{-1} discharge. Find the critical depth.

Solution

Given, discharge, $Q = 14$ m^3 s^{-1}; channel bottom width, $B = 6.0$ m; and channel side slope, $Z = 1$.
From Eq. (7.10), for critical flow,

$$\frac{Q^2 T_c}{g A_c^3} = 1$$

$$\text{Or,} \quad \frac{A_c^3}{T_c} = \frac{Q^2}{g} \tag{7.16}$$

Putting the values of Q and g (9.81 m s^{-2}), $\frac{A_c^3}{T_c} = \frac{(14 \times 14)}{9.81} = 19.98$.
For a trapezoidal channel, $A_c = B y_c + Z y_c^2$ and $T_c = B + 2 Z y_c$.
Putting A_c and T_c in Eq. (7.16), with $B = 6$ and $Z = 1$,

$$\frac{\left(6 \times y_c + 1 \times y_c^2\right)^3}{(6 + 2 \times 1 \times y_c)} = 19.98$$

By trial-and-error, $y_c = 0.79$ m.

Answer: The critical depth is 0.79 m.

Example 7.2 The discharge in a 6.2 m wide rectangular channel is 10 m^3 s^{-1}. The Manning roughness coefficient, $n = 0.014$. Determine the specific energy of the flow and the type of flow if the depth of flow is 1.0 m.

Solution

Given, channel width, $B = 6.2$ m; depth of flow, $y = 1.0$ m; discharge, $Q = 10$ m^3 s^{-1}; Manning roughness coefficient, $n = 0.014$.

The cross-sectional area of flow, $A = B\,y = 6.2 \times 1.0 = 6.2$ m^2

From the continuity equation, velocity of flow, $V = Q/A = 10/6.2 = 1.61$ m s^{-1}

Putting the values of y, V and g in Eq. (7.6), the specific energy would be

$$E = y + \frac{V^2}{2g} = 1 + \frac{1.61^2}{2 \times 9.81} = 1.13 \text{ m}$$

Similarly, putting the values of y, V and g in Eq. (7.15), Froude number would be

$$F = \frac{V}{\sqrt{gy}} = \frac{1.61}{\sqrt{(9.81 \times 1)}} = 0.51$$

Since $F < 1$, the flow in the channel is subcritical.

Answer: The specific energy of the flow is 1.13 m, and the flow is subcritical.

7.3.3 Hydraulic Jump

A hydraulic jump is defined as an abrupt rise in the water level in an open channel (Fig. 7.10). It occurs when the water flowing at a high velocity is retarded to a slow velocity. The slowing of water leads to an increase in height that changes the kinetic energy of the flow into potential energy. In the process, some of the energy is dissipated in the form of heat due to turbulence. It typically occurs below the sluice gates in a channel or when the channel bed slope changes from steep to mild. One of the significant applications of hydraulic jump is in the design of stilling basins, which are used as a terminal structure in a spillway for the energy dissipation.

Since hydraulic jump phenomenon involves substantial energy losses, the energy or specific energy equation cannot be applied. Instead, the momentum equation is used to analyse the phenomenon. Further, the following assumptions are made while deriving the mathematical expression for the hydraulic jump:

1. The channel is rectangular with a horizontal bed slope.

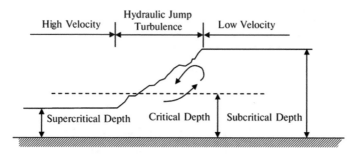

Fig. 7.10 Hydraulic jump phenomenon

Fig. 7.11 Definition sketch
to derive a mathematical
expression for the hydraulic
jump

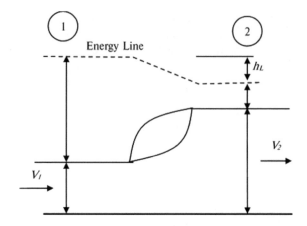

2. The pressure distribution at cross-sections 1 and 2, i.e., before and after the hydraulic jump, is hydrostatic (Fig. 7.11). Also, the velocity distribution is uniform.
3. The friction losses are negligible.

Applying the momentum equation to the control volume (Fig. 7.11), considering a unit width perpendicular to the control volume,

$$\frac{1}{2}\gamma y_1^2 - \frac{1}{2}\gamma y_2^2 = \rho q V_2 - \rho q V_1 \tag{7.17}$$

where y_1 and $y_2 =$ depth of flow at sections 1 and 2; V_1 and $V_2 =$ velocity of flow at sections 1 and 2, $q =$ discharge per unit width of the channel; $\gamma =$ specific weight of the liquid; and $\rho =$ density of the liquid.

Equation (7.17) is based on the fact that the resultant force between sections 1 and 2 in the flow direction is equal to the momentum change.

The corresponding flow depths before and after the hydraulic jump, i.e., y_1 and y_2, are called *sequent depth*.

Since $q = V_1 y_1 = V_2 y_2$ and $\gamma = \rho g$, Eq. (7.17) may be written as

$$\frac{\rho g}{2}(y_1^2 - y_2^2) = \rho\left(\frac{q^2}{y_2^2}y_2 - \frac{q^2}{y_1^2}y_1\right)$$

$$\frac{g}{2}(y_1 - y_2)(y_1 + y_2) = q^2\left(\frac{(y_1 - y_2)}{y_1 y_2}\right)$$

$$y_1 y_2(y_1 + y_2) = \frac{2q^2}{g} = \frac{2V_1^2 y_1^2}{g}$$

$$y_1^2 y_2\left(1 + \frac{y_2}{y_1}\right) = \frac{2V_1^2 y_1^2}{g} \tag{7.18}$$

Multiplying both sides of Eq. (7.18) with $\left(\frac{1}{y_1^3}\right)$, we get

$$y_1^2 y_2 \left(1 + \frac{y_2}{y_1}\right) \times \left(\frac{1}{y_1^3}\right) = \frac{2V_1^2 y_1^2}{g} \times \left(\frac{1}{y_1^3}\right)$$

$$\frac{y_2}{y_1}\left(1 + \frac{y_2}{y_1}\right) = 2\frac{V_1^2}{g y_1}$$

(7.19)

Since for rectangular channel, Froude number, $F = \frac{V}{\sqrt{gy}}$, Eq. (7.19) takes the form,

$$\left(\frac{y_2}{y_1}\right)^2 + \frac{y_2}{y_1} - 2F_1^2 = 0$$

(7.20)

Considering only the positive sign, the solution of Eq. (7.20) yields

$$\frac{y_2}{y_1} = \frac{1}{2}\left(\sqrt{1 + 8F_1^2} - 1\right)$$

(7.21)

The ratio of flow depths, (y_2/y_1), is a function of the Froude number before the hydraulic jump.

Example 7.3 The depth and velocity of flow on the upstream of a 12-m-wide sluice gate are 1.2 m, and 20 m s^{-1}, respectively. For the hydraulic jump occurring on the down-stream of the sluice gate, determine (a) the Froude number before and after the hydraulic jump, and (b) the depth and velocity after the jump.

Solution

Given: upstream flow depth, $y_1 = 1.2$ m; upstream flow velocity, $V_1 = 20$ m s^{-1}; width of the sluice gate $= 12$ m.

Hence, the upstream Froude number, $F_1 = \frac{V_1}{\sqrt{gy_1}} = \frac{20}{\sqrt{9.81 \times 1.2}} = 5.8$

Using Eq. (7.21), the conjugate depth,

$$y_2 = \frac{1}{2}y_1\left(\sqrt{1 + 8F_1^2} - 1\right) = \frac{1}{2} \times 1.2 \times \left(\sqrt{1 + 8 \times (5.8)^2} - 1\right) = 9.3 \text{ m}$$

From continuity equation, $V_2 = \frac{q}{y_2} = \frac{(1.2 \times 20)}{9.3} = 2.6$ m s^{-1}

Hence, downstream Froude number, $F_2 = \frac{V_2}{\sqrt{gy_2}} = \frac{2.6}{\sqrt{9.81 \times 9.3}} = 0.27$

Answers: (a) Froude number before the jump $= 5.8$; Froude number after the jump $= 0.27$; (b) Downstream flow depth $= 9.3$ m; downstream flow velocity $= 2.6$ m s^{-1}.

7.3.3.1 Types of Hydraulic Jump

Table 7.1 presents the classification of the hydraulic jump based on the upstream Froude number, F_1.

Table 7.1 Types of hydraulic jump

Upstream Froude Number, F_1	Type of jump	Illustration	Description
1.0–1.7	Undular		Water surface shows undulation with no sharp or abrupt rise in the water level
1.7–2.5	Weak		Water surface has a series of small rollers with the smooth downstream water surface
2.5–4.5	Oscillating		Water surface shows oscillations with an oscillating jet entering the jump from the bottom to the surface
4.5–9.0	Steady		A distinct, well-balanced steady jump is formed with considerable dissipation of energy

(continued)

Table 7.1 (continued)

Upstream Froude Number, F_1	Type of jump	Illustration	Description
>9.0	Strong		A high-velocity jet generates waves downstream with a rough surface. The rough jump causes significant dissipation of energy

7.3.3.2 Energy Dissipation in Hydraulic Jump

Considering the difference in the specific energy before and after the hydraulic jump,

$$\Delta E = E_1 - E_2 = \left(y_1 + \frac{V_1^2}{2g} \right) - \left(y_2 + \frac{V_2^2}{2g} \right)$$

$$\Delta E = (y_1 - y_2) + \left(\frac{V_1^2}{2g} - \frac{V_2^2}{2g} \right)$$

Since $q = V_1 y_1 = V_2 y_2$, substituting, $V_1 = \frac{q}{y_1}$ and $V_2 = \frac{q}{y_2} V_2 = \frac{q}{y_2}$

$$\Delta E = (y_1 - y_2) + \frac{q^2}{2g} \left(\frac{1}{y_1^2} - \frac{1}{y_2^2} \right) \tag{7.22}$$

While deriving Eq. (7.18), we have seen that $y_1 y_2 (y_1 + y_2) = \frac{2q^2}{g}$

$$\text{Thus,} \quad \frac{q^2}{2g} = \frac{1}{4} y_1 y_2 (y_1 + y_2) \tag{7.23}$$

Substituting the value of $\left(\frac{q^2}{2g} \right)$ from Eq. (7.23) into Eq. (7.22),

$$\Delta E = (y_1 - y_2) + \frac{1}{4} y_1 y_2 (y_1 + y_2) \left(\frac{1}{y_1^2} - \frac{1}{y_2^2} \right)$$

$$= (y_1 - y_2) + \frac{1}{4} y_1 y_2 (y_1 + y_2) \left(\frac{y_2^2 - y_1^2}{y_1^2 y_2^2} \right)$$

$$= (y_1 - y_2) + \frac{1}{4} y_1 y_2 (y_1 + y_2) \left(\frac{(y_2 + y_1)(y_2 - y_1)}{y_1^2 y_2^2} \right)$$

$$= (y_1 - y_2) + \frac{1}{4} \frac{(y_1 + y_2)^2 (y_2 - y_1)}{y_1 y_2}$$

$$= \frac{4 y_1 y_2 (y_1 - y_2) + (y_1 + y_2)^2 (y_2 - y_1)}{4 y_1 y_2}$$

$$= \frac{(y_2 - y_1)[-4 y_1 y_2 + (y_1 + y_2)^2]}{4 y_1 y_2}$$

$$= \frac{(y_2 - y_1)(y_2 - y_1)^2}{4 y_1 y_2}$$

$$\Delta E = \frac{(y_2 - y_1)^3}{4 y_1 y_2} \tag{7.24}$$

Based on Eqs. (7.21) and (7.24), the following inferences could be drawn:

1. If $F_1 = 1$, then $\frac{y_2}{y_1} = 1$; $\Delta E = 0$; the energy dissipation would be nil in case of critical flow.
2. If $F_1 > 1$, then $\frac{y_2}{y_1} > 1$; $\Delta E > 0$; the energy dissipation would be positive in case of supercritical flow.
3. If $F_1 < 1$, then $\frac{y_2}{y_1} < 1$; $\Delta E < 0$; the energy dissipation would be negative or in other words, there would be energy gain in case of subcritical flow. However, energy gain is practically not possible.

Example 7.4 Determine the energy dissipation for the data in Example 7.3.

Solution

From Example 7.3 data, $y_1 = 1.2$ m; $y_2 = 9.3$ m.
  Putting the values of y_1 and y_2 in Eq. (7.24),

$$\Delta E = \frac{(y_2 - y_1)^3}{4 y_1 y_2} = \frac{(9.3 - 1.2)^3}{(4 \times 1.2 \times 9.3)} = 11.9 \text{ m}$$

Answer: The energy dissipation is 11.9 m.

7.3.3.3 Length of Hydraulic Jump

Several empirical equations are available for estimating the length of the hydraulic jump (distance measured from the front face of the jump to a point on the surface immediately after the roller). Table 7.2 lists a few notable equations valid for $2 \leq F_1 \leq 20$.

Table 7.2 Empirical equations for the length of hydraulic jump

Developer (Year)	Equation
Safranez (1929)	$L = 5.2 y_2$
Smetana (1937)	$L = 6(y_2 - y_1)$
Bakhmeteff and Matzke (1936)	$L = 5(y_2 - y_1)$
Marques et al. (1997)	$L = 8.5(y_2 - y_1)$
Simoes et al. (2012)	$L = 9.52(y_2 - y_1)$

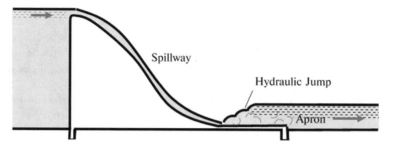

Fig. 7.12 Hydraulic jump formation on the terminal structure

7.3.3.4 Application of Hydraulic Jump for Designing Stilling Basins

An important application of the hydraulic jump is in designing stilling basins. Stilling basins are typically used as a terminal structure in a spillway for energy dissipation. Stilling basins control scour and erosion of the downstream channel.

Flow at the lower end of the spillway is usually supercritical, with subcritical flow in the downstream channel. Thus, the hydraulic jump forms near the base of the spillway. Therefore, the downstream section of the spillway must be designed so that the hydraulic jump always forms on the downstream terminal structure (Fig. 7.12).

U.S. Bureau of Reclamation has standardised the stilling basin designs (Peterka 1984).

Questions

1. Classify different types of gully control structures, and describe specific features of various structures.
2. Based on the information from the watersheds, managed under the soil management programmes, closer to your home town, describe various gully control structures.
3. Describe the types of temporary control structures used for gully erosion control.
4. With the help of the specific energy curve, derive the condition for the critical flow.
5. The discharge in a rectangular channel, having a bottom width of 2.0 m, is 7.5 m³ s⁻¹. The specific energy is estimated as 2.0 m. Determine the alternate depths and corresponding Froude numbers.

6. The discharge in a 3 m wide channel is $10 \text{ m}^3 \text{ s}^{-1}$. The specific energy is estimated as 4.0 m. Determine the critical depth, the minimum specific energy, and the alternate depths.

7. The discharge in a rectangular channel of width 6.0 m is $24 \text{ m}^3 \text{ s}^{-1}$. The Manning roughness coefficient is $0.015 \text{ m}^{-1/3} \text{ s}$. Find the Froude number and the critical depth if the channel bed slope is 1 in 200.

8. The discharge in a 2 m wide rectangular channel is $4.0 \text{ m}^3 \text{ s}^{-1}$. Determine the critical depth and the corresponding specific energy. What would be the magnitudes of the critical depth and the corresponding specific energy if the same flow passes through (i) a triangular channel having 0.5:1 side slope; (ii) a trapezoidal channel having a bottom width of 2.0 m and side slope 1.5:1.

9. Determine the length of the hydraulic jump if the downstream Froude number and flow depth are 5 and 0.50 m, respectively. Determine the magnitude of the power dissipated if the discharge is $1,600 \text{ m}^3 \text{ s}^{-1}$.

10. The depth and velocity of flow on the horizontal apron of a 30-m wide spillway are 0.2 m and 4.5 m s^{-1}. Estimate the flow depth and the Froude number before and after the jump, and the energy dissipated.

Multiple Choice Questions

Answer the questions by choosing the correct option.

1. The gradient between the lowest elevation of the upstream structure and the crest of the adjacent downstream structure is known as the

   (A) channel gradient   (B) compensation gradient
   (C) gully gradient   (D) total gradient

2. The hydraulic design of a gully control structure involves
   (A) analysing various forces acting on the structure
   (B) determining the peak flow rate for the structure
   (C) determining the dimensions of various components of the structure
   (D) determining the required strength of the various component of the structure

3. Hydraulic jump occurs when the flow changes from

   (A) supercritical to critical   (B) subcritical to critical
   (C) subcritical to supercritical   (D) supercritical to subcritical

4. The Froude number in a strong jump is greater than 9. The energy dissipation in the jump may be

   (A) less than 5% (B) 5–20%
   (C) 15–45%   (D) 75–85%

5. When the drop height exceeds 4 m, and there is a possibility of flow accumulation upstream, then the most suitable mechanical structure would be

   (A) drop spillway   (B) drop-inlet spillway

(C) chute spillway   (D) grassed waterway

6. Gabions are typically used for

(A) flow measurement   (B) flood control
(C) gully stabilisation   (D) storing runoff

7. When the critical flow condition occurs, then the

(A) specific energy is minimum   (B) specific force is minimum
(C) viscous force is minimum   (D) total force is minimum

8. In a rectangular channel, supercritical flow occurs with the Froude number of 2.0. The depth of flow is 0.63 m. The critical depth, in m, is

(A) 0.50   (B) 1.00
(C) 1.26   (D) 2.52

9. A discharge of 10 m^3 s^{-1} is flowing through a 4-m wide channel. The critical depth, in m, is

(A) 0.74   (B) 0.86
(C) 1.02   (D) 2.16

10. A hydraulic jump occurs in a horizontal rectangular channel with the sequent depths of 0.30 m and 1.50 m. The energy loss, in m, is

(A) 0.96   (B) 1.20
(C) 1.50   (D) 1.92

References

Ayres QC, Scoates WD (1939) Land drainage and reclamation, 2nd edn. McGraw-Hill Book Company, New York and London

Bakhmeteff BA, Matzke AE (1936) The hydraulic jump in terms of dynamic similarity. Trans ASCE 101:630–680

Geyik MP (1986) FAO watershed management field manual: gully control. FAO, Rome

Marques MG, Drapeau J, Verrette JL (1997) Flutuação de pressão em um ressalto hidráulico (Pressure fluctuation in a hydraulic jump). Revista Brasileira De Recursos Hídricos 2:45–52

Peterka AJ (1984) Hydraulic design of spillways and energy dissipaters. Engineering Monograph 25, Water Resources Technical Publication, US Bureau of Reclamation, Colorado

Safranez K (1929) Untersuchung iiber den Wechselsprung (Investigation on the hydraulic jump). Der Bauingenieur 10:649–651

Simões ALA, Porto RM, Schulz HE (2012) Superfície livre de escoamentos turbulentos em canais: vertedores em degraus e ressalto hidráulico (Free surface of turbulent flows in channels: stepped weirs and hydraulic jump). Revista Brasileira De Recursos Hídricos 17:125–139

Smetana J (1937) Studi Sperimentali sul Salto di Bidone Libero e Annegato (Experimental Studies on the free and submerged hydraulic jumps). L'energia Elettrica 14:829–835

Chapter 8
Drop Spillway

Abstract Drop spillway, one of the most widely used soil conservation structures, is primarily used for controlling and stabilising grades in a gully. The chapter focuses on the hydrologic, hydraulic and structural designs of drop spillways. The hydrologic design approaches for estimating the peak flow rate, i.e., the rational method, empirical or frequency factor method of frequency analysis and the hydrological or hydraulic modelling, are discussed. The hydraulic design of straight and box-inlet drop spillways under free and submerged flow conditions is presented. This chapter also includes the critical depth concept and its application in determining the dimensions of various components of the straight and box-inlet drop spillways. The structural design contains the analysis of the horizontal forces acting against the structure due to the hydrostatic pressure of the water column upstream and the earth pressure caused by the backfill. It also comprises the uplift pressure caused due to water seepage through the saturated foundation material. A detailed procedure to analyse the stability of the structure against overturning, sliding, piping, tension, and compression or contact pressure is demonstrated through a solved example.

8.1 Background

Drop spillway is one of the most commonly used gully control structures. Two significant variants of drop spillways, distinguished based on the inlet, are the *straight drop spillway* and the *box-inlet drop spillway*.

The straight drop spillway (Fig. 8.1a) has a weir in a wall as the inlet, while the box-inlet drop spillway (Fig. 8.1b) has a rectangular box, open at the top and the downstream end, as the inlet. In both these structures, flow passes through the inlet, drops to an energy dissipater and flows into the downstream channel. An apron or stilling basin may be used as the energy dissipater.

© The Author(s), under exclusive license to Springer Nature Singapore Pte Ltd. 2023 161
R. Singh, *Soil and Water Conservation Structures Design*, Water Science and Technology
Library 123, https://doi.org/10.1007/978-981-19-8665-9_8

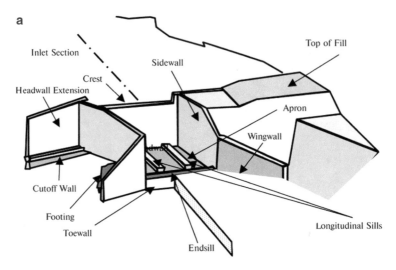

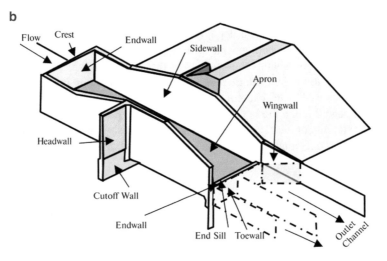

Fig. 8.1 a Straight drop spillway. **b** Box-inlet drop spillway

8.2 Functions

Drop structures are mainly used for controlling and stabilising grades in a gully. Typically, a series of drop structures are constructed along the length of the gully at fixed intervals. These structures retard the flow of water, resulting in sediment deposition and filling of the gully section. Drop structures may also be used at the gully head for passing the flow safely and protecting the gully head.

Drop structures are also used as the reservoir spillway when the drop is comparatively low and inlet and outlet structures in a tile drainage system. These may also be used to control tailwater at a spillway outlet or at the exit of grass waterways for protecting the downstream channel.

8.3 Adaptability

In general, the drop spillways may be used when the total fall or drop at a site is limited to 3.0 m. The box-inlet drop spillways are preferred when a large water flow needs to be handled in narrow channels. The long crest of the box-inlet facilitates the flow of large volumes at relatively low heads.

8.4 Advantages and Limitations

The drop spillways offer the advantages of higher stability as compared to other structures of similar discharge capacity. These are easy to construct and less susceptible to clogging due to debris.

The box-inlet drop spillways offer the additional advantage of higher discharge capacity for narrow channels.

The limitations of drop spillways include their adaptability to less than 3.0 m drop or fall. Further, these cannot be used when temporary storage is needed upstream of the structure to reduce the flow passing over the structure.

8.5 Materials of Construction

Reinforced concrete is the most widely used construction material for both straight drop and box-inlet drop spillways as it offers the advantages of longer life and low annual maintenance cost. Plain concrete, concrete blocks, or rock masonry may also be used for constructing straight drop spillways.

8.6 Drop Spillway: Components and Functions

The individual components of the drop spillway are described below, along with their functions.

1. **Headwall**: It acts as the inlet in the straight drop spillway. It is constructed across the gully width. It typically has a rectangular weir opening for flow to pass through.

2. **Headwall extension**: It is the extended portion of the headwall into the gully banks. It prevents seepage and protects the banks against erosion, besides providing structural strength to the structure.
3. **Sidewalls**: These are constructed on either side of the apron along the gully walls. The sidewalls confine the water flow within the apron and protect the gully banks against erosion.
4. **Wingwalls**: These are walls extending from the rear end of the apron into the gully banks at an inclination of about 45° from the outlet centreline. These walls protect the gully banks from eddies that may be generated as water leaves the structure.
5. **Apron**: It acts as the outlet structure of the drop spillway. It receives the flow through the inlet at a high velocity, dissipates the kinetic energy of the flow by creating a hydraulic jump, and discharges the water into the downstream channel at a safe velocity. It protects the downstream channel against erosion.
6. **Longitudinal Sills**: These sills are constructed along the apron bed to provide stability to the apron. The height of these sills ranges over 50–100 mm.
7. **End Sill**: The end sill is provided at the rear end of the apron to obstruct the water being discharged into the downstream channel and reduce its velocity. The height of the sill may range over 100–300 mm.
8. **Cutoff Wall**: It is the vertical wall constructed under the inlet. It extends into the soil foundation and acts as a key. It prevents seepage under the structure and provides structural strength to the structure against sliding.
9. **Toewall**: It is the vertical wall constructed under the downstream end of the apron. The toewall also extends into the soil foundation and provides structural strength to the structure against slippage.

8.7 Design of Drop Spillway

As discussed in Sect. 7.3.1 (Chap. 7), the design of a permanent gully control structure involves the following three phases:

1. Hydrologic design
2. Hydraulic design
3. Structural design

The hydrologic design involves determining the peak inflow rate. The hydraulic design includes determining the dimensions of various components of the structure for handling the design flow rate. The structural design involves analysing various forces acting on the structure and then providing the required strength and stability to the various components of the structure.

8.7.1 Hydrologic Design

The estimation of peak flow rate is the heart of the planning, design, and construction of drop spillways. The peak flow rate is estimated considering the structure's expected life and the extent of damage caused due to the structure's failure. Therefore, typically a return period of 25–50 years is considered while determining the flow rate. The following approaches may be adopted for estimating the peak or design flow rate:

1. The rational method
2. Frequency analysis of historical rainfall or flow data
3. Hydrological or hydraulic modelling.

8.7.1.1 Rational Method

The rational method, developed by Mulvany (1851), is popularly used for design storm analysis. Though the rational method's assumptions limit its application to small watersheds, having an area up to 12 km^2, or relatively frequent storms having a return period of 2–10 years, it is the preferred choice of the practising engineers because of its simplicity (Peel 2020). The rational formula is given by Eq. (3.19) (Chap. 3) as:

$$Q = ci\,A$$

where Q = peak runoff rate, m^3 s^{-1}; c = runoff coefficient, dimensionless; i = average rainfall intensity for a duration equal to the time of concentration of the watershed, m s^{-1}; and A = watershed area, m^2.

Section 3.2.4 (Chap. 3) includes the details of the rational method.

8.7.1.2 Frequency Analysis of Historical Rainfall or Flow Data

Frequency analysis of historical rainfall or runoff is another acceptable procedure for estimating the design flow rate. However, frequency analysis requires a minimum of 30 years of rainfall or flow data.

Frequency analysis provides the relationship between the magnitudes of extreme events and their frequency of occurrence. Frequency analysis, however, assumes that data being analysed are independent and identically distributed, and the hydrologic system producing them is stochastic and space- and time-independent (Dirk 2013).

Suppose P is the probability of occurrence of a hydrological event (rainfall or streamflow) whose magnitude is equal to or above a specified magnitude X. The recurrence interval (or return period), T, defined as the average interval between the successive occurrences of a hydrological event of magnitude equal to or greater than X, is then related to P as

$$T = \frac{1}{P} \tag{8.1}$$

Frequency analysis can be performed either by *empirical method* or by *frequency factor method*.

Empirical Method

In the empirical method, the plotting position formula calculates the probability of the precipitation or streamflow event. Weibull's formula is the most popular plotting position formula and is given below.

$$P = \frac{m}{N + 1} \tag{8.2}$$

where m = rank assigned to the data after arranging them in descending order of magnitude; and N = number of records. Thus, the highest value is assigned $m = 1$, the second-highest value $m = 2$, and the lowest value $m = N$.

After calculating P (and T) for all the events in the data series, the magnitudes of the hydrological variable are plotted against the corresponding T on a semi-log or log–log paper. The precipitation or runoff magnitude for any desired recurrence interval can be determined using the plot. This method produces good results for slight extrapolations, as errors increase with increasing extrapolation.

Frequency Factor Method

In the frequency factor method, the magnitude of the hydrological variable at the desired recurrence interval, x_T, is obtained using the following general hydrological frequency equation (Chow 1951):

$$x_T = \bar{x} + K_T \sigma \tag{8.3}$$

where $\bar{x}$ = mean of the data series; σ = standard deviation of the data series; and K_T = frequency factor for T year. recurrence interval.

The frequency factor, K_T, depends on the probability density function of x and the recurrence interval, T. K_T can be obtained for the desired recurrence interval using the equation or the standard table developed for various probability distributions like normal, log-Normal, Pearson type III, and Gumbel extreme value type I (Haan 2002).

For the normal distribution, K_T equals the standard normal variate, z, defined as

$$z = \frac{x - \mu}{\sigma} \tag{8.4}$$

where μ = population mean.

The standard normal variate z has zero mean and unit variance, and it could be obtained from the standard normal curve or cumulative normal table (Haan 2002).

For the Gumbel extreme value type-I distribution, frequency factor K_T is given as

$$K_T = -\frac{\sqrt{6}}{\pi}\left\{0.5772 + \ln\left[\ln\left(\frac{T}{T-1}\right)\right]\right\} \tag{8.5}$$

Example 8.1 The following data present peak discharge ($m^3\ s^{-1}$) of a river for 30 years. Compute the 25 year. peak flow, assuming that the data follows Gumbel extreme value type-I distribution.

1626	1538	1481	1316	1269	1248	1228	1217	1178	1098
1049	1022	1015	994	985	968	954	939	872	1065
779	829	1082	1312	1239	1123	1177	639	877	1063

Solution

Analysing the data, mean, $\bar{x} = 1106$; standard deviation, $\sigma = 220.9$

For $T = 25$ years, the frequency factor K_T for the Gumbel extreme value type-I distribution can be determined using Eq. (8.5).

$$K_T = -\frac{\sqrt{6}}{\pi}\left\{0.5772 + \ln\left[\ln\left(\frac{T}{T-1}\right)\right]\right\}$$

$$= -\frac{\sqrt{6}}{\pi}\left\{0.5772 + \ln\left[\ln\left(\frac{25}{25-1}\right)\right]\right\} = 2.044$$

Putting the values of $\bar{x}$, σ and K_T in Eq. (8.3),

$$x_T = \bar{x} + K_T\sigma = 1106 + 2.044 \times 220.9 = 1557\ m^3 s^{-1}$$

Answer: 25 year peak flow for the river $= 1557\ m^3\ s^{-1}$.

8.7.1.3 Hydrological or Hydraulic Modelling

Hydrological modelling of the catchment or the hydraulic modelling of the stream may also be used to estimate the design flow. Several hydrological and hydraulic models of varying degree of complexity may be used for this purpose. The popular hydrological models include TOPMODEL (Beven and Kirkby 1979), MIKE SHE (Refsgaard and Storm 1995), IHACRES (Jakeman et al. 1990), VIC (Liang et al. 1994), SAC-SMA (Finnerty et al. 1997), HEC-HMS (US Army Corps of Engineers 1998), and SWAT (Arnold et al. 1998). The popular hydraulic models include MIKE 11 (Havno et al. 1995) and HEC-RAS (US Army Corps of Engineers 2002). Hydrological or hydraulic models are considered superior to the rational method and statistical procedure. These take into account the physical aspects of catchment or stream characteristics while estimating the design flow estimation. However, these are data and computationally demanding. Besides, the user should have basic knowledge of modelling protocol.

8.7.2 Hydraulic Design of Straight Drop Spillway

Knowing the design flow rate, Q, and drop through the spillway, F, the hydraulic design begins with determining the length of the weir, L, and the depth of the weir, h (Fig. 8.2). In a straight drop spillway, the weir's discharge capacity may be determined using the following weir formula.

$$Q = CL\left(H + \frac{v_a^2}{2g}\right)^{3/2}$$
(8.6)

where Q = peak flow rate, m^3 s^{-1}; H = head over the crest, m; v_a = average velocity of approach, m s^{-1}; and C = weir coefficient.

The typical value of the weir coefficient, C, in SI units is 1.77. It is a lumped parameter that represents the combined effect of discharge coefficient, the gravitational constant, and the weir geometry and may be expressed as.

$$C = \frac{2}{3}C_d\sqrt{(2g)}$$
(8.7)

where C_d = discharge coefficient, dimensionless; and g = acceleration due to gravity = 9.81 m s^{-2}. The value of C_d is usually taken as 0.60.

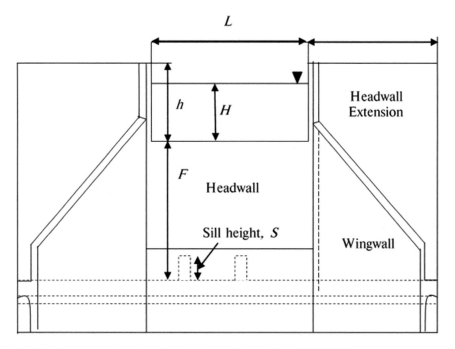

Fig. 8.2 Downstream elevation of straight drop spillway (After NRCS 2008)

The term $\left[H + \left(\frac{v^2}{2g}\right)\right]$ in Eq. (8.6) represents the energy head causing the flow over the weir crest (Fig. 8.3). The head, H, or the velocity of approach, v_a, should be measured at a distance of $3H$ or more upstream from the weir to avoid fluctuations in the head. The measurement section should also ensure a horizontal energy grade line in between the section and the weir. In Fig. 8.3, m represents the height of the weir crest from the approach channel bed. As evident, m would be positive if the channel bed is below the crest elevation at section AA.

Substituting $C = 1.77$ in Eq. (8.6), we get

$$Q = 1.77L\left(H + \frac{v_a^2}{2g}\right)^{3/2}$$ (8.8)

In practical cases, the velocity of approach, v_a, is so small that the velocity head, $(v^2/2g)$, may be ignored, especially when the water level in the approach channel remains above the weir crest. Therefore, Eq. (8.8) reduces to,

$$Q = 1.77LH^{3/2}$$ (8.9)

Equation (8.9), however, is only applicable when the free-flow condition prevails.

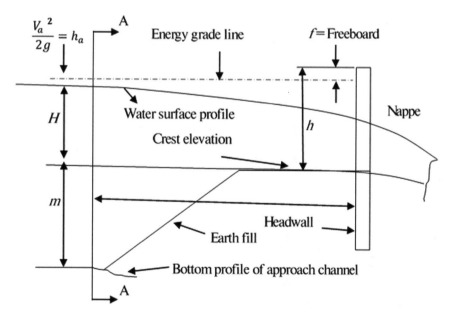

Fig. 8.3 Flow through the approach channel passing over the weir crest (After NRCS 2008)

Free Flow

Free flow occurs when the flow over the weir is independent of the water surface on the downstream end of the structure, usually referred to as the tailwater, i.e., when the tailwater surface is below the weir crest (Fig. 8.4). When the flow drops over the crest in such a condition, the lower nappe does not cling to the headwall due to airflow between the nappe and the headwall. Consequently, both upper and lower nappe surfaces are subject to atmospheric pressure.

Submerged flow

Submerged flow takes place when the tailwater surface is above the weir crest (Fig. 8.5). The elevation difference between the weir crest and the tailwater surface is termed the *depth of submergence* (H_2). Under submerged flow conditions, the weir's flow may or may not be independent of the tailwater.

As mentioned earlier, Eq. (8.9) can only be applied for the free-flow condition. Under the free-flow condition, the flow passes through critical depth near the crest, and the upstream flow becomes independent of the downstream flow conditions. Though the exact location of critical depth may vary according to the approach channel characteristics, it is assumed that the critical depth occurs at the crest. The critical depth, y_c, may be estimated using Eq. (7.14) (Sect. 7.3.2.2, Chap. 7) and replacing the discharge per unit channel width, q, with (Q/L),

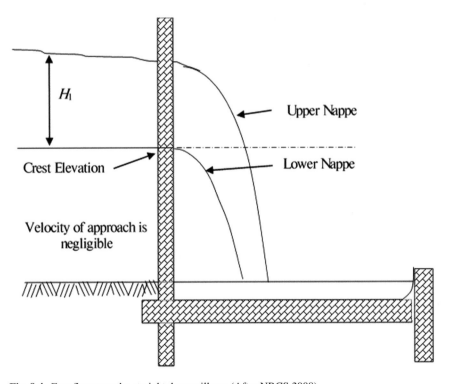

Fig. 8.4 Free flow over the straight drop spillway (After NRCS 2008)

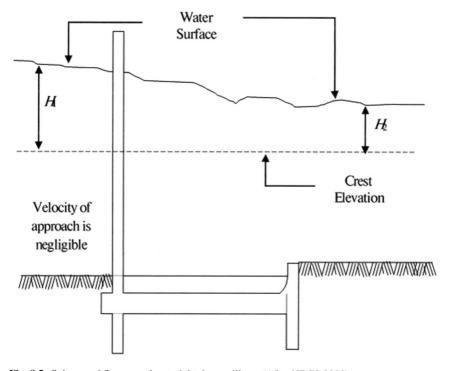

Fig. 8.5 Submerged flow over the straight drop spillway (After NRCS 2008)

$$y_c = \left(\frac{q^2}{g}\right)^{1/3} = \left(\frac{Q^2}{L^2 g}\right)^{1/3} \tag{8.10}$$

The critical depth of flow, y_c, helps determine the drop spillway's component dimensions.

Further, from Fig. 8.3, the depth of weir, h, may be expressed as

$$h = f + H + \frac{v_a^2}{2g} \tag{8.11}$$

where f = freeboard.

Moreover, the value of the ratio h/F should usually be kept lower than 0.50, and in no case it should be higher than 0.75. Also, the ratio L/h should invariably be equal to or greater than 2.0 for all rectangular weirs.

8.7.2.1 Design Cases

The following two possible design cases may be encountered:

Case I: To determine the crest length, L, and depth, h, of the weir for known discharge capacity, Q.

Case II: To determine the Q of the spillway for known L and h.

In either case, the flow may be free or submerged.

8.7.2.2   Design for Free-Flow Condition

Case I: Knowing Q, to determine L and h.

The following steps may be followed.

Step 1: Using Eq. (8.9), obtain a relationship between L and H.

$$L = \frac{Q}{1.77 H^{3/2}}$$

Step 2: Compute L using the above relationship for assumed values of H and prepare a table in the following format.

Assumed H	$H^{3/2}$	Computed L	Practical value of L	Assumed f	Computed h

Select a suitable combination of h and L based on the site condition while maintaining $(h/F) < 0.50$ and $(L/h) \geq 2.0$.

Example 8.2 Determine the crest length and height of the weir for a straight drop spillway to carry $8.0 \text{ m}^3 \text{ s}^{-1}$ at a site having a total drop of 3.0 m. The site conditions limit the weir's crest length and height to 6.0 m and 1.0 m, respectively.

Solution

Given: Discharge, $Q = 8.0 \text{ m}^3 \text{ s}^{-1}$; total drop, $F = 3.0$ m; weir crest length, $L = 6.0$ m (maximum); weir height, $h = 1.0$ m (maximum).

Step 1: Using $Q = 8.0 \text{ m}^3 \text{ s}^{-1}$, the relationship between L and H would be

$$L = \frac{Q}{1.77 H^{3/2}} = \frac{8}{1.77 H^{3/2}} = \frac{4.52}{H^{3/2}}$$

Step 2: L values are computed using the above relationship for assumed values of H, and the following table is prepared.

Assumed H (m)	$H^{3/2}$	L (m)	Practical L (m)	F (m)	H (m)	Remarks
0.60	0.46	9.7	10.0	0.15	0.75	$L >$ maximum possible value

(continued)

(continued)

Assumed H (m)	$H^{3/2}$	L (m)	Practical L (m)	F (m)	H (m)	Remarks
0.70	0.59	7.7	8.0	0.15	0.85	$L >$ maximum possible value
0.80	0.72	6.3	6.5	0.15	0.95	$L >$ maximum possible value
0.90	0.85	5.3	5.5	0.15	1.05	$h >$ maximum possible value
0.85	0.78	5.8	6.0	0.15	1.00	Both L & h are acceptable

Based on the calculation, L and h values may be taken as 6.0 m and 1.0 m, respectively [Check: $(h/F) = 0.33$ (< 0.5) and $(L/h) = 6.0$ (≥ 2.0)].

Answer: weir crest length, $L = 6.0$ m; weir height, $h = 1.0$ m.

Case II: Knowing F, L and h, to determine Q.

The solution, in this case, is simple as we need to put the known values of L and h in Eq. (8.9) to determine Q.

Example 8.3 A straight drop spillway has a crest length of 4.0 m and a weir height of 1.0 m. What is the maximum discharge capacity of the spillway?

Solution

Given: Weir crest length, $L = 4.0$ m; weir height, $h = 1.0$ m.

Assuming a freeboard, f, of 0.15 m, the flow depth, H, could be determined as.

$$H = h - f = 1.0 - 0.15 = 0.85 \, \text{m}$$

Substituting L and H in Eq. (8.9),

$$Q = 1.77 L H^{3/2} = 1.77 \times 4.0 \times (0.85)^{3/2} = 5.5 \, \text{m}^3 \, \text{s}^{-1}$$

Answer: Discharge capacity of the weir, $Q = 5.5$ m^3 s^{-1}.

8.7.2.3   Design for Submerged Flow Condition

The relationship between the free discharge and the submerged discharge is given as

$$Q_s = R Q_f \tag{8.12a}$$

or

$$q_s = R q_f \tag{8.12b}$$

where Q_S = submerged discharge; Q_f = free discharge; q_s = submerged discharge per unit length of the weir; q_f = free discharge per unit length of the weir; and R = ratio of free discharge and submerged discharge.

Figure 8.6 presents the relationship between R and submergence ratio, H_2/H_1, in graphical form (NRCS 2008). H_1 and H_2 are the upstream flow depth and the submergence depth.

Case I: To determine the crest length, L, and depth, h, of the weir for known
          discharge capacity, Q, and depth of submergence, H_2.

Since flow would only occur when $H_1 > H_2$, the solution approach is to choose h such that $H_1 > H_2$.

The calculations are done in the following table, and a suitable combination of h and L is selected based on the site conditions.

h	f	$H_1 = h - f$	$q_f = 1.77 H_1^{3/2}$	H_2/H_1	R	$q_s = R\,q_f$	$L = Q/q_s$
(1)	(2)	(3)	(4)	(5)	(6)	(7)	(8)

[Column (1): Assumed h; Column (2): Assumed f; Column (3): upstream flow depth; Column (4): Free discharge using the primary relationship; Column (5): Submergence ratio; Column (6): R from Fig. 8.6; Column (7): Submerged discharge using Eq. 8.12(b); Colum (8): Estimated L]

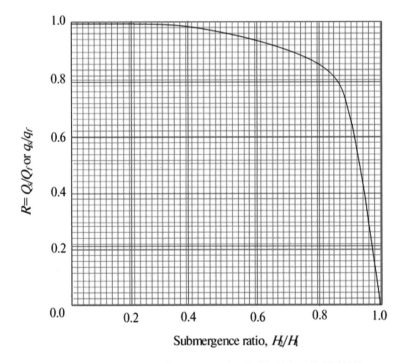

Fig. 8.6 Relationship between R and submergence ratio, H_2/H_1 (After NRCS 2008)

Example 8.4 Determine the dimensions of a straight drop spillway to carry 12 $m^3 \, s^{-1}$ under submergence. The submergence depth is 0.7 m, and the recommended freeboard is 0.23 m. The site conditions limit the crest length to 6.25 m.

Solution

Given: Submerged discharge, $Q_S = 12 \, m^3 \, s^{-1}$; depth of submergence, $H_2 = 0.7$ m; weir crest length, $L = 6.25$ m (maximum); freeboard, $f = 0.23$ m.

Following the procedure described earlier, i.e., beginning by assuming a value of h such that $H_1 \, (= h - f) > H_2$, the calculations are done in the following table.

h	H_1	q_f	H_2/H_1	R	q_s	$L = Q/q_s$	Remark
1	0.77	1.20	0.91	0.56	0.67	17.92	L > maximum value
2	1.77	4.17	0.40	0.99	4.13	2.91	L < maximum value
1.5	1.27	2.53	0.55	0.96	2.43	4.93	L < maximum value
1.4	1.17	2.24	0.60	0.94	2.11	5.70	L < maximum value
1.35	1.12	2.10	0.63	0.93	1.95	6.15	L ≈ maximum value

Thus, calculated L and h values are 6.15 m and 1.35 m, respectively. From practical consideration, L may be taken as 6.25 m.

Answer: Weir crest length, $L = 6.25$ m; weir height, $h = 1.35$ m.

Case II: To determine Q of the spillway for known L, h, and submergence condition (i.e., crest elevation and the downstream stage-discharge curve).

The following steps may be followed:

Step 1: Assume freeboard, f, and determine $H_1 \, (= h - f)$.

Step 2: Using Eq. (8.9), determine free discharge, Q_f.

Step 3: A trial-and-error procedure is used. First, assume a trial value of Q, and obtain the submergence depth, H_2, from the down- stream stage-discharge, i.e., H_2 vs Q, curve. Knowing H_2 and H_1, estimate H_2/H_1, and then read R from Fig. 8.6. Estimate Q_s using Eq. (8.12b). Continue the process until the estimated Q_s matches the assumed Q. The calculations are performed in the following table.

Trial Q	Downstream elevation	H_2	H_2/H_1	R	$Q_s = R \, Q_f$	Remarks
(1)	(2)	(3)	(4)	(5)	(6)	(7)

[Column (1): Trial Q; Column (2): From Downstream elevation vs Q curve; Column (3): Difference between elevation in Column 2 and crest elevation; Column (4): Using known H_2 and H_1; Column (5): From R vs H_2/H_1 curve (Fig. 8.6); Column (6): Submerged discharge using Eq. (8.12b); Column (7): Remarks on computed Q_s and assumed Q]

Example 8.5 A straight drop spillway has a crest length of 5.0 m and a weir depth of 1.0 m. Determine the discharge capacity of the spillway. The freeboard and crest elevations are 0.15 m and 194.8 m, respectively. The downstream stage-discharge curve is given in Fig. 8.7.

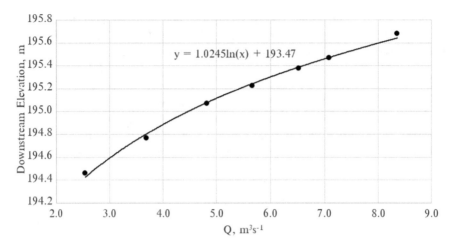

Fig. 8.7 Downstream elevation vs discharge curve (Example 8.5)

Solution

Given: Crest length, $L = 5.0$ m; weir depth, $h = 1.0$ m; Freeboard, $f = 0.15$ m; crest elevation $= 194.8$ m.

Step 1: Knowing $h = 1.0$ m and $f = 0.15$ m, upstream flow depth $H_1 = h - f = 0.85$ m.

Step 2: Free discharge, $Q_f = 1.77LH_1^{\frac{3}{2}} = 1.77 \times 5.0 \times (0.85)^{\frac{3}{2}} = 6.94$ m³ s⁻¹

Step 3: The trial-and-error procedure, described earlier, is used, and calculations are performed in the table below.

Trial Q (m³ s⁻¹)	Downstream elevation (m)	H_2 (m)	H_2/H_1	R	$Q_s = R\,Q_f$ (m³ s⁻¹)	Remarks
(1)	(2)	(3)	(4)	(5)	(6)	(7)
5	195.1	0.3	0.37	0.99	6.9	$Q_s > Q$
6	195.3	0.5	0.59	0.94	6.5	$Q_s > Q$
7	195.5	0.7	0.82	0.84	6.0	$Q_s < Q$
6.5	195.4	0.6	0.71	0.91	6.3	$Q_s < Q$
6.3	195.4	0.6	0.71	0.91	6.3	$Q_s = Q$; Okay

Answer: Discharge capacity of the weir under submergence $= 6.3$ m³ s⁻¹.

8.7.2.4 Design Dimensions of Different Components of a Straight Drop Spillway

As mentioned earlier, the critical depth of flow, y_c, is used for determining the design dimensions of various components of the drop spillway. The design dimensions of various components (Fig. 8.8) may be determined using the equations shown in Table 8.1.

Example 8.6 Determine the dimensions of various components of the straight drop spillway whose crest length and height of the weir are found in Example 8.2. The given discharge and total drop is 8.0 m³ s⁻¹ and 3.0 m, respectively.

Solution

In Example 8.2, the weir crest length, L, and weir height, h, were calculated as 6.0 m and 1.0 m, respectively. However, the given discharge, $Q = 8.0$ m³ s⁻¹, was revised to 8.32 m³ s⁻¹. With the known $Q = 8.32$ m³ s⁻¹ and $L = 6.0$ m, the critical depth, y_c, can be determined using Eq. (8.10).

Subsequently, the dimensions of various components of the drop spillway can be determined using the equations given in Table 8.1.

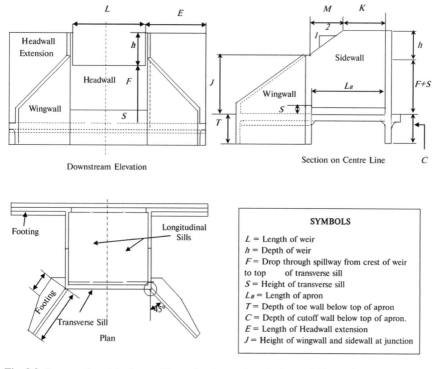

Fig. 8.8 Layout of straight drop spillway showing various design variables (After NRCS 2008)

Table 8.1 Design equations for various components of the straight drop spillway

Component	Equation	
Height of transverse sill, S	$S = \frac{y_c}{2}$	(8.13)
Height of headwall, H_b	$H_b = F + S$	(8.14)
Height of headwall extension, H_e	$H_e = F + S + h$	(8.15)
Minimum length of headwall extension, E	$E = 3h + 0.6$ or $E = 1.5F$ The higher of the two values is chosen	(8.16a) (8.16b)
Minimum length of apron, L_B	$L_B = F\left(2.28\left(\frac{h}{F}\right) + 0.52\right)$	(8.17)
Height of the sidewall and wingwall at the junction, J	$J = 2h$ or $J = [F + S + h - 0.5(L_B + 0.12)]$ or $J = (T + 0.3)$ The highest of the three values is chosen	(8.18a) (8.18b) (8.18c)
Depth of toewall (T) and the cutoff wall (C)	$T = C = 1.2$ or $T = C = 1.65\left[\frac{S + 0.4F + 0.75}{4}\right]$ The higher of the two values is chosen	(8.19a) (8.19b)
Sidewall dimension, M	$M = 2\left(F + \frac{4}{3}h - J\right)$	(8.20)
Sidewall dimension, K	$K = [(L_B + 0.12) - M]$	(8.21)

All dimensions are in m

$$y_c = \left(\frac{q^2}{g}\right)^{1/3} = \left(\frac{Q^2}{L^2 g}\right)^{1/3} = \left(\frac{(8.32)^2}{(6.0)^2 \times 9.81}\right)^{1/3} = 0.58 \text{ m}$$

Height of transverse sill, $S = \frac{y_c}{2} = \frac{0.58}{2} = 0.29$ m.
Height of headwall, $H_b = F + S = 3 + 0.29 = 3.29$ m.
Height of headwall extension: $H_e = F + S + h = 3 + 0.29 + 1 = 4.29$ m.
Minimum length of headwall extension:

$$E = 3h + 0.6 = 3 \times 1 + 0.6 = 3.6 \text{ m}$$

$$\text{Or} \quad E = 1.5F = 1.5 \times 3 = 4.5 \text{ m}$$

Since higher of the two values is taken, $E = 4.5$ m.
Minimum length of apron: $L_B = F\left(2.28\left(\frac{h}{F}\right) + 0.52\right) = 3 \times \left(2.28\left(\frac{1}{3}\right) + 0.52\right) =$ 3.84 m.
Depth of Toewall and Cutoff wall, $T = C = 1.2$ m

$$\text{Or} \quad T = C = 1.65\left[\frac{S + 0.4F + 0.75}{4}\right]$$

$$= 1.65 \times \left[\frac{0.29 + 0.4 \times 3 + 0.75}{4}\right] = 0.92 \text{ m}$$

Since higher of the two values is taken, $T = C = 1.2$ m.
Height of the sidewall and wingwall at the junction:

$$J = 2h = 2 \times 1 = 2 \text{ m}$$
$$\text{Or } J = [F + S + h - 0.5(L_B + 0.12)]$$
$$= [3 + 0.29 + 1 - 0.5(3.84 + 0.12)] = 2.31 \text{ m}$$
$$\text{Or } J = (T + 0.3) = 1.2 + 0.3 = 1.5 \text{ m}$$

Since the highest of the three values is taken, $J = 2.31$ m.
Sidewall dimension, $M = 2\left(F + \frac{4}{3}h - J\right) = 2 \times \left(3 + \frac{4}{3} \times 1 - 2.31\right) = 4.04$ m.
Sidewall dimension, $K = [(L_B + 0.12) - M] = (3.84 + 0.12) - 4.04 = -0.08 \approx 0$.

Example 8.7 Carry out the hydrologic and hydraulic design of a straight drop spillway for the free-flow condition in a gully. The gully has a catchment area of 0.6 km^2. The 10 year maximum rainfall intensity for a duration equal to the time of concentration of the catchment is 140 mm h^{-1}. The total drop in the gully bed is 2.0 m. Assume a runoff coefficient of 0.35. The site conditions limit the length of the weir crest within 5–6 m.

Solution

Given: Catchment area, $A = 0.6$ sq. km $= 6 \times 10^5$ m^2; design rainfall intensity, i $= 140$ mm h^{-1} $= 3.9 \times 10^{-5}$ m s^{-1}; total drop, $F = 2$ m; runoff coefficient $= 0.35$; range on weir crest length, $L = 5.0 - 6.0$ m.

Hydrologic Design

Using the rational formula [Eq. (3.19); Chap. 3], the design peak discharge,

$$Q = ci A = 0.35 \times 3.9 \times 10^5 \times 6 \times 10^{-5} = 8.19 \text{ m}^3\text{s}^{-1}$$

Hydraulic Design

Since Q and F are known, we need to determine L and h under the free-flow condition (Sect. 8.7.2.2).
The following steps are adopted for the design.

Step 1: Using $Q = 8.19$ m^3 s^{-1}, the relationship between L and H would be

$$L = \frac{Q}{1.77H^{3/2}} = \frac{8.19}{1.77H^{3/2}} = \frac{4.62}{H^{3/2}}$$

Step 2: L values are computed using the above relationship for assumed values of H, and a table is prepared (It is assumed that freeboard, $f = 0.15$ m).

Assumed H, m	$H^{3/2}$	L m	f m	h m	Remarks
0.60	0.46	9.93	0.15	0.75	$L >$ maximum possible value
0.70	0.59	7.88	0.15	0.85	$L >$ maximum possible value
0.80	0.72	6.45	0.15	0.95	$L >$ maximum possible value
0.90	0.85	5.41	0.15	1.05	L is within the range, but $h/F > 0.5$
0.85	0.78	5.89	0.15	1.00	L is within the range, and $h/F = 0.5$. Thus, acceptable

Based on the calculation, L and h values are 5.89 m and 1.0 m, respectively. However, from a practical consideration, L may be taken as 6.0 m [Check: $(h/F) = 0.5$ and $(L/h) = 6.0$ (≥ 2.0)]. Thus, the value of Q will be 8.32 m^3 s^{-1}.

With the known Q and L values, the critical depth, y_c, can be determined using Eq. 8.10. Subsequently, the dimensions of various components of the drop spillway can be determined using the equations given in Table 8.1.

$$y_c = \left(\frac{q^2}{g}\right)^{1/3} = \left(\frac{Q^2}{L^2 g}\right)^{1/3} = \left(\frac{(8.32)^2}{(6.0)^2 \times 9.81}\right)^{1/3} = 0.58 \text{ m}$$

Height of transverse sill, $S = \frac{y_c}{2} = \frac{0.58}{2} = 0.29$ m.
Height of headwall, $H_b = F + S = 2 + 0.29 = 2.29$ m.
Height of headwall extension: $H_e = F + S + h = 2 + 0.29 + 1 = 3.29$ m.
Minimum length of headwall extension $E = 3h + 0.6 = 3 \times 1 + 0.6 = 3.6$ m

$$\text{Or } E = 1.5F = 1.5 \times 2 = 3.0 \text{ m}$$

Since higher of the two values is taken, $E = 3.6$ m.
Minimum length of apron: $L_B = F\left(2.28\left(\frac{h}{F}\right) + 0.52\right) = 2 \times \left(2.28\left(\frac{1}{2}\right) + 0.52\right) = 3.32$ m.
Depth of toewall and the cutoff wall, $T = C = 1.2$ m

$$\text{Or } T = C = 1.65\left[\frac{S + 0.4F + 0.75}{4}\right] = 1.65 \times \left[\frac{0.29 + 0.4 \times 2 + 0.75}{4}\right] = 0.76 \text{ m}$$

Since higher of the two values is taken, $T = C = 1.2$ m.
Height of the sidewall and wingwall at the junction:

$$J = 2h = 2 \times 1 = 2.0 \text{ m}$$
$$\text{Or } J = [F + S + h - 0.5(L_B + 0.12)]$$
$$= [2 + 0.29 + 1 - 0.5(3.32 + 0.12)] = 1.57 \text{ m}$$
$$\text{or } J = (T + 0.3) = 1.2 + 0.3 = 1.5 \text{ m}$$

Since the highest of the three values is taken, $J = 2.0$ m.

Sidewall dimension, $M = 2(F + \frac{4}{3}h - J) = 2 \times (2 + \frac{4}{3} \times 1 - 2.0) = 2.67$ m.
Sidewall dimension, $K = [(L_B + 0.12) - M] = (3.32 + 0.12) - 2.67 = 0.77$ m.

8.7.3 Hydraulic Design of Box-Inlet Drop Spillway

8.7.3.1 Design for Free-Flow Condition

The box-inlet drop spillway has a rectangular box, open at the top and downstream, as the inlet (Fig. 8.1b). Free flow takes place when the flow through the spillway is independent of the tailwater. However, unlike a straight drop spillway, the flow through a box-inlet spillway may be controlled either by the crest of the box-inlet or the opening in the headwall (Blaisdell and Donnelly 1951). The crest of the box-inlet controls the flow until the box-inlet is flooded out. Once the box-inlet is flooded out at higher flows, the opening in the headwall controls the flow. The box-inlet dimensions, i.e., its depth and length, govern the shift of control from the crest to the headwall opening.

Therefore, the following two possible design cases may be encountered:

Case I: When the crest of the box-inlet controls the flow.
Case II: When the opening of the headwall controls the flow.

8.7.3.2 Case I: When the Crest of the Box-Inlet Controls the Flow

When the crest controls the box-inlet drop spillway flow, the flow enters over the crest and two sides. The discharge capacity of the spillway may be determined using the rectangular weir formula [Eq. (8.6)]. As mentioned earlier, in practical cases, the velocity of approach, v_a, is so small that the velocity head, $(v_a^2/2g)$, may be ignored. Therefore, Eq. (8.6) may be expressed as:

$$Q = CLH^{3/2} \tag{8.22}$$

Crest length, L, in Eq. (8.22) is taken as the sum of the lengths of the three inflow sides of the box-inlet, i.e., $L = 2B + W$, where B is the inside length of the box-inlet and W is the width of the box-inlet (Fig. 8.9).

The typical value of the discharge coefficient, C, in SI units is 1.89 when $B/W = 1.0$, $W_C/L > 3.0$ and $H/W > 0.6$. Here W_C is the approach channel width. However, the value of C needs to be corrected to reflect the effects of various factors like head over the spillway, shape of the box-inlet, width of the approach channel, and dike position.

Correction in C due to Head
The value of C in Eq. (8.22) is 1.89 when $H/W > 0.6$. Thus, the value of C needs to be corrected when $H/W < 0.6$. For $H/W < 0.6$, the value of C may be corrected by

Fig. 8.9 Box-inlet drop
spillway (After Blaisdell and
Donnelly 1951)

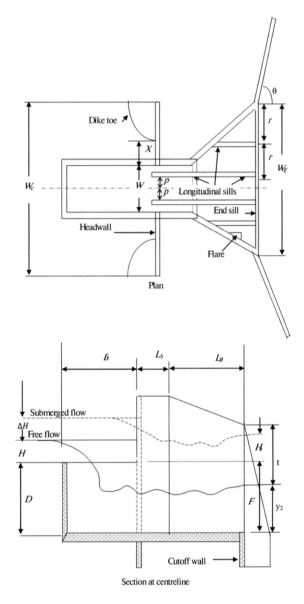

multiplying it with the correction factor, $C_{\text{CORRECTION\_}H}$, read from Fig. 8.10 (valid
for $0 < H/W < 1.5$).

Correction in C due to Box-Inlet Shape

The value of C in Eq. (8.22) is 1.89 when $\frac{B}{W} = 1.0$. Thus, C needs to be corrected
when $\frac{B}{W} \neq 1.0$. The value of C may be corrected by multiplying it with the correction
factor, $C_{\text{CORRECTION\_Shape}}$, read from Fig. 8.11 (valid for $0 < B/W < 4.0$).

Fig. 8.10 Correction in discharge coefficient C due to head (After Blaisdell and Donnelly 1951)

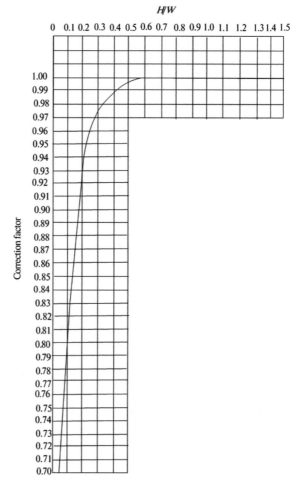

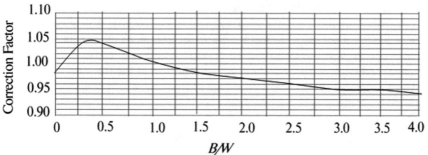

Fig. 8.11 Correction in discharge coefficient C due to box-inlet shape (After Blaisdell and Donnelly 1951)

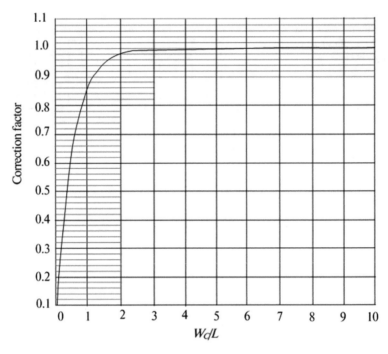

Fig. 8.12 Correction in discharge coefficient C due to approach channel width (After Blaisdell and Donnelly 1951)

Correction in C due to Approach Channel Width

The value of C in Eq. (8.22) is 1.89 when $W_C/L > 3.0$. Thus, the value of C needs to be corrected when $W_C/L < 3.0$. For $W_C/L < 3.0$, the value of C may be corrected by multiplying it with the correction factor, $C_{\text{CORRECTION\_Width}}$, read from Fig. 8.12 (valid for $0.4 < W_C/L < 10$).

Correction in C due to Dike Position

The dike position affects the discharge coefficient because it may obstruct the flow to the box-inlet adjacent to the headwall. Table 8.2 presents the correction factor, $C_{\text{CORRECTION\_Dike}}$, as a function of X/H, X being the distance between the dike's toe and the box-inlet crest (Fig. 8.9), for $B/W = 2$, 1 or 0.5.

Therefore, the discharge capacity of the spillway may be determined using the modified form of Eq. 8.22 expressed as:

$$Q = C\big(C_{\text{CORRECTION\_}H} \times C_{\text{CORRECTION\_Shape}}$$
$$\times C_{\text{CORRECTION\_Width}} \times C_{\text{CORRECTION\_Dike}}\big)LH^{3/2} \qquad (8.23)$$

Subsequently, for known Q, like in the straight drop spillway, a relationship between L and H is obtained using Eq. (8.22). L is computed for assumed values of H, and the combination of L and h best suited for the site conditions is selected.

Table 8.2 Correction for dike effect (After Blaisdell and Donnelly 1951)

B/W	X/H					
	0.0	0.7	1.4	2.9	5.7	∞
	Correction factor					
2.0	0.84	0.85	0.93	0.97	0.99	1.0
1.0	0.73	0.75	0.88	0.95	0.98	1.0
0.5	0.60	0.62	0.82	0.92	0.98	1.0

For a known Q value, L and h could be obtained using the trial-and-error procedure as in Case I.

8.7.3.3 Case II: When the Opening of the Headwall Controls the Flow

When the opening of the headwall controls the flow, the discharge capacity of the spillway may be determined using the modified form of the weir formula as follows:

$$Q = CW(H - H_C)^{3/2} \tag{8.24}$$

In Eq. (8.24), H is measured at the box-inlet crest. Since the effective opening of the headwall is $(H + D)$, D being the depth of the box-inlet, H needs to be corrected. Therefore, the head correction H_C is needed. Similarly, C also needs to be corrected to consider the relative increase in the depth D/W of the box.

Variation of C with the Relative Depth of the Box

The value of C in Eq. (8.24) increases as D/W increases. The value of C may be taken from Fig. 8.13.

Head Correction, H_C

The relative head correction, H_C/D, may be obtained as a function of B/D from Fig. 8.14 when D/W is between 1/4 and 1.0. When D/W is between 1/8 and 1.0, Fig. 8.15 may be used. However, for using Fig. 8.15, any two of B/D, B/W or D/W should be known.

8.7.3.4 Design Dimensions of Different Components of a Box-Inlet Drop Spillway

As in the straight drop spillway, the critical depth of flow, y_c, is used for determining the design dimensions of various components of the box-inlet drop spillway. The critical depth may be determined by replacing the crest length, L, in Eq. (8.10) by the width of the box-inlet, W, as follows:

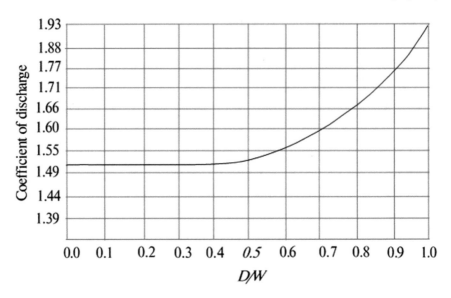

Fig. 8.13 Coefficient of discharge for different D/W (After Blaisdell and Donnelly 1951)

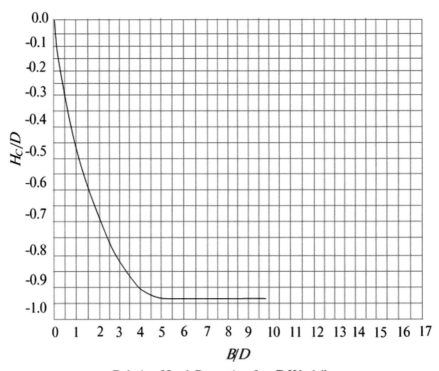

Relative Head Correction for $D/W \geq 1/4$

Fig. 8.14 Relative head correction for $D/W \geq 1/4$ (After Blaisdell and Donnelly 1951)

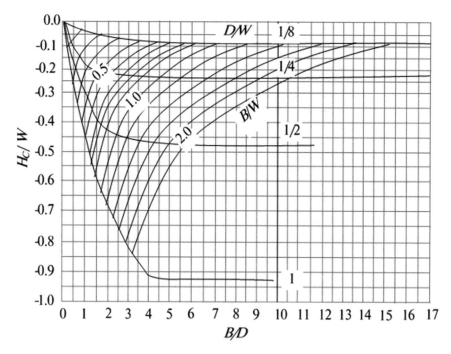

Fig. 8.15 Relative head correction for D/W between 1/8 and 1.0 (After Blaisdell and Donnelly 1951)

$$y_c = \left(\frac{Q^2}{W^2 g}\right)^{1/3}$$

(8.25a)

However, Eq. (8.25a) is only valid for the straight section of the outlet, which has a length of L_S (Fig. 8.9).

Since the exit of the stilling basin section of the outlet, having a length of L_B, may be wider than the straight section, the critical depth may be determined using the following relationship:

$$y_{ce} = \left(\frac{Q^2}{W_E^2 g}\right)^{1/3}$$

(8.25b)

where y_{ce} and W_E are the critical depth and the width at the stilling basin exit.

The design dimensions of various components (Fig. 8.9) may be determined using the equations shown in Table 8.3.

Table 8.3 Design equations for various components of the box-inlet drop spillway

Component	Equation	
Minimum length of the straight section of the outlet, L_S	$L_S = y_c\left(\frac{0.2}{(B/W)} + 1.0\right)$	(8.26)
The flare of the stilling basin sidewalls	1 in ∞ (parallel extension of the straight section walls) to 2:1 (H:V)	
Minimum length of the stilling basin, L_B	$L_B = \frac{(2B+W)}{2\left(\frac{B}{W}\right)}$	(8.27)
Minimum tailwater depth over the basin floor, y_2	When $W_E < 11.5y_{ce}$, $y_2 = 1.6y_{ce}$	(8.28a)
	When $W_E > 11.5y_{ce}$, $y_2 = y_{ce} + 0.052W_E$	(8.28b)
Height of the end sill, S	$S = \frac{y_2}{6}$	(8.29)
Number, location and the height of the longitudinal sills	Longitudinal sills are not required when the stilling basin walls are parallel to the straight walls When $W_E < 2\,W$, only two longitudinal sills are required The sills could be placed at a distance p of $W/6$ to $W/4$ from the centerline When $W_E > 2\,W$, two additional longitudinal sills may be used. These sills should be placed on either side of the centerline, between the centre sills and the sidewalls at the exit of the stilling basin The height of the longitudinal sills should be equal to S	
Minimum height of the sidewall above the water surface at the stilling basin exit, t	$t = \frac{y_2}{3}$	(8.30)
Wingwall	Wingwall should be triangular in elevation, and its top should have a downward slope of 45° with the horizontal Wingwall should flare in the plan at an angle, θ, of 60° with the outlet centerline	

All dimensions are in m

8.7.3.5 Submergence Effect

In the case of the box-inlet drop spillway, the discharge Q, dimensions B, D, W and W_E, and the tailwater depth above the box-inlet crest, H_t, should be known for determining the submergence effect.

The submergence effect is usually considered in terms of an increase in the free-flow head, H, due to the submergence. The increase in H due to submergence could be referred from the submergence curves that relate a different combination of B/W, D/W and W_E/W and ratio $Q/W^{5/2}$ (Blaisdell and Donnelly 1951).

Example 8.8 Determine the dimensions of a box-inlet drop spillway to carry 6.0 $m^3 s^{-1}$ at a site having a total drop of 1.5 m. The approach channel has a trapezoidal section with a bottom width of 12 m and a side slope of 3:1. The crest of the box-inlet and the exit channel is at elevation 30 m and 28.5 m, respectively. The exit channel has a bottom width of 3.0 m and a side slope of 2:1. The depth of flow in the exit channel is 1.0 m.

Solution

Given: discharge, $Q = 6.0 \ m^3 \ s^{-1}$; total drop at the site = 1.5 m; bottom width of the approach channel = 12 m; Side slope of the approach channel = 3:1; elevation of the box-inlet crest = 30 m; elevation of the exit channel = 28.5 m; Bottom width of the exit channel = 3.0 m; side slope of the exit channel = 2:1; depth of flow in the exit channel = 1.0 m.

Referring to Fig. 8.9, and considering the given data, we find that,
Depth of the box-inlet, $D = 1.5$ m.
Width of the approach channel, $W_C = 12 + 2 \times 3 \times H = 12 + 6H.$

For initiating the hydraulic design, the values of L and H are determined first, assuming that the crest of the box-inlet controls the flow. Thus, using Eq. (8.6) with $C = 1.89$, the relationship between L and H may be expressed as:

$$H = \left(\frac{Q}{1.89L}\right)^{2/3} = \left(\frac{6}{1.89L}\right)^{2/3} = \frac{2.16}{L^{2/3}}$$

The values of H for various assumed values of L are tabulated below.

Assumed L (m)	3	6	9
H (m)	1.04	0.65	0.5

Considering H and L values of 0.65 m and 6.0 m, respectively, and assuming a square box-inlet,

$$B = 2.0 \ m; \ W = 2.0 \ m; \ B/W = 1.0$$

The next step would be to determine the head, assuming that the opening of the headwall controls the flow. In the final design, the greater of the two heads are used, i.e., the head when the crest or the headwall opening controls the flow.

When the opening of the headwall controls the flow, the discharge capacity of the spillway is determined using Eq. (8.24):

$$Q = CW(H - H_C)^{3/2}$$

However, the value of C needs to be referred from Fig. 8.13 for the known relative depth D/W. The head correction H_C needs to be obtained from Fig. 8.14 or Fig. 8.15 for known values of B/D and D/W.

In our case, $D/W = 1.5/2 = 0.75$, therefore, from Fig. 8.14, $C = 1.62$.

Since D/W is 0.75, i.e., in between 1/4 and 1.0, from Fig. 8.15, for $B/D = 1.33$, $H_C/D = -0.63$. Hence, $H_C = -0.63 \times 1.5 = -0.95$.

Substituting known values of Q, C, W and H_C in Eq. (8.24),

$$H = \left(\frac{Q}{CW}\right)^{2/3} + H_C = \left(\frac{6}{1.62 \times 2}\right)^{2/3} - 0.95 = 0.56 \text{ m}$$

Since the head obtained here is less than the previously determined value of 0.65, the control would be at the box-inlet crest, and $H = 0.65$ m would be used in the subsequent design.

Since the control is at the crest of the box-inlet, the value of C in Eq. (8.6) needs to be corrected, and the corrected H needs to be determined using Eq. (8.23).

$$Q = C(C_{\text{CORRECTION\_}}H \times C_{\text{CORRECTION\_}}\text{Shape} \times C_{\text{CORRECTION\_}}\text{Width}$$
$$\times C_{\text{CORRECTION\_}}\text{Dike})LH^{3/2}$$

Correction in C due to Head: The value of C needs to be corrected when $H/W < 0.6$. Since in this case, $H/W = 0.65/2 = 0.33$, using Fig. 8.11, $C_{\text{CORRECTION\_}H} = 0.98$.

Correction in C due to Box-Inlet Shape: The value of C in Eq. (8.22) is 1.89 when $B/W = 1.0$. Since in this case, $B/W = 1.0$, $C_{\text{CORRECTION\_Shape}} = 1.0$.

Correction in C due to Approach Channel Width: The value of C in Eq. (8.22) is 1.89 when $W_C/L > 3.0$. Since in this case, $W_C/L = (12 + 5\frac{1}{2}0.65)/6 = 2.55$, using Fig. 8.13, $C_{\text{CORRECTION\_Width}} = 0.995$.

Correction in C due to Dike Position: Since the dike position is not specified, $C_{\text{CORRECTION\_Dike}} = 1.0$.

Putting the known values in Eq. (8.23),

$$6 = 1.89 \times (0.98 \times 1.0 \times 0.995 \times 1.0) \times 6 \times H^{3/2}; \ H = 0.67 \text{ m}$$

Thus, the finalised dimensions of the spillway until this stage are:

$B = 2.0$ m; $W = 2.0$ m; $D = 1.5$ m; $W_C = 15.4$ m; $L = 6.0$ m and $H = 0.67$ m.

The dimensions of various other components may be decided based on the critical depths at the box-inlet crest and the stilling basin exit. Using Eqs. 8.25a–8.25b, the critical depths may be determined as follows:

Critical depth at the crest of the box-inlet, $y_c = \left(\frac{Q^2}{W^2 g}\right)^{1/3} = \left(\frac{6^2}{2^2 \times 9.81}\right)^{1/3} = 0.97$ m.

Critical depth at the exit of the stilling basin, $y_{ce} = \left(\frac{Q^2}{W_E^2 g}\right)^{1/3} = \left(\frac{6^2}{3^2 \times 9.81}\right)^{1/3} = 0.74$ m.

The width at the exit of the stilling basin, W_E, is assumed equal to the exit channel's bottom width, i.e., 3.0 m.

The design dimensions of various components are determined using the equations in Table 8.3.

Minimum length of the straight section of the outlet, L_S, $L_S = y_c\left(\frac{0.2}{(B/W)} + 1.0\right) = 0.97 \times \left(\frac{0.2}{1} + 1.0\right) = 1.16$ m.

Minimum length of the stilling basin, L_B, $L_B = \frac{(2B+W)}{2\left(\frac{B}{W}\right)} = \frac{(2\times2+2)}{2\times\left(\frac{2}{2}\right)} = 3.0$ m.

For determining the minimum tailwater depth over the basin floor, y_2, we need to find whether $W_E < 11.5 y_{ce}$ or $> 11.5 y_{ce}$. Since $W_E = 3.0$ and $y_{ce} = 0.74$, $W_E < 11.5 y_{ce}$. Hence, Eq. (8.28a) would be used to determine y_2.

$$y_2 = 1.6 y_{ce} = 1.6 \times 0.74 = 1.18 \text{ m}$$

It is worth mentioning that the problem statement specifies the depth of flow in the exit channel as 1.0 m, with the elevation of the exit channel at 28.5 m. Since the calculated tailwater depth is more than the natural depth of flow in the channel, the exit channel's elevation must be lowered by 0.18 m to 28.32 or 28.3 m to maintain the naturally available tailwater depth.

Height of the end sill, S, $S = \frac{y_2}{6} = \frac{1.18}{6} = 0.2$ m.

Number, location and the height of the longitudinal sills.

Since $W_E = 3.0$ and $W = 2.0$, $W_E < 2 W$; only two longitudinal sills would be required. These longitudinal sills may be located at a distance p of $W/5$ from the centreline (assumed in between $W/6$ and $W/4$); thus, $p = \frac{W}{5} = 0.4$ m.

The height of the longitudinal sills = the height of the end sill = 0.2 m.

Minimum height of the sidewall above the water surface at the stilling basin exit, t,

$$t = \frac{y_2}{3} = \frac{1.18}{3} = 0.4 \text{ m}$$

The wingwalls may have an angle of $45°$ with the outlet centreline. The tops of wingwalls should be sloped to fit the fill. The wingwalls may finish where their tops coincide with the ground surface.

8.7.4 Structural Design of Straight Drop Spillway

As described earlier, the structural design involves analysing various forces acting on the structure and then providing the required strength and stability to the various components of the structure. The interaction among the water, the foundation soil and the material used for constructing the structure governs the stability of the structure. The structural failure may occur due to piping, unstable fill, sliding, overturning, undermining, uplift, and lateral scour.

The functions of various components of a drop spillway are listed in Sect. 8.6. The functions of the structural components are listed below from the stability point of view:

1. **Headwall extension**: It provides a stable fill and prevents piping around the structure to protect the gully banks against erosion.
2. **Sidewalls**: The sidewalls provide a stable fill and protect the gully banks against erosion.
3. **Wingwalls**: The wingwalls provide a stable fill and prevent scour of the fill and gully banks.
4. **Cutoff Wall**: It prevents piping under the structure, reduces uplift pressure and resists sliding.
5. **Toewall**: It prevents piping under the structure and prevents undermining of the apron.

The structural stability also depends on the depth of the groundwater table at the construction site. If the groundwater depth is more, the fill around the structure remains dry, minimising the piping and uplift problems.

A drop structure may fail due to overturning, sliding, piping, compression, or tension. Therefore, the design of the structure has to be thoroughly checked for its ability to withstand failure against any of these five causes. While analysing the structural stability, we need to begin by considering the horizontal forces acting against the headwall. The headwall needs to withstand the hydrostatic pressure due to the water column upstream and the earth pressure caused by the backfill. The magnitude of the earth pressure is governed by the backfill material characteristics like permeability, cohesion, angle of internal friction, weight, void ratio, and moisture content.

The headwall may have any of the following load conditions:

Case I: No earth fill against the headwall (Fig. 8.16a); thus, only hydrostatic pressure acts against the headwall.
Case II: Gully graded full to crest elevation (Fig. 8.16b); thus, no standing water behind the headwall and only saturated soil pressure act against the headwall, besides the pressure due to water flow above the crest.
Case III: Compacted earth-fill berm constructed to crest elevation (Fig. 8.16c); thus, both hydrostatic pressure and saturated soil pressure act against the headwall, besides the pressure due to water flow above the crest.

In addition to the horizontal forces acting against the headwall, the uplift pressure caused due to water seepage through the saturated foundation material also acts on the structure. Uplift pressure reduces the self-weight of the structure. The uplift pressure is generally considered equal to the head of the water on the upstream of the headwall, i.e., assuming the tailwater depth to be zero. Its line of action is vertically upward through the centre of the structure.

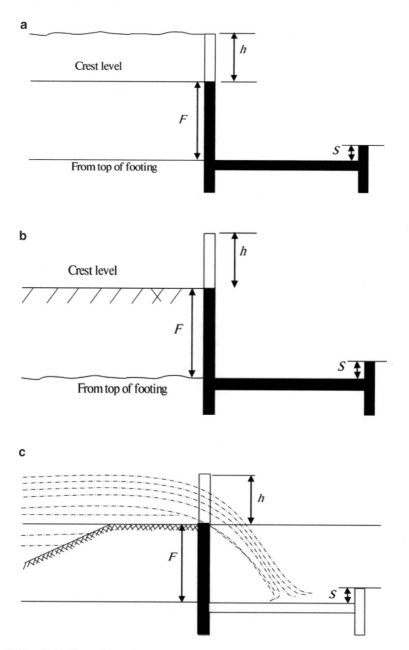

Fig. 8.16 **a** Backfill condition: Case I—No backfill against headwall. **b** Backfill condition: Case II—Gully graded full to crest elevation. **c** Backfill condition: Case III—Compacted earth-fill berm constructed to crest elevation

8.7.4.1   Safety of the Structure Against Overturning

We need to check the structure for safety against overturning about the toe of the structure. The hydrostatic pressure and the lateral earth pressure acting against the wall and the uplift pressure produce the *turning moment* (*TM*), which must be resisted by the *resisting moment* (*RM*) produced by the weight of the structure.

The factor of safety against overturning (*FS$_O$*) is given as the ratio of the total resisting moment to the total turning moment about the toe and expressed as

$$FS_O = \frac{\sum RM}{\sum TM} \qquad (8.31)$$

For the structure to be safe against overturning, *FS$_o$* should be more than 1.5.

Let us consider Case III for analysis, i.e., headwall having a compacted earth-fill berm constructed to crest elevation on its upstream and water flowing above the crest. Three types of pressure act against the headwall: (1) hydrostatic pressure due to flowing water of depth H; (2) lateral pressure due to water of depth F; (3) lateral pressure due to submerged weight of soil of depth F.

Figure 8.17 presents the forces acting on a typical straight drop structure.

The horizontal forces acting on the headwall are:

1. Hydrostatic force due to flowing water of depth H,

$$F_1 = \gamma_w H F \qquad (8.32)$$

where γ_w = unit weight of water, N m^{-3}.
    F_1 acts at the height of ($F/2$) from the base.
2. Lateral force due to water of depth F,

$$F_2 = \frac{1}{2}\gamma_w F^2 \qquad (8.33)$$

F_2 acts at the height of ($F/3$) from the base.
3. Lateral force due to the submerged weight of soil of depth F,

$$F_3 = \frac{1}{2}K_a\gamma_{sub}F^2 \qquad (8.34)$$

where γ_{sub} = submerged unit weight of the backfill, N m^{-3}; and K_a = coefficient of active earth pressure, given as

$$K_a = \frac{(1 - \sin\varphi)}{(1 + \sin\varphi)} \qquad (8.35)$$

where φ = angle of internal friction of the backfill, degrees.
    F_3 acts at the height of ($F/3$) from the base.

Fig. 8.17 Forces acting on a typical straight drop structure

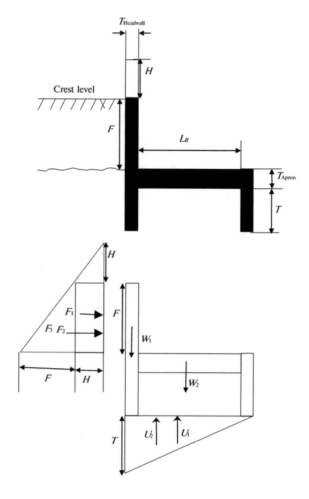

The differential head between the water elevations on upstream and downstream of the structure causes the uplift pressure. In cases where tailwater depth is zero, the upstream head causes the uplift pressure. The uplift pressure is typically assumed to act over the entire base of the structure. The uplift pressure is considered to have two components, given as.

1. Uplift force due to the head equal to the depth of the toewall, T,

$$U_1 = (T \times (L_B + T_{\text{Headwall}}))\gamma_w \qquad (8.36)$$

where T_{Headwall} = thickness of the headwall, m.

U_1 acts at a distance of $((L_B + T_{\text{Headwall}})/2)$ from the toe.

2. Uplift force due to the head equal to $(F - T)$,

$$U_2 = \frac{1}{2}\gamma_w (F - T)(L_B + T_{\text{Headwall}}) \tag{8.37}$$

U_2 acts at a distance of $(\frac{2}{3}(L_B + T_{\text{Headwall}}))$ from the toe.

Horizontal forces F_1, F_2 and F_3, and uplift pressures U_1 and U_2 cause the turning moment.

The weight of the structure, having the following two components, causes the resisting moment:

1. Weight of the headwall,

$$W_1 = F \times T_{\text{Headwall}} \times \gamma_{\text{material}} \tag{8.38}$$

where $\gamma_{\text{material}}=$ unit weight of the construction material, N m^{-3}.

W_1 acts at a distance of $(L_B + (T_{\text{Headwall}}/2))$ from the toe.

2. Weight of the stilling basin,

$$W_2 = (L_B + T_{\text{Headwall}}) \times T_{\text{Apron}} \times \gamma_{\text{material}} \tag{8.39}$$

where $T_{\text{Apron}} =$ apron thickness, m.

W_2 acts at a distance of $((L_B + T_{\text{Headwall}})/2)$ from the toe.

Substituting moments due to various forces (Eqs. 8.32–8.37) and weights (Eqs. 8.38–8.39) in Eq. (8.31), the factor of safety against overturning may be determined as follows:

$$FS_O = \frac{\sum RM}{\sum TM}$$
$$= \frac{[\{W_1 \times (L_B + (T_{\text{Headwall}}/2))\} + \{W_2 \times ((L_B + T_{\text{Headwall}})/2)\}]}{[\{F_1 \times (F/2)\} + \{F_2 \times (F/3)\} + \{F_3 \times (F/3)\} + \{U_1 \times (((L_B + T_{\text{Headwall}})/2)\} + \{U_2 \times (\frac{2}{3}((L_B + T_{\text{Headwall}})))\}]} \tag{8.40}$$

As mentioned earlier, FS_o should be more than 1.5 for the structure to be safe against overturning.

8.7.4.2   Safety of the Structure Against Sliding

We need to check the structure for safety against sliding. The horizontal forces acting on the structure in the downstream direction may cause the failure of the structure due to sliding. Usually, two forces help the structure resist the sliding: (i) the frictional resistance of the foundation and (ii) the cohesive resisting force of the foundation material.

The factor of safety against sliding (FS_S) is given as the ratio of the total resisting forces to the total horizontal forces and expressed as

$$FS_S = \frac{\sum \text{Resisting Forces}}{\sum \text{Horizontal Forces}} = \frac{(f \sum V + CA)}{\sum H} \tag{8.41}$$

where $f = \tan \emptyset =$ coefficient of friction; $\sum V =$ sum of all vertical forces $= (W_1 + W_2 - U_1 - U_2)$; $C =$ cohesive resistance of the foundation material; $A =$ area of the plane of sliding $= [F \times (L_B + T_{\text{Headwall}})]$; and $\sum H =$ sum of all horizontal forces $= (F_1 + F_2 + F_3)$.

Equation (8.41) may be expanded as

$$FS_S = \frac{\left(f \sum V + CA\right)}{\sum H} = \frac{\left[\{\tan \varphi (W_1 + W_2 - U_1 - U_2)\} + \{C(F(L_B + T_{\text{Headwall}}))\}\right]}{(F_1 + F_2 + F_3)} \tag{8.42}$$

For the structure to be safe against sliding, FS_S should be more than 1.1.

8.7.4.3 Safety of the Structure Against Piping

Piping refers to removing foundation material due to seepage of water below the structure (Fig. 8.18). Since piping may lead to structural failure due to foundation instability, we need to check the structure for safety against piping.

Since most of the structural failures due to piping occur along the line of least resistance to the flow, i.e., the line of contact between the structure and the foundation, the *Line of Creep Theory* (Bligh 1910; Lane 1935) is usually adopted to analyse the safety of the structure against piping. As per the theory, the weighted creep distance is the sum of all the steep contact lengths and one-third of all the contact lengths flatter than 45° between the upstream and tailwater head along the contact surface of the structure and the foundation.

The weighted creep ratio, C_W, for a structure and the foundation is given as

$$C_w = \frac{\text{Weighted creep distance}}{\text{Differential head between the upstream and tailwater head}} = \frac{\left(\frac{1}{3} \sum L_H + \sum L_V\right)}{H_{Diff}} \tag{8.43}$$

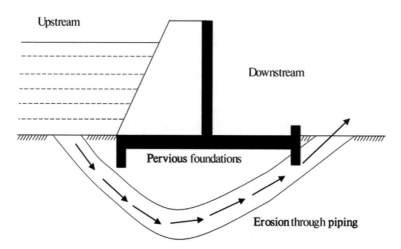

Fig. 8.18 Piping through the foundation of the straight drop structure

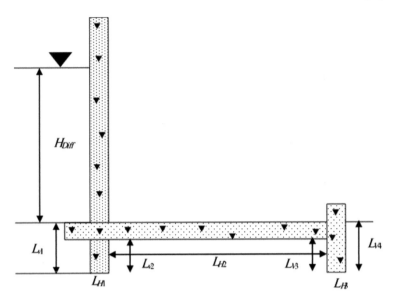

Fig. 8.19 Determination of $\sum L_H$ and $\sum L_V$ for calculating the weighted creep ratio

where $\sum L_H$ = sum of all horizontal or flat contact lengths; $\sum L_V$ = sum of all vertical or steep contact lengths; and H_{Diff} = head causing the flow.

Figure 8.19 shows the horizontal and vertical contact lengths used for determining $\sum L_H$ and $\sum L_V$.

$$\text{From the figure, } \sum L_H = L_{H_1} + L_{H_2} + L_{H_3} \tag{8.44a}$$

$$\sum L_V = L_{V_1} + L_{V_2} + L_{V_3} + L_{V_4} \tag{8.44b}$$

In Fig. 8.19, L_{V_1} and L_{V_4} are equal to the depth of the Toewall, T, while L_{V_2} and L_{V_3} are equal to $(T - T_{\text{Apron}})$. Similarly, L_{H_2} is equal to L_B, while L_{H_1} and L_{H_3} are equal to the thickness of the cutoff wall and toewall

The computed value of C_w obtained from Eq. (8.43) is compared with the weighted creep ratio's recommended value for different foundation materials (Table 8.4). The computed C_w should be equal to or greater than C_w's recommended value for a stable structure.

8.7.4.4 Safety of the Structure Against Tension

We need to check the structure for safety against tension that may develop due to irregular forces. Safety against tension is especially needed when brick or stone masonry is used as the construction material. In such cases, it is essential to ensure

Table 8.4 Recommended values of C_w for different foundation materials (After NRCS 2008)

Foundation material	C_w
Very fine sand and silt	8.5
Fine sand	7.0
Clean sand or sand and gravel mixture	6.5
Medium sand	6.0
Well-graded mixture of sand, silt, and less than 15% clay	5.5
Coarse sand	5.0
Clean gravel	5.0
Well-graded mixture of sand, silt, and more than 15% clay	4.0
Medium gravel	3.5
Soft clay	3.0
Coarse gravel, including cobbles	3.0
Boulders with some cobbles and gravel	2.5
Firm clay	2.3
Medium clay	2.0
Hard clay	1.8
Very hard clay or hardpan	1.6

that no tension develops in the base of the structure. The resultant force acting on the structure must pass through the middle third of the base of the structure to ensure the safety of the structure against tension.

The position of the resultant force is determined using the following relationship:

$$Z = \frac{\sum M}{\sum V} \tag{8.45}$$

where Z = position at which the resultant force acts on the structure (Fig. 8.21); $\sum M$ = sum of moments of all horizontal and vertical forces acting on the structure = $(\sum RM - \sum TM)$; and $\sum V$ = sum of all vertical forces = $(W_1 + W_2 - U_1 - U_2)$.

The eccentricity is given as : $e = \dfrac{d}{2} - Z$ (8.46)

where e = eccentricity, defined as the longitudinal distance between the centroid of the base area and the point of application of the resultant vertical force; and d = base length, i.e., the distance between the upstream and the downstream edges of the base area that includes the length of the apron and widths of the headwall and end sill (Fig. 8.20).

When $\left(Z > \frac{d}{2}\right)$ or $\left(e \leq \frac{d}{6}\right)$, no tension would develop in the base of the structure, and the structure would be safe against tension.

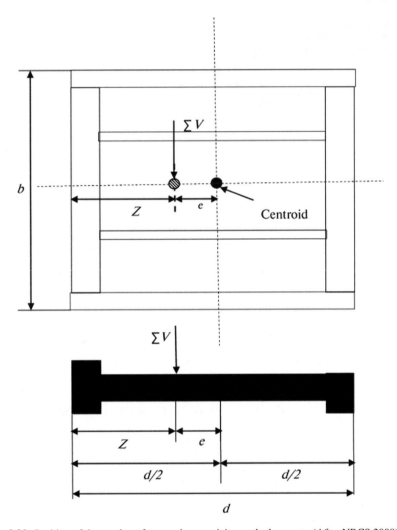

Fig. 8.20 Position of the resultant force and eccentricity on the base area (After NRCS 2008)

8.7.4.5   Safety of the Structure Against Compression or Contact Pressure

We need to check the structure for safety against compression or contact pressure. The resultant force acting on the foundation of the structure due to all horizontal and vertical forces acting on the structure needs to be determined to ensure the safety of the structure against compression or contact pressure. The resultant force acts vertically downward and is compressive. The resultant force is transmitted into the foundation by the direct contact of the foundation material and the structure. Hence, it is also called contact pressure.

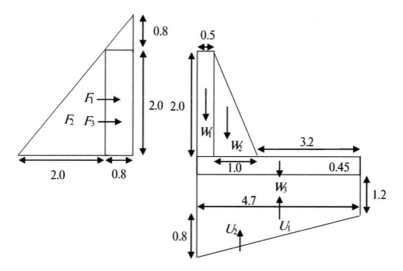

Fig. 8.21 Forces acting on the straight drop structure (Example 8.9)

The contact pressure may be determined using the following relationship:

$$P = \frac{\sum V}{A}\left(1 \pm \frac{6e}{d}\right) \tag{8.47}$$

where P = contact pressure; A = base area = $(b \times d)$; and b = base width that includes the width of the apron and sidewalls (Fig. 8.20).

Equation (8.47) results in two values of P, which belong to the upstream (with '+ve' sign) and the downstream (with '−ve' sign) edges of the base.

The structure is considered safe against compression or contact pressure when both P values are positive.

8.7.4.6 Apron Thickness

The apron should be thick enough to withstand the uplift pressure exerted on it. However, as a thumb rule, the thickness of the plain concrete apron may be considered as a function of F and referred to in Table 8.5 (Sharda et al. 2007).

Table 8.5 Thickness of concrete apron for different values of F (After (Sharda et al. 2007)

Drop, F (m)	0.50–0.75	1.0–2.0	2.0– 3.0
Apron thickness (m)	0.20	0.25	0.30

Table 8.6 Wall thickness for stone or brick masonry drop structure After (Sharda et al. 2007)

	Headwall	Sidewall	Wingwall and Headwall Extension
Minimum top width (m)	0.45	0.30	0.30
Drop, F (m)	Recommended bottom width (m)		
0.5	0.45	0.30	0.30
1.0	0.67	0.55	0.40
1.5	1.00	0.82	0.60
2.0	1.33	1.10	0.80
2.5	1.67	1.37	1.00
3.0	2.00	1.85	1.20

For masonry structures, the apron thickness should be 1.5 times the table value

8.7.4.7 Wall Thickness

Table 8.6 presents the minimum top and bottom widths of the headwall, sidewall, wingwall, and headwall extension (Sharda et al. 2007). The recommended values are for stone or brick masonry structures. The bottom width is considered as a function of the drop, F.

Example 8.9 Carry out the hydraulic and structural design of a straight drop spillway for carrying 6.0 m³ s⁻¹ under the free-flow condition in a gully. The total drop in the gully bed is 2.0 m. The site conditions limit the length of the weir crest within 4–5 m.

Brick masonry, having a unit weight of 22 kN m⁻³, is used as a construction material. The submerged unit weight of the backfill material is 19 kN m⁻³, and its angle of internal friction is 25°. The foundation material is very hard clay. The cohesive resistance of the foundation material is 5.0 kPa, and its recommended creep ratio is 1.6. Take the unit weight of water as 10 kN m⁻³.

Solution

Given: Design discharge, $Q = 6$ m³ s⁻¹; total drop, $F = 2.0$ m; range on weir crest length, $L = 4-5$ m; unit weight of brick masonry, $\gamma_{material} = 22$ kN m⁻³; unit weight of water, $\gamma_w = 10$ kN m⁻³.

Backfill: Submerged unit weight of the backfill material, $\gamma_{sub} = 19$ kN m⁻³; Angle of internal friction, $\varphi = 25°$.

Foundation: Cohesive resistance of the foundation material, $C = 5.0$ kPa; creep ratio, $C_w = 1.6$.

Hydraulic Design

Since Q and F are known, we need to determine L and h under the free-flow condition described under Sect. 8.7.2.2.

The following steps are adopted for the design.

Step 1: Using $Q = 6.0$ m³ s⁻¹, the relationship between L and H would be

$$L = \frac{Q}{1.77 H^{3/2}} = \frac{6}{1.77 H^{3/2}} = \frac{3.39}{H^{3/2}}$$

Step 2: L values are computed using the above relationship for assumed values of H, and the following table is prepared (It is assumed that freeboard, $f = 0.15$ m).

Assumed H (m)	$H^{3/2}$	L (m)	f (m)	h (m)	Remarks
0.7	0.59	5.8	0.15	0.85	$L >$ maximum possible value
0.8	0.72	4.7	0.15	0.95	L within acceptable limit; $h/F < 0.5$
0.9	0.85	4.0	0.15	1.05	L within acceptable limit; $h/F > 0.5$
1.0	1.00	3.4	0.15	1.15	$L <$ minimum possible value

Based on the calculation, L and h values may be taken as 4.7 m and 0.95 m, respectively [Check: $(h/F) < 0.5$ and $(L/h) = 3.8 \, (> 2.0)$].

With the known Q and L values, the critical depth, y_c, can be determined using Eq. (8.10). Subsequently, the dimensions of various components of the drop spillway can be determined using the equations given in Table 8.1.

$$y_c = \left(\frac{q^2}{g}\right)^{1/3} = \left(\frac{Q^2}{L^2 g}\right)^{1/3} = \left(\frac{(6)^2}{(4.7)^2 \times 9.81}\right)^{1/3} = 0.55 \text{ m}$$

Height of transverse sill, $S = \frac{y_c}{2} = \frac{0.55}{2} = 0.28$ m.
Height of headwall, $H_b = F + S = 2.0 + 0.28 = 2.28$ m.
Height of headwall extension: $H_e = F + S + h = 2.0 + 0.28 + 0.95 = 3.23$ m.
Minimum length of headwall extension:

$$E = 3h + 0.6 = 3 \times 0.95 + 0.6 = 3.45 \text{ m}$$

$$\text{Or} \quad E = 1.5F = 1.5 \times 2.0 = 3.0 \text{ m}$$

Since higher of the two values is taken, $E = 3.45$ m.
Minimum length of apron: $L_B = F\left(2.28\left(\frac{h}{F}\right) + 0.52\right) = 2.0 \times \left(2.28\left(\frac{0.95}{2.0}\right) + 0.52\right) = 3.2$ m.
The length of the apron, L_B, is taken as 3.2 m.
Depth of toewall and the cutoff wall, $T = C = 1.2$ m

$$\text{Or } T = C = 1.65\left[\frac{S + 0.4F + 0.75}{4}\right]$$
$$= 1.65 \times \left[\frac{0.28 + 0.4 \times 2.0 + 0.75}{4}\right] = 0.75 \text{ m}$$

Since higher of the two values is taken, $T = C = 1.2$ m.

Height of the sidewall and wingwall at the junction:

$$J = 2h = 2 \times 0.95 = 1.90 \text{ m}$$
$$\text{Or } J = [F + S + h - 0.5(L_B + 0.12)]$$
$$= [2 + 0.28 + 0.95 - 0.5(3.2 + 0.12)] = 1.57 \text{ m}$$
$$\text{Or } J = (T + 0.3) = 1.2 + 0.3 = 1.5 \text{ m}$$

Since the highest of the three values is taken, $J = 1.90$ m.

Sidewall dimension, $M = 2\left(F + \frac{4}{3}h - J\right) = 2 \times \left(2.0 + \frac{4}{3} \times 0.95 - 1.90\right) = 2.73$ m.

Sidewall dimension, $K = [(L_B + 0.12) - M] = (3.2 + 0.12) - 2.73 = 0.59$ m.

Structural Design

Since brick masonry is used as the construction material, the apron thickness should be 0.45 m (1.5 times the recommended value from Table 8.5). Similarly, the top and bottom widths of the headwall should be 0.45 m and 1.33 m, respectively (Table 8.6). The top width of the headwall ($Headwall_{TopW}$) and bottom width of the headwall ($Headwall_{BotW}$) are taken as 0.50 m and 1.50 m, respectively. Figure 8.21 shows the drop structure with various forces acting on it.

From the figure, the horizontal forces acting on the headwall are:

1. Hydrostatic force due to flowing water of depth H,

$$F_1 = \gamma_w H F = 10 \times 0.8 \times 2 = 16 \text{ kN}$$

F_1 acts at the height of $(F/2)$ or 1.0 m from the base.

2. Lateral force due to water of depth F,

$$F_2 = \frac{1}{2}\gamma_w F^2 = \frac{1}{2} \times 10 \times \left(2^2\right) = 20 \text{ kN}$$

F_2 acts at the height of $(F/3)$ or 0.67 m from the base.

3. Lateral force due to the submerged weight of soil of depth F,

$$F_3 = \frac{1}{2}K_a\gamma_{sub}F^2 = \frac{1}{2} \times \left(\frac{(1 - \sin 25)}{(1 + \sin 25)}\right) \times 19 \times \left(2^2\right) = 15.39 \text{ kN}$$

F_3 acts at the height of $(F/3)$ or 0.67 from the base.

Similarly, the uplift pressure has two components, given as.

1. Uplift force due to the head equal to the depth of the Toewall, T,

$$U_1 = (T \times (L_B + Headwall_{BotW}))\gamma_w = (1.2 \times 4.7) \times 10 = 56.4 \text{ kN}$$

U_1 acts at a distance of $((L_B + Headwall_{BotW})/2)$ or 2.35 m from the toe.

2. Uplift force due to the head equal to $(F - T)$,

$$U_2 = \frac{1}{2}\gamma_w(F - T)(L_B + \text{Headwall}_{\text{BotW}}) = \frac{1}{2} \times 10 \times (2.0 - 1.2) \times 4.7 = 18.8 \text{ kN}$$

U_2 acts at a distance of $\left(\frac{2}{3}(L_B + \text{Bottom width of headwall})\right)$ or 3.13 m from the toe.

The weight of the structure has the following three components:

1. Weight of the rectangular portion of the headwall,

$$W_1 = F \times \text{Headwall}_{\text{HTopW}} \times \gamma_{\text{material}} = 2 \times 0.5 \times 22 = 22 \text{ kN}$$

W_1 acts at a distance of $(L_B + \text{Bottom width of the triangular portion of headwall} + (\text{Headwall}_{\text{TopW}}/2)$ or 4.45 m from the toe.
2. Weight of the triangular portion of the headwall,

$$W_2 = \frac{1}{2}F(\text{Headwall}_{\text{BotW}} - \text{Headwall}_{\text{TopW}})\gamma_{\text{material}} = \frac{1}{2} \times 2 \times (1.5 - 0.5) \times 22 = 22 \text{ kN}$$

W_2 acts at a distance of $\left(L_B + \frac{2}{3}(\text{Headwall}_{\text{BotW}} - \text{Headwall}_{\text{TopW}})\right)$ or 3.87 m from the toe.
3. Weight of the stilling basin,

$$W_3 = (L_B + \text{Headwall}_{\text{BotW}}) \times T_{\text{Apron}} \times \gamma_{\text{material}} = 4.7 \times 0.45 \times 22 = 46.53 \text{ kN}$$

W_3 acts at a distance of $((L_B + \text{Bottom width of headwall})/2)$ or 2.35 m from the toe.

Table 8.7 presents the analysis of forces acting on the structure.

Safety against Overturning

From Table 8.7,

Sum of turning moments, $\sum TM = (\text{Sum of moments due to } F_1 - F_3 \text{ and } U_1 - U_2)$ = 231.04 kN m.

Sum of resisting moments, $\sum RM = (\text{Sum of moments due to } W_1 - W_3) = 292.31$ kN m.

Substituting $\sum RM$ and $\sum TM$ in Eq. (8.31),

$$FS_O = \frac{\sum RM}{\sum TM} = \frac{292.31}{231.04} = 1.27$$

Since calculated $FS_o < 1.5$, the structure is not safe against overturning.

Hence, we increase the apron thickness to 0.90 m. Also, the top and bottom widths of the headwall are revised to 0.45 m and 0.90 m, respectively.

Note that the final apron thickness and headwall dimensions were arrived at after several iterations while ensuring safety against various failures.

The revised analysis of forces acting on the structure is presented in Table 8.8.

Thus, the revised sum of turning moments, $\sum TM = 185.28$ kN m.

Table 8.7   Analysis of forces acting on the structure

Force	Volume of component per unit length (m^3)	Unit weight (kN m^{-3})	Force per unit length (kN)	Lever arm length (m)	Moment (kN m)	Type of moment
F_1	(0.8 × 2.0)	10	16	1.0	16.00	Turning
F_2	(0.5 × 2.0 × 2.0)	10	20	0.67	13.33	Turning
F_3	(0.5 × 2.0 × 2.0)	19	15.39	0.67	10.26	Turning
U_1	(1.2 × 4.7)	10	56.4	2.35	132.54	Turning
U_2	[0.5 × (2.0–1.2) × 4.7]	10	18.8	3.13	58.91	Turning
W_1	(2.0 × 0.5)	22	22	4.45	97.90	Resisting
W_2	[0.5 × 2.0 × 1.5–0.5)]	22	22	3.87	85.07	Resisting
W_3	(3.0 + 1.7) × 0.45	22	46.53	2.35	109.35	Resisting

Table 8.8   Revised analysis of forces acting on the structure

Force	Volume of component per unit length (m^3)	Unit weight (kN m^{-3})	Force per unit length (kN)	Lever arm length (m)	Moment (kN m)	Type of moment
F_1	(0.8 × 2.0)	10	16	1.0	16.00	Turning
F_2	(0.5 × 2.0 × 2.0)	10	20	0.67	13.33	Turning
F_3	(0.5 × 2.0 × 2.0)	19	15.39	0.67	10.26	Turning
U_1	(1.2 × 4.1)	10	49.2	2.05	100.86	Turning
U_2	[0.5 × (2.0–1.2) × 4.1]	10	16.4	2.73	44.83	Turning
W_1	(2.0 × 0.45)	22	19.8	3.875	76.73	Resisting
W_2	[0.5 × 2.0 × (0.9–0.45)]	22	9.9	3.5	34.65	Resisting
W_3	(3.2 + 0.9) × 0.9	22	81.18	2.05	166.42	Resisting

The revised sum of resisting moments, $\sum RM = 277.79$ kN m.
Substituting $\sum RM$ and $\sum TM$ in Eq. 8.31,

$$FS_O = \frac{\sum RM}{\sum TM} = \frac{277.79}{185.28} = 1.5$$

Since calculated $FS_o = 1.5$, the structure is now safe against overturning.

Safety against Sliding

From Eq. (8.42), $FS_S = \frac{(f \sum V + CA)}{\sum H} = \frac{[\{\tan \varphi (W_1 + W_2 + W_3 - U_1 - U_2)\} + \{C(F(L_B + T_{\text{Headwall}}))\}]}{(F_1 + F_2 + F_3)}$.

From Table, $W_1 + W_2 + W_3 - U_1 - U_2 = 45.28$ and $F_1 + F_2 + F_3 = 51.39$.
Also, $F(L_B + T_{\text{Headwall}}) = 2.0 \times (3.2 + 0.9) = 8.2$ m^2; $f = \tan \varphi = \tan 25 = 0.467$; and $C = 5.0$ kPa.

Substituting the values in Eq. (8.42), $FS_S = 1.21$.
Since calculated $FS_S > 1.1$, the structure is safe against sliding.

Safety against Piping

From Eq. (8.43), weighted creep ratio, $C_w = \frac{(\frac{1}{3} \sum L_H + \sum L_V)}{H}$.
Referring to Fig. 8.20, $\sum L_H = L_{H_1} + L_{H_2} + L_{H_3} = 0.5 + 4.1 + 0.5 = 5.1$ m

$$\sum L_V = L_{V_1} + L_{V_2} + L_{V_3} + L_{V_4} = 1.2 + (1.2 - 0.9) + (1.2 - 0.9) + 1.2 = 3.0 \, \text{m}$$

Substituting $\sum L_H$ and $\sum L_V$ in Eq. (8.43), with $H = 2.8$,

$$C_w = \frac{(\frac{1}{3} \sum L_H + \sum L_V)}{H} = \frac{\left(\frac{5.1}{3}\right) + 3.0}{2.8} = 1.68$$

Since the calculated $C_W >$ recommended C_W value of 1.6, the structure is safe against piping.

Safety against Tension

From Eq. (8.45), the position of the resultant force, $Z = \frac{\sum M}{\sum V}$.
From Table 8.7, $\sum M = \left(\sum RM - \sum TM\right) = 277.79 - 185.28 = 92.51$ and $\sum V = 45.28$,

Hence, $Z = \frac{\sum M}{\sum V} = \frac{92.51}{45.28} = 2.04$ m.
Base length $d = 4.1$, thus, Z < (d/2); hence, structure is safe against tension.
Thus, using Eq. (8.46), eccentricity, $e = \frac{d}{2} - Z = \frac{4.1}{2} - 2.04 = 0.01$ m.
Thus, $\left(e \leq \frac{4.1}{6} \text{ or } 0.68\right)$, no tension would develop.
Hence, the structure is safe against tension.

Safety against Compression or Contact Pressure

From Eq. (8.47), the contact pressure, $P = \frac{\sum V}{A}\left(1 \pm \frac{6e}{d}\right)$.
Substituting $\sum V = 45.28$, $A = b \times d = (5.7 \times 4.1)$, and $e = 0.01$,

$$P = \frac{\sum V}{A}\left(1 \pm \frac{6e}{d}\right) = \frac{45.28}{(5.7 \times 4.1)} \times \left(1 \pm \frac{6 \times 0.01}{4.1}\right) = 1.96 \, \text{or} \, 1.92 \, \text{kN m}^{-2}$$

Since both P values are positive, the structure is safe against compression or contact pressure.

Since the structure is safe against various failures, the design dimensions of various components are acceptable.

Questions

1. Describe various types of drop structures along with their specific functions and conditions under which they are preferred.
2. Based on the information from the watersheds closer to your hometown, managed under the soil management programmes, describe various existing drop structures.
3. Draw a straight drop spillway showing its structural components. Describe the function of various components.
4. Determine the crest length and height of the weir for a straight drop spillway to carry 6.0 m s^{-1} at a site having a total drop of 2.5 m. The site conditions limit the weir's crest length and height to 5.0 m and 1.5 m, respectively.
5. The following data present peak discharge (m^3 s^{-1}) of a river for 30 years. Compute the 25 year peak flow, assuming that the data follows Gumbel extreme value type-I distribution.

1428	1758	1701	1536	1489	1468	1448	1437	1408	1285
1269	1242	1235	1214	1205	1188	1164	1159	1092	1038
999	996	1193	1159	929	1102	1039	1617	926	1183

6. Determine the dimensions of a straight drop spillway to carry 8.0 m s^{-1} under submergence. The submergence depth is 0.5 m, and the recommended freeboard is 0.2 m. The site conditions limit the crest length to 6.0 m.
7. A straight drop spillway has a crest length of 4.0 m and a weir depth of 1.0 m. Determine the discharge capacity of the spillway. The freeboard and crest elevations are 0.15 m and 194.5 m, respectively. The downstream stage-discharge curve is given by the following relationship: $Downstream\,Elevation = 1.02\ln(Q) + 193.5$ m; Q is in m^3 s^{-1}.
8. Determine the dimensions of a box-inlet drop spillway to carry 8.0 m s^{-1} at a site having a total drop of 2.0 m. The approach channel has a trapezoidal section with a bottom width of 12 m and a side slope of 3:1. The crest of the box-inlet and the exit channel is at elevation 30 m and 28 m, respectively. The exit channel has a bottom width of 4.0 m and a side slope of 2:1. The depth of flow in the exit channel is 1.5 m.
9. Carry out the hydrologic and hydraulic design of a straight drop spillway for the free-flow condition in a gully. The gully has a catchment area of 0.8 sq. km. The 25 yr. maximum rainfall intensity for a duration equal to the time of concentration of the catchment is 120 mm h^{-1}. The total drop in the gully bed is 2.0 m. Assume a runoff coefficient of 0.30. The site conditions limit the length of the weir crest within 4–6 m.

10. Design a straight drop spillway for carrying 8.0 m³ s⁻¹ under the free-flow condition and carry out its structural analysis. The total drop in the gully bed is 2.0 m. The site conditions limit the length of the weir crest within 4–6 m. Brick masonry, having a unit weight of 19 kN m⁻³, is used as a construction material. The submerged unit weight of the backfill material is 17 kN m⁻³, and its angle of internal friction is 25°. The foundation material is medium clay. The cohesive resistance of the foundation material is 5.0 kPa, and its recommended creep ratio is 2.0. Take the unit weight of water as 9.8 kN m⁻³.

Multiple Choice Questions
Answer the questions by choosing the correct option.

1. In a straight drop spillway, the height of the transverse sill is equal to

   A. 2×(critical depth)
   B. 1×(critical depth)
   C. 0.5×(critical depth)
   D. 0.25×(critical depth)

2. Box-inlet drop spillways are preferred

   A. when large water flow needs to be handled in wide channels
   B. large water flow needs to be handled in narrow channels
   C. flow from a reservoir has to be removed
   D. flow has to pass through a wide downstream channel

3. In a box-inlet drop spillway, the opening in the headwall controls the flow when

   A. the flow is free
   B. the flow is submerged
   C. the depth of flow is more than half the depth of box-inlet
   D. the box-inlet is flooded out

4. The thickness of the apron in a straight drop spillway is determined based on

   A. horizontal and vertical forces
   B. frictional forces
   C. uplift pressure
   D. eccentricity

5. In a drop structure, the cut off wall is provided to

   A. dissipate the kinetic energy of the flow
   B. provide stability to the gully wall
   C. guide the flow safely to the downstream end
   D. impart structural strength against sliding

6. Longitudinal sills are provided in the stilling basin to

A. provide stability to the stilling basin
B. provide structural strength against sliding
C. provide structural strength against slippage
D. protect the structure against siltation

7. The self-weight of a structure, having top and bottom width of 1.0 m and 2.0 m, is 2700 kN. If the eccentricity is 0.3 m, then the maximum stress at the base, in kN m^{-2}, is

A. 675   B. 1775
C. 1925   D. 2880

8. The horizontal and vertical forces on a drop spillway are 42 kN and 35.8 kN, respectively. The area of the plane of sliding is 8.0 m^2. The angle of internal friction and cohesive resistance of the foundation material is 25° and 5.0 kPa. The factor of safety against sliding is

A. 0.55   B. 0.74
C. 1.35   D. 1.47

9. For the structure to be safe against tension, the following criterion must be satisfied:

A. The resultant force must always pass through the mid-point of the base
B. The resultant force must pass through the middle third of the base
C. The resultant force must pass through the upstream extremity of the middle third of the base
D. The resultant force must pass through the downstream extremity of the middle third of the base

10. According to the weighted creep theory, the weighted creep distance is the sum of

A. 1/3rd of the horizontal creep and the vertical creep
B. 1/3rd of the vertical creep and the horizontal creep
C. 1/3rd of the horizontal creep and 2/3rd of the vertical creep
D. 1/3rd of the vertical creep and 2/3rd of the horizontal creep

References

Arnold JG, Srinivasan R, Muttiah RS, Williams JR (1998) Large-area hydrologic modelling and assessment: Part I. Model development. J Am Water Resour Asso 34:73–89

Beven KJ, Kirkby MJ (1979) A physically based variable contributing area model of basin hydrology. Hydrol Sci Bull 24:43–69

Blaisdell FW, Donnelly CA (1951) Hydraulic design of the box inlet drop spillway. USDA Soil Conservation Service Technical Paper 106. SCS, Washington DC

Bligh WG (1910) The practical design of irrigation works. Lond Const 162–205

Chow VT (1951) A general formula for hydrologic frequency analysis. Trans AGU 32:231–237

Dirk R (2013) Frequency analysis of rainfall data. In: College on soil physics—30th anniversary (1983–2013). The Abdus Salam International Centre for Theoretical Physics, pp 244–10

Finnerty BD, Smith MB, Seo DJ, Koren V, Moglen GE (1997) Space-time sensitivity of the Sacramento model to radar-gage precipitation inputs. J Hydrol 203:21–38

Haan CT (2002) Statistical methods in hydrology, 2nd edn. Iowa State Press, Ames, Iowa

Havno K, Madsen MN, Dørge J (1995) MIKE11—a generalized river modelling package. In: Singh VP (ed) Computer models of watershed hydrology. Water Resources Publications, Englewood

Jakeman AJ, Littlewood IG, Whitehead PG (1990) Computation of the instantaneous unit hydrograph and identifiable component flows with application to two small upland catchments. J Hydrol 117:275–300

Lane EW (1935) Security from under seepage masonry dams on earth foundations. Trans ASCE 100:1234–1351

Liang X, Lettenmaier DP, Wood EF, Burges SJ (1994) A simple hydrologically based model of land surface water and energy fluxes for GCMs. J Geophys Res 99:14415–14428

Mulvany T (1851) On the use of self-registering rain and flood gauges in making observations of the relation of rainfall and flood discharges in given catchment. Trans Inst Civil Eng Ireland 4:18–33

Natural Resources Conservation Service (NRCS) (1984) Structures, Chapter 6: In: Engineering field manual. USDA Natural Resources Conservation Service, Washington DC

Natural Resources Conservation Service (NRCS) (2008) National engineering handbook, Section 11: Drop spillways. USDA Natural Resources Conservation Service, Washington DC

Peel MC, McMahon TA (2020) Historical development of rainfall-runoff modeling. WIREs Water 7. https://doi.org/10.1002/wat2.1471

Refsgaard JC, Storm B (1995) MIKE SHE. In: Singh VP (ed) Computer models of watershed hydrology. Water Resources Publications, Englewood

Sharda VN, Juyal GP, Prakash C, Joshi BP (2007) Training manual: soil conservation and watershed management, Vol-II (soil water conservation engineering). Central Soil and Water Conservation Research and Training Institute (CSWCRTI), Dehradun

US Army Corps of Engineers (1998) HEC-HMS hydrologic modeling system: User's Manual, US Army Corps of Engineers, Hydrologic Engineering Centre, Davis, California

US Army Corps of Engineers (2002) HEC-RAS river analysis systems: Hydraulic reference manual, version 3.1, US Army Corps of Engineers, Hydrologic Engineering Centre, Davis, California

Chapter 9
Drop-Inlet Spillway

Abstract Drop-inlet spillway, a widely used soil conservation structure, is preferred for sites providing substantial temporary storage above the inlet, especially in gullies having more than 3 m fall or drop. The chapter focuses on the hydraulic design of two general types of drop-inlet spillways, the first having a circular or rectangular box-type flat crest and the second having a standard or funnel-shaped crest, the latter popularly known as 'morning glory' or 'glory hole' spillway. It discusses the typical head-discharge relationships of the structure, controlled by its various components, besides the composite head-discharge relationship. The pressure distribution in various components of a drop-inlet spillway, essential for determining the hydraulic loading to ensure safety against cavitation, is discussed. The chapter mainly focuses on designing the standard-crested and the flat-crested drop-inlet spillways under specific discharge and pressure conditions. The design includes computing the water surface profile in the conduit and developing the composite head-discharge relationship. The complex computations involved in the design are demonstrated through solved examples.

9.1 Background

A drop-inlet spillway is one in which flow enters through a vertical flaring funnel, with its top as the lip of the spillway, and flows to the downstream channel through a horizontal or near horizontal circular conduit (Fig. 9.1). The drop-inlet spillway is also known as the *shaft spillway* or *morning glory* (or *glory hole*) spillway.

The drop-inlet spillway may be of two general types, the first having a circular or rectangular box-type flat crest (Fig. 9.2a) and the second having a standard or funnel-shaped crest (Fig. 9.2b). The latter, i.e., the drop-inlet spillway having the standard or funnel-shaped crest, is referred to as *morning glory* or *glory hole* spillway.

The flat-crested spillway has a weir section, a free-falling section, a vertical shaft section, and a conduit as its components. The water first approaches the crest on a flat slope and then begins the free-fall through the free-falling section, which follows the path of the free-falling jet. Finally, the free-falling jet reaches the vertical shaft section, which is fully filled with water and connected to the conduit through an

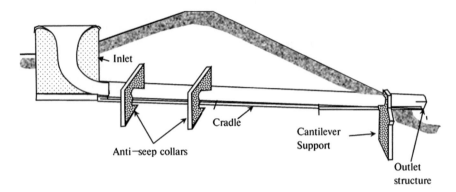

Fig. 9.1 Drop-inlet spillway

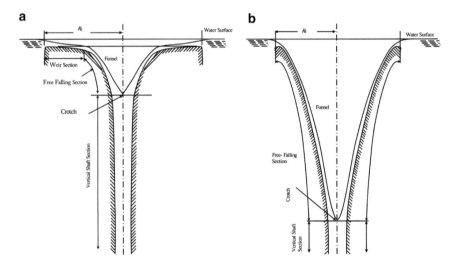

Fig. 9.2 **a** A flat-crested drop-inlet spillway showing various components (R_S represents the crest radius of the spillway), **b** a standard-crested drop-inlet spillway showing various components

elbow. The weir section, the free-falling section, and the vertical shaft section are collectively referred to as the inlet or the riser. It may be noted that in the flat-crested spillway, the water first approaches the crest on a flat slope and then begins the free-fall.

There is no weir section in the standard-crested spillway, and the free-falling section begins from the crest; i.e., the water begins its free-fall immediately upon leaving the crest. The free-falling section approaches the vertical shaft section, connected to the conduit through an elbow. Thus, the inlet or the riser in the standard-crested spillway consists of the free-falling section and the vertical shaft section.

9.1.1 Standard-Crested and Flat-Crested Drop-Inlet Spillway

The genesis of the *standard crest* used in the modern structures' design is the realisation that a partial vacuum underneath the water spilling over the structure, say a dam spillway, may result in an additional overturning force on the structure. Therefore, the crest and the downstream face of the structure should mimic the shape of the lower nappe. Consequently, filling the area below the lower nappe with masonry was found appropriate for obtaining the standard crest (Fig. 9.3). The same principle is adopted for obtaining the shape of the standard-crested drop-inlet spillway, except that the crest is circular (Fig. 9.4).

The standard-crested spillway offers the advantage of a smaller diameter crest. Therefore, it is preferred if the spillway is a tower, i.e., protruding from the upstream face of the dam. However, the flat-crested spillway has a smaller funnel; therefore, it is preferred where the spillway is excavated in the rock. In cases where the height of the free-fall section for the standard-crested spillway is greater than the available vertical distance to the elbow, a flat-crested spillway is usually adopted.

9.2 Functions

Drop-inlet spillways are generally used for removing excess water from a reservoir or similar water storage structures. The water storage structure may also be constructed in gullies by building earth dams across them. The water storage structures created in gullies, with drop-inlet spillways as the principal spillway, help control, and stabilise grades.

Drop-inlet spillways are also used as a principal spillway for farm ponds, an inlet for irrigation and drainage systems, and the downstream end of the water

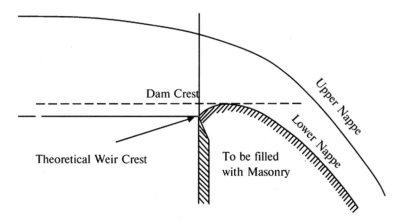

Fig. 9.3 Nappe for sharp-crested weir with a vertical upstream face

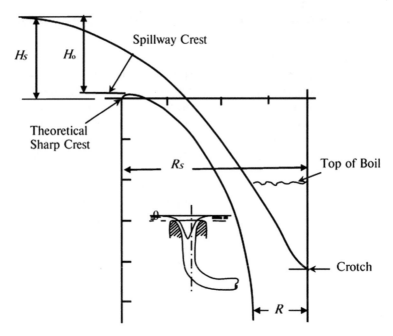

Fig. 9.4 A typical standard-crested drop-inlet spillway (H_o and H_S represent head over the spillway crest and theoretical sharp-crest, respectively, and R represents the radius of the vertical shaft)

disposal systems. These also help in flood prevention when used in conjunction with reservoirs.

9.3 Adaptability

Drop-inlet spillway is effective in controlling gully heads having more than 3 m fall or drop. It is especially suited for sites providing substantial temporary storage above the inlet. The morning glory structures are preferred in narrow gullies where the space is limited. Drop-inlet spillways may also be adapted for relatively low heads when used as a road culvert.

9.4 Advantages and Limitations

Since drop-inlet spillways are used at sites providing temporary storage, the spillway capacity and the construction cost are reduced. In addition, the reduced spillway discharge helps reduce the downstream channel capacity. The lower spillway discharge also helps in channel grade stabilisation and flood prevention.

Morning glory spillways offer the advantage of achieving the design capacity at relatively low heads. Therefore, these spillways perform ideally at sites that limit the peak spillway discharge.

The limitations of drop-inlet spillways include their adaptability to locations where earthen dams or embankments could be constructed. Another major limitation of these spillways is their susceptibility to clogging due to debris at low heads. Also, to attain the full inlet capacity of the spillway, a substantial head or stage is needed in the reservoir, which may result in unreasonably high embankments.

9.5 Materials of Construction

Plain concrete, reinforced concrete, a corrugated or smooth metal pipe may be used as the construction material for drop-inlet spillways. Usually, similar construction material is used for both the riser or inlet and the conduit.

9.6 Drop-Inlet Spillway: Components and Functions

As mentioned earlier, a drop-inlet spillway typically consists of the following components:

1. Inlet or Riser
2. Conduit
3. Outlet or Terminal Structure.

The individual components of the drop-inlet spillway are described below, along with their functions.

Inlet or Riser: The inlet or riser may consist of a weir section, a free-falling section, and a vertical shaft section.

The weir section consists of a circular converging section with a slightly rounded upstream edge or a rectilinear flat crest weir with a slightly rounded upstream edge (Fig. 9.2a). The rounded upstream edge is adopted to minimise the head loss. The weir section is followed by an open portion of a freely falling jet section. As mentioned earlier, there is no weir section in the standard-crested spillway, and the free-falling jet section begins from the crest itself. The section follows the lower nappe of the free-falling jet and continues until the crotch, i.e., the point at which the upper nappe meets the centre line of the spillway. The crotch is accompanied by a boil formed just above it due to energy conversion. At the crotch, a closed portion of the freely falling jet section begins and continues downward to the elevation at which the water acquires a final vertical velocity. A closed cylindrical vertical shaft section begins at this elevation, connected to a horizontal or nearly horizontal circular conduit using a connecting elbow.

The design of the inlet or riser includes the determination of the size and shape of the weir section or the freely falling jet section and the size of the vertical shaft to handle the design flow and transfer it to the conduit.

The inlet should have safety provisions against debris. Typically, *trash racks*, in the form of a cage atop the crest, are provided to check debris' entry. Similarly, an *antivortex device* may also be provided to control the vortex formation. Vortex formation occurs due to the non-uniform approach flow to the crest section. Consequently, the air is sucked into the inlet and replaces a portion of water, resulting in a decreased flow, accompanied by randomly and erratically varying head and discharge. Typically, a horizontal circular plate is used as the *antivortex device* at the inlet entrance.

Conduit: The horizontal or nearly horizontal circular conduit passes through or under an earthen embankment and conveys the flow from the vertical shaft to the outlet or the terminal structure. An elbow bend may form the transition from the circular vertical shaft section to the circular conduit.

The design of the conduit includes the determination of its size and slope. The slope of the conduit governs the flow through the conduit, besides influencing the energy dissipation through head loss due to friction. The inside diameter of the conduit should be adequate to handle the design flow, besides permitting access for inspection, repair, and maintenance.

The conduit is typically equipped with *antiseep collars* to grip the conduit for firm placement inside the embankment and prevent seepage through the joints. Also, a masonry or concrete *cradle* is provided to prevent uneven settlement of the conduit.

Outlet or Terminal Structure: The outlet or the terminal structure is usually provided in the drop-inlet spillway to dissipate the flow's kinetic energy. It usually consists of either a hydraulic jump stilling basin or an impact basin.

The hydraulic jump stilling basins, discussed in detail in Sect. 10.6.2.4 (Chap. 10), are ordinarily preferred as these are easy to maintain and facilitate access to the conduit for inspection and maintenance. However, the adaptation of the hydraulic jump stilling basins depends on the tailwater condition. On the contrary, the performance of the impact basin is independent of the tailwater. In the impact basin, energy dissipation occurs as the flow from the conduit strikes a vertical hanging baffle provided at the entrance of the basin.

9.7   Design of Drop-Inlet Spillway

As discussed in Sect. 7.3.1 (Chap. 7), a permanent gully control structure design involves the hydrologic design, hydraulic design, and structural design. As described earlier, the hydrologic design involves determining the peak flow rate. The hydraulic design includes determining the dimensions of various structural components for handling the design flow rate. Finally, the structural design involves analysing various forces acting on the structure and then providing the required strength and stability to the various components of the structure.

The hydrologic design concepts and approaches discussed in Sect. 8.7.1 (Chap. 8) are also applicable to the hydrologic design of the drop-inlet spillways. A return period of 25–50 years is considered. Any of the three approaches, i.e., the rational method, frequency analysis of historical rainfall or flow data, or hydrological or hydraulic modelling, may be adopted for determining the inflow to the reservoir. The design flow rate of the drop-inlet spillway is subsequently determined based on the upstream and downstream site conditions and the reservoir storage requirement.

The hydraulic design of the drop-inlet spillway is presented in the following sections.

9.7.1   Hydraulic Design of Drop-Inlet Spillway

9.7.1.1   Head-Discharge Relationship

The head-discharge relationship of a drop-inlet spillway may be controlled by various components of the structure as follows:

1. The crest section, resulting in weir control or orifice control at the crest
2. The vertical shaft, resulting in short-tube control
3. The conduit, resulting in pipe control.

9.7.1.2   Weir Control by Crest Section

For small heads, flow over the crest section is controlled by the crest discharge characteristics. As mentioned earlier, the crest section is followed by an open portion of a freely falling jet section until the crotch. The weir flow governs the head-discharge relationship of the crest until the crotch. During the weir control, a drawdown occurs in the inlet (Fig. 9.5).

Fig. 9.5 Weir control at the drop-inlet crest section

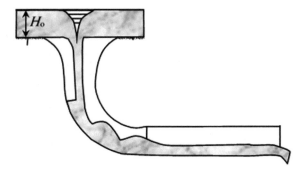

The flow over the crest section may be determined using the weir formula given in Sect. 8.7.2 (Chap. 8):

$$Q = CLH_o^{3/2} \tag{9.1}$$

where Q = flow rate, m³ s⁻¹; L = weir crest length, m; H_o = head over the crest, m; and C = weir coefficient, dimensionless.

It may be noted that the velocity head is neglected in Eq. (9.1).

If L is expressed in terms of the spillway radius, R_S, then Eq. (9.1) becomes

$$Q = C(2\pi R_s)H_o^{3/2} \tag{9.2}$$

where RS = spillway radius, m.

The process of obtaining the C for a particular case is discussed later in Sect. 9.7.1.8 .

Equation (9.2) is used for both standard-crested and flat-crested drop-inlet spillways. The changes in the value of the coefficient of discharge, C, take care of changes in the flow characteristics of the crests.

9.7.1.3 Orifice Control by Crest Section

After the crotch and boil form during the weir control, submergence begins to affect the weir flow. As the head increases, the crotch and boil disappear, depicting only a slight depression of the water surface over the crest. The head-discharge relationship is then controlled by the crest section acting as an orifice. Usually, no air passes through the spillway when the control is an orifice, except the air sucked in through vortices, and the vertical shaft is partly full (Fig. 9.6). The discharge formula for orifice flow at the crest is

$$Q = CA_o\sqrt{2gH_o} \tag{9.3}$$

Fig. 9.6 Orifice control at the drop-inlet crest section

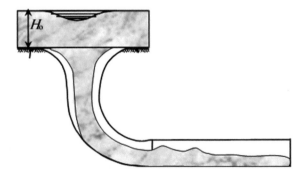

Fig. 9.7 Short-tube control
by the vertical shaft

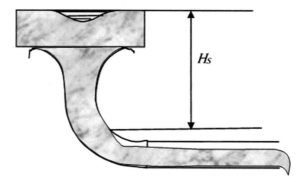

where $A_O=$ horizontal area of the vertical shaft, m^2.

However, it may be noted that even for the orifice control, the weir formula in Eq. (9.2) is used in place of Eq. (9.3). The changes in the value of C take care of changes from weir flow to the orifice flow.

9.7.1.4 Short-Tube Control by Vertical Shaft

The short-tube control occurs when the vertical shaft is full, but the conduit is only partly full (Fig. 9.7). Like in the orifice flow, there may be a slight depression of the water surface over the inlet with no air passing through the spillway.

The discharge equation for short-tube control by the vertical shaft is (Blaisdell 1952),

$$Q = CA_oH_s^{1/2} \tag{9.4}$$

where $H_S =$ distance from the headwater surface to a point in the vertical shaft where the water separates from the vertical shaft.

9.7.1.5 Pipe Control by Conduit

The head-discharge relationship is governed by pipe flow equations when the conduit is full (Fig. 9.8). As evident from the figure, head H_T causes the flow. When the conduit exit is not submerged, H_T equals the elevation difference between the head-water surface and the point at which the hydraulic grade line cuts the centre of the conduit at the plane of its exit. However, when the conduit exit is submerged, H_T equals the elevation difference between the headwater and tailwater surfaces.

The head H_T causing flow may be expressed as follows:

$$H_T = h_e + h_f + h_o \tag{9.5}$$

Fig. 9.8 Pipe control by the conduit

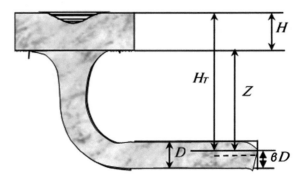

where h_e = entrance head loss; h_f = friction head loss; and h_o = head loss at the outlet.

The entrance and outlet losses are usually expressed in terms of the velocity head as,

$$h_e = K_e \frac{V_P^2}{2g} \tag{9.6a}$$

$$h_0 = K_o \frac{V_P^2}{2g} \tag{9.6b}$$

where K_e = head loss coefficient for the entrance to the crest, dimensionless; K_o = head loss coefficient for the exit of the conduit, dimensionless; and V_P = flow velocity in the conduit, m s^{-1}.

The friction head loss is typically determined by the Darcy–Weisbach equation, given as,

$$h_f = f \frac{l}{D} \frac{V_P^2}{2g} = K_c \frac{V_P^2}{2g} l \tag{9.6c}$$

where f = Darcy–Weisbach friction coefficient, dimensionless; l = length of the conduit, m; D = diameter of the conduit, m; and K_c = friction head loss coefficient, dimensionless.

Substituting the values of h_e, h_o, and h_f from Eqs. (9.6a)–(9.6c) in Eq. (9.5),

$$H_T = [K_e + K_o + K_c l] \frac{V_P^2}{2g} \tag{9.7}$$

Straub and Morris (1950) reported that the value of K_o was nearly 1.0 and recommended K_e and K_c values based on laboratory experiments. Therefore, Eq. (9.7) reduces to,

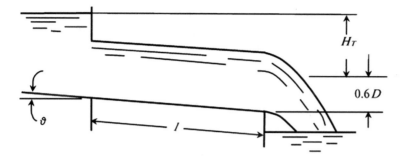

Fig. 9.9 Pipe control by the conduit

$$H_T = [1 + K_e + K_c l]\frac{V_P^2}{2g} \tag{9.8}$$

The discharge equation for pipe control by the conduit is,

$$Q = \frac{A\sqrt{2gH_T}}{\sqrt{(1 + K_e + K_c l)}} \tag{9.9}$$

where A = cross-sectional area of the conduit, m².

The value of H_T is usually approximated as the vertical distance between the headwater surface and a point located at $0.6D$ above the downstream invert (Fig. 9.9).

The relationship between the Manning roughness coefficient, n, and Darcy–Weisbach friction coefficient, f, is given as follows:

$$n = R^{1/6}\sqrt{\frac{f}{8g}} \tag{9.10a}$$

Or,

$$f = \frac{125n^2}{D^{1/3}} \tag{9.10b}$$

Straub and Morris (1950) experimentally determined the values of the Manning roughness coefficient, n, and the entrance head loss coefficient, Ke, for smooth concrete and plain corrugated metal pipes.

The pipe control may also be analysed considering the concept of the *neutral slope*, which is also referred to as friction or normal slope (NRCS 1972). The neutral slope is defined as that slope of the conduit at which the friction head loss is equal to the conduit slope, i.e., when the conduit is parallel to the hydraulic gradient.

Referring to Fig. 9.9,

$$\sin \theta = \frac{h_f}{l} \tag{9.11}$$

where θ = angle conduit makes with the horizontal.
  Substituting h_f from Eq. (9.6c) in Eq. (9.11),

$$\sin \theta = K_c \frac{V_P^2}{2g} \tag{9.12}$$

Hence, neutral slope,

$$S_n = \tan \theta = \frac{K_c \frac{V_P^2}{2g}}{\sqrt{1 - \left(K_c \frac{V_P^2}{2g}\right)^2}} \tag{9.13}$$

Example 9.1 A drop-inlet spillway is expected to operate under a maximum head of 3.0 m and handle a peak discharge of 5.0 m³ s⁻¹. The length of the concrete conduit is 15.0 m. Determine the minimum uniform diameter and the slope of the conduit flowing full to carry the discharge. Assume the head loss coefficient for the entrance to the crest, $K_e = 0.15$, and the friction head loss coefficient, $K_c = 0.04$.

Solution

Given: Discharge, $Q = 5.0$ m³ s⁻¹; Head causing flow, $H_T = 3.0$ m; Length of the conduit, $l = 15.0$ m; Head loss coefficient for the entrance to the crest, $K_e = 0.15$; Friction head loss coefficient, $K_c = 0.04$. Conduit flows full at the downstream end.
  The pipe flow occurs when the conduit slope is less than the neutral slope.
  The discharge for pipe control by the conduit is given by Eq. (9.9):

$$Q = \frac{A\sqrt{2g\,H_T}}{\sqrt{(1 + K_e + K_c l)}}$$

Table 9.1 presents the head loss coefficient, K_c, as a function of n and D for circular pipes flowing full.
  Substituting the given values of Q, H_T, K_e, K_c, and l in Eq. (9.9), and taking $g = 9.81$ m s⁻², we get,

$$A = \frac{Q\sqrt{(1 + K_e + K_c l)}}{\sqrt{2g\,H_T}} = \frac{5 \times \sqrt{(1 + 0.15 + 0.04 \times 15)}}{\sqrt{2 \times 9.81 \times 3}} = 0.86\,\text{m}^2$$

Since $A = (\pi/4)\,D^2$,

$$D = \sqrt{\frac{4A}{\pi}} = \sqrt{\frac{(4 \times 0.86)}{3.14}} = 1.05\,\text{m}$$

Table 9.1 Head loss coefficient, K_c for circular pipes flowing full (After NRCS 1972)

Pipe inside diameter, mm	Flow area, mm²	Head loss coefficient, K_c				
		Manning roughness coefficient, n				
		0.010	0.013	0.016	0.020	0.025
13	133	4.071	6.881	10.423	16.286	25.447
25	491	1.702	2.877	4.358	6.6810	10.641
51	2043	0.658	1.112	1.685	2.632	4.113
76	4536	0.387	0.653	0.990	1.546	2.416
102	8171	0.261	0.441	0.669	1.045	1.632
127	12,668	0.195	0.329	0.499	0.780	1.218
152	18,146	0.153	0.259	0.393	0.614	0.959
203	32,365	0.104	0.176	0.267	0.417	0.652
254	50,671	0.0774	0.131	0.198	0.309	0.484
305	73,062	0.0606	0.102	0.155	0.242	0.379
381	114,009	0.0451	0.0761	0.115	0.180	0.282
457	164,030	0.0354	0.0598	0.0950	0.141	0.221
533	223,123	0.0288	0.0487	0.0737	0.115	0.180
610	292,247	0.0241	0.0407	0.0616	0.0962	0.150
762	456,037	0.0179	0.0302	0.0458	0.0715	0.112
914	656,119	0.0140	0.0237	0.0359	0.0561	0.0877
1219	1,167,071	0.00956	0.0162	0.0245	0.0382	0.0597
1524	1,824,147	0.00710	0.0120	0.182	0.0284	0.0444

Hence, the conduit diameter is taken as 1.05 m.

The neutral slope, S_n, of the conduit, is given by Eq. (9.13):

$$S_n = \frac{K_c \frac{V_P^2}{2g}}{\sqrt{1 - \left(K_c \frac{V_P^2}{2g}\right)^2}}$$

The velocity of flow in the conduit,

$$V_P = \frac{Q}{A} = \frac{5}{0.86} = 5.8 \, \text{m s}^{-1}$$

Substituting K_c, V_P and g in Eq. (9.13),

$$S_n = \frac{K_c \frac{V_P^2}{2g}}{\sqrt{1 - \left(K_c \frac{V_P^2}{2g}\right)^2}} = \frac{0.04 \times \frac{5.8^2}{(2 \times 9.81)}}{\sqrt{1 - \left(0.04 \times \frac{5.8^2}{2 \times 9.81}\right)^2}} = 0.069$$

Answer: The conduit diameter is 1.05 m, and its slope should be less than the neutral slope, 0.069, to flow full.

Example 9.2 A drop-inlet spillway is designed for a gully having a catchment area of 50 ha. The rainfall intensity of duration equal to the time of concentration for a 25-year return period storm is 120 mm h^{-1}. The upstream water surface elevation is 28.6 m, and the elevation of the centre of the conduit at its outlet is 25.1 m. The conduit is 26 m long, with $K_e = 0.5$ and $K_c = 0.015$. Determine the conduit diameter to discharge 50% of peak flow. Take runoff coefficient, $c = 0.30$.

Solution

Given: Gully catchment area, $A_{catchment} = 50$ ha $= 50 \times 10^4$ m^2; Rainfall intensity of duration equal to time of concentration, $i = 120$ mm h$^{-1} = 3.33 \times 10^{-5}$ m s^{-1}; Elevation of the upstream water surface $= 28.6$ m; Elevation of the centre of the conduit at its outlet $= 25.1$ m; Length of the conduit, $l = 26$ m; $K_e = 0.5$; $K_c = 0.015$. Runoff coefficient, $c = 0.30$.

The peak rate of runoff from the gully catchment can be determined using the rational method, given by Eq. (3.19):

$$Q_{catchment} = c i A_{catchment}$$

Putting the known values of c, i and $A_{catchment}$ in Eq. (3.19),

$$Q_{catchment} = c i A_{catchment} = 0.3 \times \left(3.33 \times 10^{-5}\right) \times \left(50 \times 10^4\right) = 5 \, \text{m}^3\text{s}^{-1}$$

Hence, conduit discharge $Q = 50\%$ of $5 = 2.5$ m^3 s^{-1}

Head causing flow through the conduit, $H_T =$ Elevation difference between the upstream water surface and the conduit centre at the outlet $= 28.6 - 25.1 = 3.5$ m.

(It may be noted that while calculating H_T, the conduit centre at the outlet is taken against the recommended point located at 0.6 times the conduit diameter above the downstream invert.)

The discharge for pipe control by the conduit is given by Eq. (9.9):

$$Q = \frac{A\sqrt{2g H_T}}{\sqrt{(1 + K_e + K_c l)}}$$

Substituting the given values of Q, H_T, K_e, K_c, and l in Eq. (9.9), and taking $g = 9.81$ m s^{-2}, we get,

$$A = \frac{Q\sqrt{(1 + K_e + K_c l)}}{\sqrt{2g H_T}} = \frac{2.5 \times \sqrt{(1 + 0.5 + 0.015 \times 26)}}{\sqrt{2 \times 9.81 \times 3.5}} = 0.41 \, \text{m}^2$$

Since $A = (\pi/4)D^2$,

$$D = \sqrt{\frac{4A}{\pi}} = \sqrt{\frac{(4 \times 0.41)}{3.14}} = 0.73 \, \text{m} \approx 0.75 \, \text{m}$$

Answer: The conduit diameter may be taken as 0.75 m.

9.7.1.6 Composite Head-Discharge Relationship

Figure. 9.10a presents the head-discharge curves representing the equations for four types of flow described earlier. It may be noted that a different origin is used for each curve; i.e., the origins are shifted downward so that the vertical distances are always represented downward from the headwater surface. Thus, in the case of weir and orifice, the origin is at O_1, shifting downward to O_2 in the short-tube. The distance between O_1 and O_2 represents the effective height of the vertical shaft. Similarly, in the pipe flow curve, the origin shifts to O_3 from O_1 by Z when the conduit exit is not submerged (Fig. 9.8).

It may be noticed that the dashed portions of the head-discharge curves in Fig. 9.10a are hypothetical and do not govern the head-discharge relationship. Thus, only the solid portions of the curves represent the head-discharge relationship.

Figure 9.10b presents the composite head-discharge relationship for a drop-inlet spillway. It is evident from Fig. 9.10b that for heads between the ordinates of 'a' and 'g', the weir control by the crest section prevails. In between 'g' and 'm', orifice control by the crest would govern the flow, and the pipe control would be for heads above the ordinate of 'n'. The short-tube control would prevail in-between 'm' and 'n'.

It may be noted that the level of the headwater surface would depend on the inflow to the reservoir and outflow through the spillway. For a constant inflow rate to the reservoir, the headwater surface level will stabilise in due course if the weir or the pipe controls the flow. On the contrary, if the orifice and short-tube controls are also present, the headwater surface level and the spillway discharge may never stabilise. Therefore, the drop-inlet spillway design would be ideal when the orifice and the short-tube cannot control the head-discharge relationship. In the case of the ideal design, it would be possible to confidently predict the spillway discharge for a given headwater surface level.

Example 9.3 Determine and draw the discharge curve up to 1.5 m head over crest for a standard-crested drop-inlet spillway having a crest radius of 1.0 m. The inlet crest elevation is 30 m, and the invert elevation at the downstream end is 24.6 m. The 1.2 m diameter concrete pipe is 50 m long, with $K_e = 0.5$ and $K_c = 0.015$. Assume that pipe flows full and controls the flow. Take the coefficient of discharge for the inlet as 0.80.

Solution

Given: Crest radius, $R_S = 1.0$ m; Conduit diameter, $D = 1.2$ m; Elevation of the inlet crest $= 30$ m; Elevation of the downstream invert $= 24.6$ m; Length of the conduit, $l = 50$ m; $K_e = 0.5$; $K_e = 0.015$; Coefficient of discharge for inlet, $C = 0.80$.

The discharge over the crest section is given by Eq. (9.2):

$$Q = C(2\pi R_s)H_o^{3/2}$$

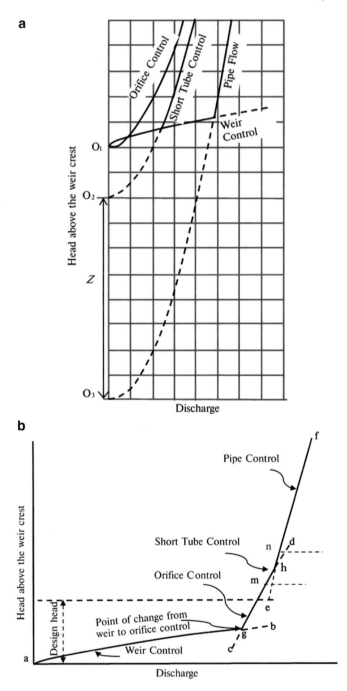

Fig. 9.10 **a** Typical head-discharge curves for different flow controls (After USBR 1987), **b** Composite head-discharge curve for a drop-inlet spillway (After USBR 1987)

Substituting the known values of C and R_S in Eq. (9.2), the Q-H relationship can be expressed as follows:

$$Q = C(2\pi R_s)H_o^{3/2} = (0.8 \times 2 \times 3.14 \times 1)H_o^{3/2}$$

Or,

$$Q = 5.024H_o^{3/2} \tag{9.14}$$

Q -H relationship for the crest section can be determined for different values of H_o (up to 1.5 m).

The discharge for pipe control by the conduit is given by Eq. (9.9):

$$Q = \frac{A\sqrt{2g H_T}}{\sqrt{(1 + K_e + K_c l)}}$$

The cross-sectional area of conduit, $A = \frac{\pi}{4}D^2 = \left(\frac{3.14}{4}\right) \times (1.2)^2 = 1.13\,\text{m}^2$

Substituting the known values of A, K_e, K_c, and l in Eq. (9.9), and taking $g = 9.81\,\text{m s}^{-2}$, we get,

$$Q = \frac{A\sqrt{2g H_T}}{\sqrt{(1 + K_e + K_c l)}} = \frac{\left(1.13 \times \sqrt{2 \times 9.81}\right)H_T^{1/2}}{\sqrt{(1 + 0.5 + 0.015 \times 50)}}$$

Or,

$$Q = 3.34H_T^{1/2} \tag{9.15}$$

Since H_T is usually approximated as the vertical distance between the headwater surface and a point located at $0.6D$ above the downstream invert, the relationship between H_T and H_o can be expressed as,

$$H_T = H_o + [\text{Crest elevation} - (\text{Downward invert elevation} + 0.6D)]$$
$$= H_o + [30 - (24.6 + 0.6 \times 1.2)] = H_o + 4.68$$

Substituting H_T in Eq. (9.15) by $(4.68 + H_o)$, the head-discharge relationship for the conduit is as follows:

$$Q = 3.34(4.68 + H_o)^{1/2} \tag{9.16}$$

Now, the discharge curve for the drop-inlet spillway can be developed using Eqs. (9.14) and (9.16) for different values of H_o, as shown in Table 9.2.

The discharge curve is plotted in Fig. 9.11.

Table 9.2 Head-discharge relationship

Head on the crest	Crest control	Pipe control
H_o (m)	Q (m³ s⁻¹)	Q (m³ s⁻¹)
0.2	0.45	7.37
0.4	1.27	7.52
0.6	2.33	7.67
0.8	3.59	7.81
1.0	5.02	7.96
1.25	7.02	8.13
1.50	9.23	8.30

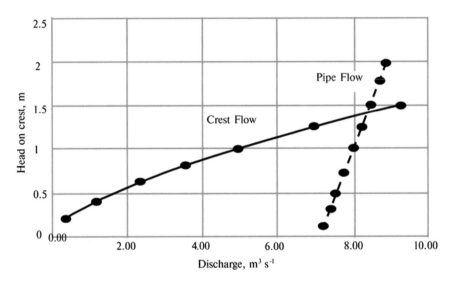

Fig. 9.11 Discharge curve for the drop-inlet spillway

9.7.1.7 Design Approaches

The design approaches adopted for standard-crested, and flat-crested drop-inlet spillways are described in the following sections.

9.7.1.8 Standard-Crested Drop-Inlet Spillway

Known: Discharge, Q; Head, H_o; Crest and outlet invert elevations; Horizontal length of the conduit, l.

To find: Minimum radius of the crest and the vertical shaft; Minimum uniform conduit diameter under specific discharge and pressure conditions.

The following two cases may arise:

Case I: The crest radius must be minimised, and subatmospheric pressures along the crest may be accepted.

Case II: The subatmospheric pressures along the crest should be minimised, and the crest radius is flexible.

The following steps may be followed to accomplish the appropriate design.

Step 1: Determine the crest radius, R_S.

As discussed earlier, we use Eq. (9.2) to determine the discharge capacity of the crest section:

$$Q = C(2\pi R_s)H_o^{3/2}$$

A trial-and-error procedure is adopted to determine R_S, i.e., assuming R_S and computing the discharge with known Q and H_o.

For finding the appropriate value of C, we need to assume P/R_S and R_S values. We may assume a large P/R_S, say, $P/R_S = 2.0$, in Case I, where we need to minimise R_S (it may be noted that for the same P, R_S will be lower when P/R_S is large). However, in Case II, where we need to minimise the subatmospheric pressures along the crest, a small P/R_S, say, $P/R_S = 0.15$, may be considered.

We then assume a reasonable value of R_S and calculate H_o/R_S.

Now for known P/R_S and H_o/R_S, we obtain C from Fig. 9.12. It may be noted that the C values obtained from the figure are valid only if the crest profile and transition shape conform to the free-falling jet and if the lower nappe surface is fully aerated.

The known values of C, R_S and H_o are substituted in Eq. (9.2), and Q is calculated. If the calculated Q matches the desired Q, the assumed value of R_S is considered final, and the design proceeds. Otherwise, we assume a different value of R_S and repeat the process.

Step 2: Determine the crest profile.

With R_S finalised in Step 1, we use Fig. 9.13 to obtain H_S/H_o for known P/R_S and H_o/R_S.

Tables 9.3, 9.4 and 9.5 present the coordinates to define the shape of the lower surface of the nappe for different H_S/R_S and P/R_S values (USBR 1987).

The X–Y coordinates of the lower nappe of the crest profile for the portions above and below the weir crest are interpreted, tabulated, and subsequently plotted. While interpreting and plotting the crest profile for the portion above the weir crest, the maximum value of Y, obtained by subtracting H_o from H_S, should be considered.

Figure 9.14 presents the comparison of lower nappe shapes for a circular weir for different values of H_S/R_S. As evident, the profiles for larger H_S/R_S fall inside the profiles for lower H_S/R_S, implying that if H_S/R_S exceeds 0.20 or 0.30, subatmospheric pressure may occur. Therefore, in design Case II, where the subatmospheric pressures need to be minimised, H_S/R_S or H_o/R_S should be as low as possible.

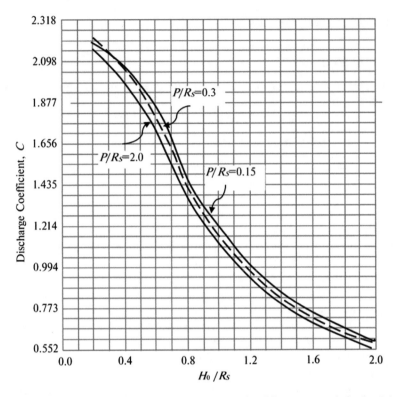

Fig. 9.12 Discharge coefficient, C, as a function of H_o/R_S for different approach depths, P (After USBR 1987)

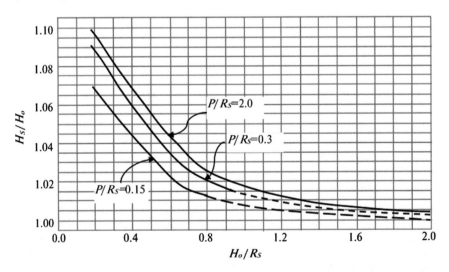

Fig. 9.13 H_s/H_o relationship as a function of H_o/R_s for different approach depths, P (After USBR 1987)

Table 9.3 Coordinates of lower nappe surface for different H_S/R_S when $P/R_S = 2.0$ (After USBR 1987)

$\frac{X}{H_S}$	$\frac{Y}{H_S}$ for portion of the profile above the weir crest								
0.000	0.0000	0.0000	0.0000	0.0000	0.0000	0.0000	0.0000	0.0000	0.0000
0.010	0.0150	0.0133	0.0128	0.0122	0.0116	0.0112	0.0104	0.0095	0.0070
0.020	0.0280	0.0250	0.0236	0.0225	0.0213	0.0202	0.0180	0.0159	0.0090
0.030	0.0395	0.0350	0.0327	0.0308	0.0289	0.0270	0.0231	0.0198	0.0085
0.040	0.0490	0.0435	0.0403	0.0377	0.0351	0.0324	0.0268	0.0220	0.0050
0.050	0.0575	0.0506	0.0471	0.0436	0.0402	0.0368	0.0292	0.0226	
0.100	0.0860	0.0762	0.0705	0.0642	0.0570	0.0482	0.0264	0.0089	
0.200	0.1105	0.0938	0.0819	0.0688	0.0521	0.0292			
0.300	0.1105	0.0850	0.0668	0.0446	0.0174				
0.400	0.0970	0.0620	0.0250	0.0060					
0.500	0.0700	0.0250							
0.600	0.0320								

$\frac{H_S}{R_S}$	0.00	0.20	0.30	0.40	0.50	0.60	0.8	1.00	2.00
$\frac{Y}{H_S}$	$\frac{X}{H_S}$ for portion of the profile below the weir crest								
0.000	0.668	0.554	0.487	0.413	0.334	0.262	0.158	0.116	0.048
−0.020	0.705	0.592	0.526	0.452	0.369	0.293	0.185	0.145	0.074
−0.040	0.742	0.627	0.563	0.487	0.400	0.320	0.212	0.165	0.088
−0.060	0.777	0.660	0.596	0.519	0.428	0.342	0.232	0.182	0.100
−0.080	0.808	0.692	0.628	0.549	0.454	0.363	0.250	0.197	0.110
−0.100	0.838	0.772	0.657	0.577	0.478	0.381	0.266	0.210	0.118
−0.200	0.978	0.860	0.790	0.698	0.575	0.459	0.326	0.260	0.144
−0.300	1.100	0.976	0.900	0.797	0.648	0.518	0.368	0.296	0.160
−0.400	1.207	1.079	1.000	0.880	0.706	0.562	0.400	0.322	0.168
−0.500	1.308	1.172	1.087	0.951	0.753	0.598	0.427	0.342	0.173
−1.000	1.713	1.564	1.440	1.189	0.899	0.710	0.508	0.402	0.188
−2.000	2.032	2.126	1.891	1.381	1.025	0.810	0.572		
−3.000	2.778	2.559	2.119	1.468	1.086	0.853			
−4.000		2.914	2.201	1.500					
−5.000		3.178	2.227						
−6.000		3.405	2.232						

However, for avoiding subatmospheric pressures along the crest while maintaining the design head requirement, the crest radius is appropriately increased. Figure 9.15 presents the plot of R_S'/R_S as a function of H_o/R_S for determining the increased radius, R_S'. Thus, in Case II, the increased radius, R_S', is used, and the crest shape is subsequently adjusted.

Step 3: Determine the vertical shaft radius, R.

The next step in the design is to find the transition shape to carry the desired Q at the given H_o. The continuity equation for the flow emerging as a circular jet from a horizontal orifice may be used to derive the appropriate equation. If the radius of the jet is R, then from continuing equation,

Table 9.4 Coordinates of lower nappe surface for different H_S/R_S when $P/R_S = 0.30$ (After USBR 1987)

$\frac{X}{H_S}$	$\frac{Y}{H_S}$ for portion of the profile above the weir crest					
0.000	0.0000	0.0000	0.0000	0.0000	0.0000	0.0000
0.010	0.0130	0.0130	0.0120	0.0115	0.0110	0.0100
0.020	0.0245	0.0240	0.0225	0.0195	0.0180	0.0170
0.030	0.0340	0.0330	0.0300	0.0270	0.0240	0.0210
0.040	0.0415	0.0390	0.0365	0.0320	0.0285	0.0240
0.050	0.0495	0.0455	0.0420	0.0370	0.0325	0.0245
0.100	0.0740	0.0660	0.0575	0.0500	0.0395	0.0190
0.200	0.0885	0.0745	0.0575	0.0435	0.0180	
0.300	0.0780	0.0580	0.0340	0.0050		
0.400	0.0495	0.0240				
0.500	0.0090					
$\frac{H_S}{R_S}$	0.20	0.30	0.40	0.50	0.60	0.80
$\frac{Y}{H_S}$	$\frac{X}{H_S}$ for portion of the profile below the weir crest					
0.000	0.519	0.455	0.384	0.310	0.238	0.144
−0.020	0.560	0.495	0.423	0.345	0.272	0.174
−0.040	0.598	0.532	0.458	0.376	0.300	0.198
−0.060	0.632	0.567	0.491	0.406	0.324	0.220
−0.080	0.664	0.600	0.522	0.432	0.348	0.238
−0.100	0.693	0.631	0.552	0.456	0.368	0.254
−0.200	0.831	0.763	0.677	0.558	0.451	0.317
−0.300	0.953	0.880	0.779	0.634	0.510	0.362
−0.400	1.060	0.981	0.867	0.692	0.556	0.396
−0.500	1.156	1.072	0.938	0.745	0.595	0.424
−1.000	1.549	1.430	1.180	0.892	0.707	0.504
−2.000	2.120	1.892	1.380	1.022	0.810	0.569
−3.000	2.557	2.133	1.464	1.081	0.852	
−4.000	2.911	2.200	1.499			
−5.000	3.173	2.223				
−6.000	3.400					

$$Q = area \times velocity = \left(\pi R^2\right)\left(\sqrt{2g H_a}\right) \qquad (9.17a)$$

where H_a = distance between the headwater surface and any horizontal plane in the vertical shaft (Fig. 9.16).

Substituting $\pi = 3.14$ and taking $g = 9.81$ m s^{-2}, we get,

$$R = 0.268 \frac{Q^{1/2}}{H_a^{\frac{1}{4}}} \qquad (9.17b)$$

Assuming a total head loss of $0.1 H_a$ to account for the friction and jet contraction losses and putting $0.9 H_a$ in place of H_a in Eq. (9.17b), we get.

Table 9.5 Coordinates of lower nappe surface for different H_S/R_S when $P/R_S= 0.15$ (After USBR 1987)

$\dfrac{X}{H_S}$	$\dfrac{Y}{H_S}$ for portion of the profile above the weir crest					
0.000	0.0000	0.0000	0.0000	0.0000	0.0000	0.0000
0.010	0.0120	0.0115	0.0110	0.0105	0.0100	0.0090
0.020	0.0210	0.0195	0.0185	0.0170	0.0160	0.0140
0.030	0.0285	0.0265	0.0250	0.0225	0.0200	0.0165
0.040	0.0345	0.0325	0.0300	0.0265	0.0230	0.0170
0.050	0.0450	0.0375	0.0345	0.0300	0.0250	0.0170
0.100	0.0590	0.0535	0.0465	0.0375	0.0255	0.0065
0.200	0.0670	0.0560	0.0295	0.0200		
0.300	0.0520	0.0330	0.0100			
0.400	0.0210					
$\dfrac{H_S}{R_S}$	0.20	0.30	0.40	0.50	0.60	0.80
$\dfrac{Y}{H_S}$	$\dfrac{X}{H_S}$ for portion of the profile below the weir crest					
0.000	0.454	0.392	0.325	0.253	0.189	0.116
−0.020	0.499	0.437	0.369	0.292	0.189	0.149
−0.040	0.540	0.478	0.407	0.328	0.259	0.174
−0.060	0.579	0.516	0.443	0.358	0.286	0.195
−0.080	0.615	0.550	0.476	0.386	0.310	0.213
−0.100	0.650	0.584	0.506	0.412	0.331	0.228
−0.200	0.795	0.729	0.639	0.516	0.413	0.293
−0.300	0.922	0.843	0.741	0.594	0.474	0.342
−0.400	1.029	0.947	0.828	0.656	0.523	0.381
−0.500	1.128	1.040	0.902	0.710	0.567	0.413
−1.000	1.525	1.420	1.164	0.878	0.696	0.498
−2.000	2.104	1.879	1.372	1.013	0.810	0.560
−3.000	2.550	2.105	1.457	1.077	0.840	
−4.000	2.904	2.180	1.487			
−5.000	3.169	2.207				
−6.000	3.396					

$$R = 0.275 \frac{Q^{1/2}}{H_a^{\frac{1}{4}}} \qquad (9.18)$$

Equation (9.18) represents the shape of the jet and is used for the transition design, i.e., for determining the minimum size of the vertical shaft. The equation may be used to compute the vertical shaft profile. The vertical shaft profile values are then plotted alongside the crest profile, and a smooth curve is drawn through the crest and transition profiles to obtain the final shape of the crest and transition.

Figure 9.16 presents the distinctive crest and vertical shaft profiles depicted by lines 'abc' and 'mnp', respectively. As evident, points 'b' and 'n' overlap and lines 'abp' represent the final shape of crest and transition. Section A-A represents the control section, below which free flow must prevail. As mentioned earlier, profile 'abc' represents the minimum shaft size required to handle the desired flow. If a

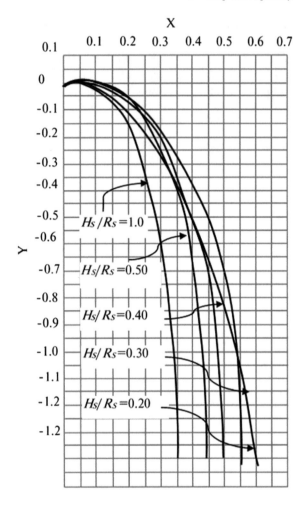

Fig. 9.14 Comparison of lower nappe profile for various H_S/R_S (After USBR 1987)

larger transition is chosen, provisions should be made for adequate aeration at the control; otherwise, the structure may face cavitation and severe vibration.

Step 4: Determine the conduit diameter, D.

As discussed earlier, the conduit controls flow capacity when the conduit is full, and the conduit slope is less than the neutral slope or the hydraulic gradient. The point 'h' in Fig. 9.10b represents the head at which the conduit flows full. If the head is more than that at point 'n', the conduit will flow full under pressure.

However, for heads less than 'n', the conduit will flow partly full, and the transition will control the flow.

Typically, a uniform diameter conduit is selected. Further, to prevent subatmospheric pressure along the length, the conduit diameter is chosen such that it would never flow full beyond the inlet transition. On the contrary, the conduit diameter is

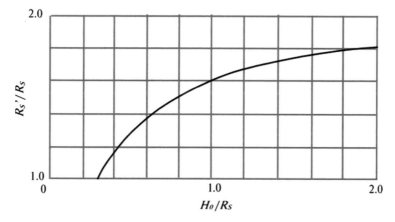

Fig. 9.15 Increased crest radius for minimising subatmospheric pressure along the crest (After USBR 1987)

selected to not flow more than 75 per cent full (in terms of area) at the downstream end.

The following procedure is followed for the conduit design:

(1) A trial conduit diameter D is selected, and Eq. (9.18) is used to determine H_a, or identify the throat location.
(2) The conduit length between the throat and the outlet end is determined.
(3) The friction head loss is determined using the Darcy–Weisbach equation (Eq. 9.6c) or Manning equation.

    While determining the friction head loss, we need to know the flow condition in the conduit, i.e., whether it flows full or partially full over its length.

    Table 9.6 can be used to obtain the area and hydraulic radius for partially full conduits along with the Manning roughness coefficients. For example, if the conduit flows 75% full, then $A/D^2 = 0.5888$. The values of d/D and r/D, corresponding to A/D^2, can be obtained from Table 9.6. Here, d and r represent the depth of flow and the hydraulic radius. For the assumed D, hydraulic radius r can be determined, and subsequently, the Manning equation can be used to determine the friction slope and the head loss.
(4) The downstream invert elevation of the conduit is determined using the following relationship:

Invert elevation $=$ Throat elevation $+$ pressure head at throat

$\qquad\qquad\qquad\quad$ $-$velocity head in the conduit$-$friction losses in the conduit

$\qquad\qquad\qquad\quad$ $-$depth at the outlet $\qquad\qquad\qquad\qquad\qquad\qquad$ (9.19)

If the calculated invert elevation matches the given invert elevation, the assumed conduit diameter is accepted. Otherwise, the process is repeated by assuming a different conduit diameter.

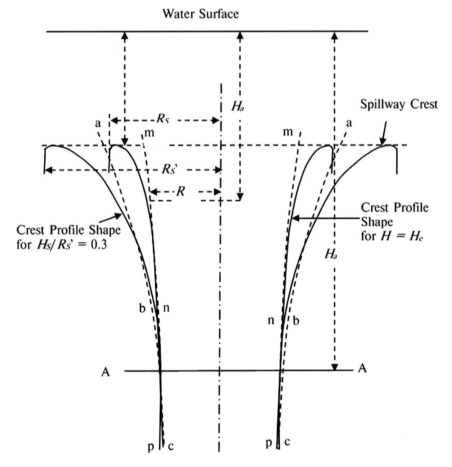

Fig. 9.16 Comparison of transition profiles for different flow conditions (After USBR 1987)

Water Surface Profile Computation

Once the conduit diameter is finalised, the water surface profile through the conduit is computed based on Bernoulli's theorem to verify the approximate solution obtained here. The procedure adopted to compute the water surface profile is presented in detail in Example 9.4.

Head-Discharge Computation

Figure 9.17 may be used to determine the discharge coefficients for partial heads of H_e on the crest for developing the discharge-head relationship. It may be noted that for adopting coefficients in Fig. 9.17, H_e/R_S must be limited to 0.4 to avoid subatmospheric pressures affecting the discharge computation.

For the lower range of heads (i.e., $H_e/R_S \leq 0.4$), C is first assumed as 2.07 for $H_e/R_S = 0.3$ (from Fig. 9.12 for $H_o/R_S = 0.3$) and then multiplied with the correction

Table 9.6 Uniform flow in circular conduits flowing partially full (After USBR 1987)

$\dfrac{d}{D}$	$\dfrac{A}{D^2}$	$\dfrac{r}{D}$	$\dfrac{Qn}{D^{8/3}s^{1/2}}$	$\dfrac{Qn}{d^{8/3}s^{1/2}}$	$\dfrac{d}{D}$	$\dfrac{A}{D^2}$	$\dfrac{r}{D}$	$\dfrac{Qn}{D^{8/3}s^{1/2}}$	$\dfrac{Qn}{d^{8/3}s^{1/2}}$
1	2	3	4	5	1	2	3	4	5
0.05	0.0147	0.0325	0.00222	6.55	0.51	0.4027	0.2531	0.239	1.442
0.06	0.0192	0.0389	0.00328	5.95	0.52	0.4127	0.2562	0.247	1.415
0.07	0.0242	0.0451	0.00455	5.47	0.53	0.4227	0.2592	0.255	1.388
0.08	0.0294	0.0513	0.00604	5.09	0.55	0.4426	0.2649	0.271	1.336
0.09	0.0350	0.0575	0.00775	4.76	0.56	0.4526	0.2676	0.279	1.311
0.10	0.0409	0.0635	0.00967	4.49	0.57	0.4625	0.2703	0.287	1.286
0.11	0.0470	0.0695	0.01181	4.25	0.58	0.4724	0.2728	0.295	1.262
0.12	0.0534	0.0755	0.01417	4.04	0.59	0.4822	0.2753	0.303	1.238
0.13	0.0600	0.0813	0.01674	3.86	0.63	0.5212	0.2842	0.335	1.148
0.14	0.0668	0.0871	0.01952	3.69	0.64	0.5308	0.2862	0.343	1.126
0.15	0.0739	0.0929	0.02250	3.54	0.65	0.5404	0.2882	0.350	1.105
0.16	0.0811	0.0985	0.0257	3.41	0.66	0.5799	0.2900	0.358	1.084
0.17	0.0885	0.1042	0.0291	3.28	0.67	0.5594	0.2917	0.366	1.064
0.18	0.0961	0.1097	0.0327	3.17	0.68	0.5687	0.2933	0.373	1.044
0.19	0.1039	0.1152	0.0365	3.06	0.69	0.5780	0.2948	0.380	1.024
0.20	0.1118	0.1206	0.0406	2.96	0.70	0.5872	0.2962	0.388	1.004
0.21	0.1199	0.1259	0.0448	2.87	0.71	0.5964	0.2975	0.395	0.985
0.22	0.1281	0.1312	0.0492	2.79	0.72	0.6054	0.2987	0.402	0.965
0.23	0.1365	0.1364	0.0537	2.71	0.73	0.6143	0.2998	0.409	0.947
0.24	0.1449	0.1416	0.0585	2.63	0.74	0.6231	0.3008	0.416	0.928
0.25	0.1535	0.1466	0.0634	2.56	0.75	0.6319	0.3017	0.422	0.910
0.26	0.1623	0.1516	0.0686	2.49	0.76	0.6406	0.3024	0.429	0.891
0.27	0.1711	0.1566	0.0739	2.42	0.77	0.6489	0.3031	0.435	0.873
0.28	0.1800	0.1614	0.0793	2.36	0.78	0.6573	0.3036	0.441	0.856
0.29	0.1890	0.1662	0.0849	2.30	0.79	0.6655	0.3039	0.447	0.838
0.30	0.1982	0.1709	0.0907	2.25	0.80	0.6736	0.3042	0.453	0.821
0.31	0.2074	0.1756	0.0966	2.20	0.81	0.6815	0.3043	0.458	0.804
0.32	0.2167	0.1802	0.1027	2.14	0.82	0.6893	0.3043	0.463	0.787
0.33	0.2260	0.1847	0.1089	2.09	0.83	0.6969	0.3041	0.468	0.770
0.34	0.2355	0.1891	0.1153	2.05	0.84	0.7043	0.3038	0.473	0.753
0.35	0.2450	0.1935	0.1218	2.00	0.85	0.7115	0.3033	0.477	0.736
0.36	0.2546	0.1978	0.1284	1.958	0.86	0.7186	0.3026	0.481	0.720
0.37	0.2642	0.2020	0.1351	1.915	0.87	0.7254	0.3018	0.485	0.703
0.38	0.2739	0.2062	0.1420	1.875	0.88	0.7320	0.3007	0.488	0.687
0.39	0.2836	0.2102	0.1490	1.835	0.89	0.7384	0.2995	0.491	0.670
0.40	0.2934	0.2142	0.1561	1.797	0.90	0.7445	0.2980	0.494	0.654
0.43	0.3229	0.2258	0.1779	1.689	0.93	0.7612	0.2921	0.498	0.604
0.44	0.3328	0.2295	0.1854	1.655	0.94	0.7662	0.2895	0.498	0.588
0.45	0.3428	0.2331	0.1929	1.622	0.95	0.7707	0.2865	0.498	0.571
0.47	0.3627	0.2401	0.208	1.559	0.97	0.7785	0.2787	0.494	0.535
0.48	0.3727	0.2435	0.216	1.530	0.98	0.7817	0.2735	0.489	0.517
0.49	0.3827	0.2468	0.224	1.500	0.99	0.7841	0.2666	0.483	0.496
0.50	0.3927	0.2500	0.232	1.471	1.00	0.7854	0.2500	0.463	0.463

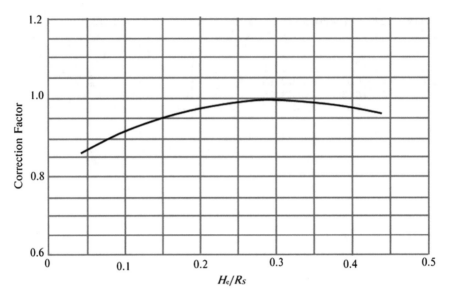

Fig. 9.17 Discharge coefficients for other than design head (After USBR 1987)

factor from Fig. 9.15. Finally, the discharge is determined using Eq. (9.2). For the higher ranges of the head, the discharge is determined using Eq. (9.18). For the highest value of H_e, the highest H_a value is determined first as described in Step 4 for the chosen conduit diameter, and then the rest of the values are back-calculated.

The head-discharge curves are plotted independently for both lower and higher ranges of heads. An approximate transition curve is used at the intersection of the curves to represent the actual conditions.

The head-discharge curves are plotted independently for both lower and higher ranges of heads. An approximate transition curve is used at the intersection of the curves to represent the actual conditions.

9.7.1.9   Flat-Crested Drop-Inlet Spillway

Known: Discharge, Q; Head, H; Crest and outlet invert elevations; Horizontal length of the conduit, l.

To find: Minimum radius of the crest and the vertical shaft; Minimum uniform conduit diameter under specific discharge and pressure conditions.

Figure 9.18 presents the definition sketch of a flat-crested drop-inlet spillway. The detailed design of the flat-crested drop-inlet spillway has been given in Kurtz (1925). The design steps are summarised here.

Step 1: Determine the crest radius, R_S.

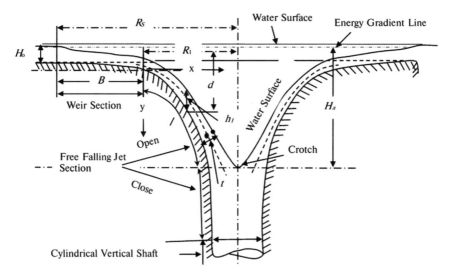

Fig. 9.18 Definition sketch of a flat-crested drop-inlet spillway (After Kurtz 1925; Redrawn with permission from ASCE)

As discussed earlier, we may use Eq. (9.2) in the following form to determine the discharge capacity of the crest section:

$$Q = C(2\pi R_S)H^{3/2}$$

For a flat-crested weir, the theoretical value of the discharge coefficient, C, is 1.7. The practical value of C may be referred to from Fig. 9.19 for a known value of H/B, B being the width of the flat crest.

With known Q and H, and an assumed value of C, the crest radius, R_S, can be determined from Eq. (9.2). The head loss, h_f, at the upstream edge of the weir can be estimated as,

$$Q = C_{assumed}H^{3/2} = C_{theoretical}(H - h_f)^{3/2}$$

With the known water surface elevation, the elevation of the energy gradient can be obtained by subtracting h_f from the water surface elevation and energy gradient line plotted.

Step 2: Determine the limiting radius to locate the downstream end of the weir crest.

The distance d to the weir floor, measured down from the energy gradient elevation, can be determined for any radius R_1 using the following equation:

$$d = (H - h_f)\left(\frac{R_s}{R_1}\right)^{2/3} \tag{9.20}$$

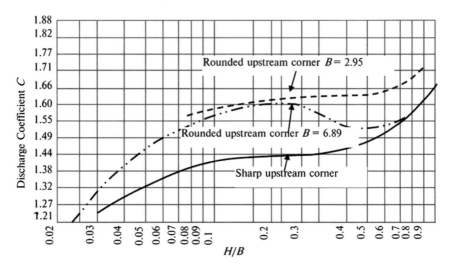

Fig. 9.19 Coefficient of discharge, C, as a function of H/B (After Kurtz 1925; Redrawn with permission from ASCE)

Also, the depth of water corresponding to any d can be determined using the following equation:

$$h_1 = 2d/3 \tag{9.21}$$

Equations (9.20) and (9.21) can be used to determine d and h_1 for various values of R_1, and the floor of the weir and the water surface plotted.

The spillway weir section should end at a particular value of R_1, and the water allowed to continue as a freely falling jet in the vertical shaft. The value of the limiting value of R_1 for the downstream end of the weir should be as low as possible while keeping the angle of inclination of the upper nappe with the vertical at the crotch to a minimum. The limiting value is obtained using a trial-and-error procedure based on the principle that at the end of the weir section, the vertical component of the velocity and the velocity head would be negligible.

For the assumed limiting value of R_1 for the downstream end of the weir, the tangent S of the water surface will be determined. The vertical velocity head will then be $[S^2(d - h_1)]$. The assumed limiting value of R_1 will be accepted if the vertical velocity head is negligible.

The width of the weir crest will be $(R - R_1)$.

Step 3: Determine the centre line and thickness of the free-falling jet.

The centre line of the freely falling jet can be determined by the following equation (Kurtz 1925):

$$y + 0.36h_1 = \frac{(x + 0.36h_1)^2}{4.56h_1} \tag{9.22}$$

where h_1 is the depth of water at the downstream end of the weir section, and x and y are the coordinates of the centre line (Fig. 9.18).

However, Eq. (9.22) may be considered only a close approximation as the centre line of the jet may be slightly steeper than the theoretical value (duPont 1937; Camp and Howe 1939).

The thickness of the jet at any point on the centre line can be determined by the following equation (Kurtz 1925):

$$t = \frac{Q}{2\pi R_1 \sqrt{2g(y + 1.5h_1)}} \tag{9.23}$$

This calculated thickness is placed normal to the centre line of the jet, with one-half on either side of the centre line until the crotch is reached.

Step 4: Determine the radius of the vertical shaft.

As defined earlier, the vertical velocity will be zero at the crotch or the vortex of the parabolic centre line of the free-falling jet. From Eq. (9.22), the vortex could be located at $y = -0.36h_1$ (Kurtz 1925).

Therefore, the total vertical velocity head is $(y + 0.36h_1)$.

The net vertical velocity head is given as,

$$h_v = y + 0.36h_1 - f_1 - f_2 \tag{9.24}$$

where f_1 and f_2 are the friction losses between the crest and the crotch and between the crotch and the point in the vertical shaft.

Therefore, the radius R_1 of the vertical shaft below the crotch can be determined using the following relationship:

$$R_1 = \sqrt{\frac{Q}{\pi \sqrt{2g(y + 0.36h_1 - f_1 - f_2)}}} \tag{9.25}$$

Using Eq. (9.25), the radius of the vertical shaft can be determined for different y values. As evident, R_1 decreases with increasing y. The vertical shaft may be determined in this fashion until a suitable conduit size is reached.

Step 5: Determine the conduit diameter, D.

The procedure described in Step 4 of the standard-crested drop-inlet spillway for determining the conduit diameter is also used for the flat-crested drop-inlet spillway.

Example 9.4 Design a standard-crested drop-inlet spillway that will carry a discharge of 56 m^3 s^{-1} under a maximum head of 3 m. The horizontal length of the conduit is 76 m. The crest elevation is 30 m, and outlet invert elevation is 16 m.

Determine the minimum radius of the overflow crest and the transition shaft radius. Also, find the minimum uniform conduit diameter, with the condition that the conduit must not flow more than 75 per cent full at the downstream end. Develop the discharge-water surface elevation relationships for minimising the crest radius. Subatmospheric pressures along the crest may be accepted.

Solution:

Condition: The crest radius must be minimised, and subatmospheric pressures along the crest may be accepted.

Given: Discharge, $Q = 56 \, \text{m}^3 \, \text{s}^{-1}$; Head over the crest, $H_o = 3.0 \, \text{m}$; Crest elevation $= 30.0 \, \text{m}$; Outlet invert elevation $= 16.0 \, \text{m}$; Length of the conduit, $l = 76.0 \, \text{m}$; Conduit must not flow more than 75 per cent full at the downstream end.

The design is done following the steps described for Case I in Sect. 9.7.1.8.

Step 1: Determine the crest radius, R_S.

As described earlier, we need first to find the appropriate value of the discharge coefficient, C, to use in Eq. (9.2).

Let us assume $P/R_S = 2.0$ (as we need to minimise R_S) and $R_S = 2.0 \, \text{m}$.

Thus,

$$\frac{H_o}{R_S} = \frac{3}{2} = 1.5$$

From Fig. 9.12, for $P/R_S = 2.0$ and $H_o/R_S = 1.5$, $C = 0.745$.

Now, putting known values of C, R_S, and H_o in Eq. (9.2), the discharge capacity of the crest section:

$$Q = C(2\pi R_s)H_o^{3/2} = 0.745 \times (2 \times 3.14 \times 2) \times (3)^{3/2} = 48.6 \, \text{m}^3 \text{s}^{-1}$$

The calculated value of Q is less than the desired Q of $56 \, \text{m}^3 \, \text{s}^{-1}$. Hence, we need to assume a larger value of R_S.

Let us assume $R_S = 2.15 \, \text{m}$.

Thus,

$$\frac{H_o}{R_s} = \frac{3}{2.15} = 1.4$$

From Fig. 9.12, for $P/R_S = 2.0$ and $H_o/R_S = 1.4$, $C = 0.801$.

Now, once again putting known values of C, R_S, and H_o in Eq. (9.2), the discharge capacity of the crest section:

$$Q = C(2\pi R_s)H_o^{3/2} = 0.801 \times (2 \times 3.14 \times 2.15) \times (3)^{3/2} = 56.2 \, \text{m}^3 \text{s}^{-1}$$

The calculated value of Q matches the desired Q,

Hence, $R_S = 2.15 \, \text{m}$.

Table 9.7 Crest profile for the portion above the weir crest

X (m)		0.00	0.06	0.09
Y (m)		0.00	0.05	0.05

Table 9.8 Crest profile for the portion below the weir crest

X (m)	0.27	0.35	0.41	0.45	0.49	0.52
Y (m)	0.00	−0.06	−0.12	−0.18	−0.24	−0.30
X (m)	0.65	0.73	0.79	0.83	0.96	
Y (m)	−0.61	−0.91	−1.22	−1.52	−3.04	

Step 2: Determine the crest profile.

With $R_S = 2.15$ m (finalised in Step 1), we use Fig. 9.13 to obtain H_S/H_o for known P/R_S and H_o/R_S.

From Fig. 9.13, for $P/R_S = 2.0$ and $H_o/R_S = 1.4$, $H_S/H_o = 1.013$.

Hence, $H_S = 1.013 \times 3 = 3.04$ m; $H_S/R_S = 3.04/2.15 = 1.41$.

From Fig. 9.4, the crest profile portion above the crest,

$$y = H_S - H_o = 3.04 - 3 = 0.04 \text{ m}$$

Using Table 9.3 (for $P/R_S = 2.0$), the X–Y coordinate of the lower nappe of the crest profile for the portion above the weir crest is computed (interpolated between $H_S/R_S = 1.0$ and 2.0) and presented in Table 9.7.

Similarly, the profile of the crest shape, which conforms to the lower nappe surface below the crest, is also interpreted and tabulated (Table 9.8).

Subsequently, the lower nappe is plotted (Fig. 9.20). Please note that while plotting, the maximum value of Y for the portion above the weir crest should be considered (0.04 m calculated earlier).

Step 3: Determine the vertical shaft radius, R.

The next step in the design is to find the transition shape to carry 56 m^3 s^{-1} at the given $H_o = 3.0$ m. Equation (9.22) is used for this purpose. Thus, putting $Q = 56$ m^3 s^{-1} in Eq. (9.22), we get,

$$R = 0.275 \frac{Q^{1/2}}{H_a^{\frac{1}{4}}} = 0.275 \times \frac{(56)^{1/2}}{H_a^{\frac{1}{4}}} = \frac{2.06}{H_a^{\frac{1}{4}}}$$

Points on the transition are computed and tabulated (Table 9.9).

The points on the transition are plotted in Fig. 9.20, on which the crest profile has been plotted. A smooth curve is drawn through the crest and transition profiles to obtain the final shape of the crest and transition.

Step 4: Determine the conduit diameter, D.

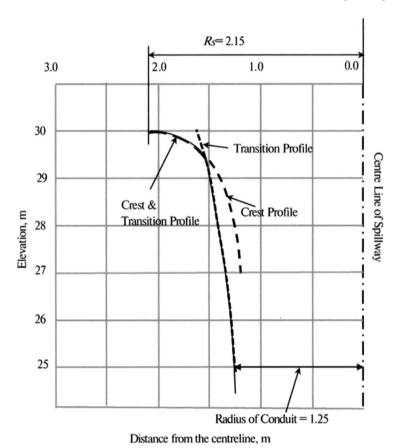

Fig. 9.20 Crest and transition shapes for the drop-inlet spillway

Table 9.9 Points on the transition

Elevation of section (m)	H_a (m)	$R = 2.06/H_a^{1/4}$ (m)
30	3	1.57
29	4	1.46
28	5	1.38
27	6	1.32
26	7	1.27
25	8	1.22

The final step in the drop-inlet spillway design is to determine the minimum uniform conduit diameter while ensuring that the conduit does not flow more than 75 per cent full.

Let us assume a conduit diameter of 2.5 m (radius of 1.25 m). The throat elevation can be determined using the final form of Eq. (9.22) obtained by inputting the value of Q in Step 3:

$$R = \frac{2.06}{H_a^{1/4}}$$

Or,

$$H_a = \left(\frac{2.06}{R}\right)^4 = \left(\frac{2.06}{1.25}\right)^4 = 7.38 \text{ m}$$

The throat location is 33 - 7.38 = 25.62 m.

With the known conduit length of 76 m from the throat transition to the downstream end of the conduit, friction losses are approximated, assuming the conduit flows 75% full for its entire length.

The velocity of flow in the conduit, $V_P = \frac{Q}{A} = \frac{56}{[0.75 \times \frac{3.14}{4} \times (2.5)^2]} = 15.21 \text{ m s}^{-1}$.

Velocity head in the conduit, $h_v = \frac{V_P^2}{2g} = \frac{(15.21)^2}{2 \times 9.81} = 11.8 \text{ m}$.

From Table 9.6, for 75% full flow, i.e., for $(A/D^2) = 0.5888$, interpolating column (2), we get $d/D = 0.702$ (d being the depth of flow), and $r/D = 0.2964$ (r being the hydraulic radius).

Therefore, depth of flow, $d = 0.702 \times D = 0.702 \times 2.5 = 1.75 \text{ m}$

hydraulic radius, $r = 0.2964 \times D = 0.2964 \times 2.5 = 0.741 \text{ m}$

Assuming Manning roughness coefficient, $n = 0.014$, the energy slope gradient by Manning equation,

$$s = \left(\frac{V_P n}{r^{2/3}}\right)^2 = \left(\frac{15.21 \times 0.014}{(0.741)^{2/3}}\right)^2 = 0.067$$

Hence, head loss due to friction in the conduit,

$$h_f = s \times l = 0.067 \times 76 = 5.09 \text{ m}$$

The downstream invert elevation of the conduit is determined using Eq. (9.19):

Invert elevation = Throat elevation + pressure head at throat
$\qquad$ − velocity head in the conduit
$\qquad$ − friction losses in the conduit − depth at the outlet
$\qquad$ = 25.62 + 0.9 × (33.0−25.62)−11.8−5.09−1.75 = 13.62 m

It may be noted that the pressure head is assumed as 90% of the elevation difference between the water surface and the throat elevation,

However, the calculated invert elevation of 13.62 m is lower than the given elevation of 16.0 m. Hence, the conduit diameter has to be increased.

Assuming a conduit diameter of 2.6 m (radius of 1.30 m) and repeating the steps, we get,

$$R = \frac{2.06}{H_a^{1/4}}$$

Or,

$$H_a = \left(\frac{2.06}{R}\right)^4 = \left(\frac{2.06}{1.30}\right)^4 = 6.31\,\text{m}$$

The throat location is $33 - 6.31 = 26.69$ m.

The velocity of flow in the conduit, $V_P = \frac{Q}{A} = \frac{56}{[0.75 \times \frac{3.14}{4} \times (2.6)^2]} = 14.07\,\text{m s}^{-1}$.

The velocity head in the conduit, $h_v = \frac{V_P^2}{2g} = \frac{(14.07)^2}{2 \times 9.81} = 10.09\,\text{m}$.

From Table 9.6, for an $(A/D^2) = 0.5888$, interpolating column (2), we get $d/D = 0.702$ (d being the depth of flow), and $r/D = 0.2964$ (r being the hydraulic radius).

Therefore,
$$d = 0.702 \times D = 0.702 \times 2.6 = 1.82\,\text{m}$$
$$r = 0.2964 \times D = 0.2964 \times 2.6 = 0.77\,\text{m}$$

Assuming Manning roughness coefficient, $n = 0.014$, the energy slope gradient by Manning equation,

$$s = \left(\frac{V_P n}{r^{2/3}}\right)^2 = \left(\frac{14.07 \times 0.014}{(0.77)^{2/3}}\right)^2 = 0.055$$

Hence, head loss due to friction in the conduit, $h_f = s \times l = 0.055 \times 76 = 4.17\,\text{m}$. The downstream invert elevation of the conduit is determined using Eq. (9.19):

Invert elevation = Throat elevation + pressure head at throat

− velocity head in the conduit − friction losses in the conduit

− depth at the outlet

$= 26.69 + 0.9 \times (33.0 - 26.69) - 10.09 - 4.17 - 1.82 = 16.29\,\text{m}$

Now, the calculated invert elevation is close to the given invert elevation of 16 m. Hence, the conduit diameter is acceptable.

Water Surface Profile Computation

The water surface profile through the 2.6 m diameter conduit is computed to verify the obtained solution.

Table 9.10 presents the computed water surface profile. As evident from the table, the final value of d/D for the outlet end is 0.72 with an A/D^2 value of 0.6054. Thus, the conduit will be 77% full, which is higher than the given upper limit of 75%. However, we have earlier accepted the invert elevation of 16.29 m against the given value of 16.0 m. Hence, the design is accepted. However, one may revise the entire calculation first to obtain the invert elevation of 16.0 m and then check the design.

Head-Discharge Computations

Table 9.9 presents the head-discharge computation. The discharge curve is plotted in Fig. 9.21.

Example 9.5 Design a flat-crested drop-inlet spillway that will carry a discharge of 750 m³ s⁻¹ under a maximum head of 2 m. The horizontal length of the conduit is 125 m. The crest elevation is 30.0 m, and outlet invert elevation is 2.0 m.

Determine the minimum radius of the overflow crest, width of the crest, and the transition shaft radius. Also, find the minimum uniform diameter of the conduit flowing full to carry the discharge.

Solution:

Given: Discharge, $Q = 750$ m³ s⁻¹; Head over the crest, $H = 2.0$ m; Crest elevation $= 30.0$ m; Outlet invert elevation $= 2.0$ m; Length of the conduit, $l = 125.0$ m; Conduit flows full at the downstream end.

The design may be accomplished following the steps described in Sect. 9.7.1.9.
Step 1: Determine the crest radius, R_S.
The discharge capacity of the crest section is determined using Eq. (9.2),

$$Q = C(2\pi R_s)H^{3/2}$$

For a flat-crested weir, the theoretical discharge coefficient, C, is 1.7.
However, the practical value of C depends on the ratio H/B. Let us arbitrarily assume $C = 1.6$.
Putting Q, H, and C in Eq. (9.2),

$$R_S = \frac{Q}{C(2\pi)H^{3/2}} = \frac{750}{(1.6 \times 2 \times 3.14)(2)^{3/2}} = 26.4 \text{ m}$$

The head loss, h_f, at the upstream edge of the weir can be estimated as,

$$Q = CH^{3/2} = C(H - h_f)^{3/2}$$

Or,

$$1.6 \times (2)^{\frac{3}{2}} = 1.7 \times (2 - h_f)^{\frac{3}{2}}$$

Table 9.10 Water surface profile computation in the conduit Known: $Q = 56$ m³ s⁻¹; $D = 2.6$ m; $n = 0.014$

Stn	Delta l (m)	Trial d/D	d (m)	A/D^2	A (m²)	V_P (ms⁻¹)	h_v (m)	r/D	r (m)	s	Ave s	Delta h_l (m)	Sum delta h_l (m)	$d + h_v +$ sum delta h_l (m)	Invert Ele (m)	Datum Ele (m)	Remarks
1	2	3	4	5	6	7	8	9	10	11	12	13	14	15	16	17	18
1	0	1	2.6	0.7854	5.31	10.55	5.67	0.2500	0.65	0.0387					26.72	32.36	
2	5.5	0.60	1.560	0.492	3.33	16.84	14.45	0.2776	0.722	0.0858	0.0623	0.343	0.343	16.35	16.4	32.75	High
		0.61	1.586	0.5016	3.39	16.52	13.90	0.2799	0.728	0.0817	0.0602	0.331	0.331	15.82		32.22	OK
3	36.5	0.65	1.690	0.5404	3.65	15.33	11.98	0.2882	0.749	0.0677	0.0747	2.725	3.057	16.72	16.2	32.92	High
		0.652	1.695	0.5483	3.71	15.11	11.63	0.2886	0.754	0.0656	0.0667	2.433	2.764	16.09		32.29	OK
4	34	0.75	1.950	0.6319	4.27	13.11	8.76	0.3017	0.784	0.0466	0.0557	1.893	4.898	15.61	16	31.61	Low
		0.72	1.872	0.6054	4.09	13.68	9.54	0.2987	0.777	0.0514	0.0581	1.976	4.980	16.40		32.40	OK

Col (2): Segments over the conduit length; Col (3): Assumed d/D; Col (4): Value from Table 9.6 corresponding to Col (3); Col (5): Value from Table 9.6 corresponding to Col (3); Col (6): Col (4) × D^2; Col (7): Q/Col (6); Col (8): Col (7)²/2 g; Col (9): Value from Table 9.6 corresponding to Col (3); Col (10): Col (9) × D; Col (11): Friction slope from Manning equation; Col 12: Average of Col (11); Col (13): Col (12) × Col (2); Col (14): Accumulated Col (13); Col (15): Col (4) + Col (8) + Col (14); Col (16): First value assuming a maximum elevation difference of 0.4 m over the conduit length (almost equally divided) with downstream end at the given invert elevation (here, 16 for downstream end and 16.2 and 16.4 at 34 m and 70.5 m from the downstream end); Col (17): First value is Col (16) + Col (8); this has to be maintained throughout; Rest of the values are Col (15) + Col (16); Calculations are done following an iterative procedure until the calculated value of datum elevation for the assumed d/D matches with the first value

Table 9.11 Head-discharge computation Known: $R_S = 2.15$ m; $D = 2.6$ m

Head on the crest	Crest control			Throat control	
H_e (m)	H_e/R_S	C	Q (m³ s⁻¹)	H_a (m)	Q (m³ s⁻¹)
(1)	(2)	(3)	(4)	(5)	(6)
0.2	0.09	1.90	2.30		
0.4	0.19	2.03	6.93		
0.6	0.28	2.07	12.99		
0.8	0.37	2.01	19.40		
0.9	0.42	1.97	22.67		
1.0				4.28	46.23
1.5				4.78	48.86
2.0				5.28	51.35
2.5				5.78	53.73
3.0				6.28	56.00

Col (1): Assumed head on the crest (maximum value limited to 3.0 m); Col (2): Col (1)/R_S; Col (3): C assumed as 2.07 for $H_e/R_S = 0.3$ (from Fig. 9.12 for $H_0/R_S = 0.3$, and then multiplying with the correction factor from Fig. 9.17; Col (4): Weir discharge using Eq. (9.2) for head in Col (1); Col (5): $H_a = 6.28$ m (determined in Step 4 for conduit diameter $D = 2.6$ m) is used for $H_e = 3.0$ m, and rest of the values are back-calculated; Col (6): Throat discharge using Eq. (9.18) for head in Col (5)

Note that H_e/R_S must be limited to 0.4 to avoid subatmospheric pressures affecting the discharge computation

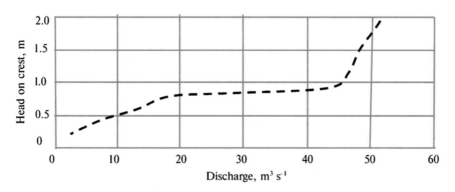

Fig. 9.21 Discharge curve for standard-crested drop-inlet spillway

Hence, $h_f = 2 - \left[\left(\frac{1.6}{1.7}\right) \times (2)^{\frac{3}{2}}\right]^{\frac{2}{3}} = 0.08$ m.

The water surface elevation is 32.0 m (as per the problem statement),
Hence, the elevation of the energy gradient $= 32 - 0.08 = 31.92$ m.
The water surface elevation and the energy gradient line can be plotted (Fig. 9.22).

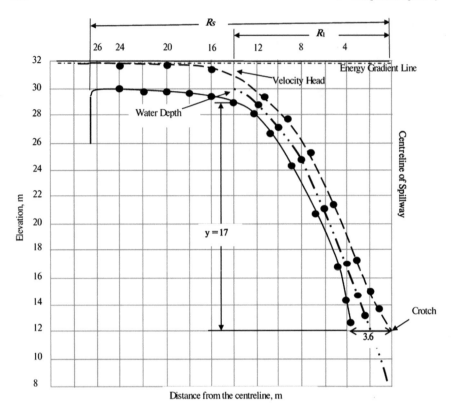

Fig. 9.22 Designed flat-crested spillway

Step 2: Determine the limiting radius to locate the downstream end of the weir crest.

As discussed earlier, the distance d to the weir floor, measured down from the energy gradient elevation, can be determined for any radius R_1 using Eq. (9.20). Also, the depth of water corresponding to any d can be determined using Eq. (9.21). Table 9.12 presents the values of d and h_1 for various values of R_1.

The floor of the weir and the water surface can now be plotted in Fig. 9.22.

Since the weir section of the spillway should end at a particular value of R_1, let us assume the limiting value of $R_1 = 14.0$ m.

At $R_1 = 14.0$ m, $d = 2.9$ m and $h_1 = 2.0$ m.

Thus, water surface is $(d - h_1)$ or $(2.9 - 2.0) = 0.9$ m below the energy gradient. The difference between energy gradient and water surface elevations represents the velocity head.

At this point, the tangent of the water surface or the average slope of the water filament, S, is approximately 0.1.

Thus, the vertical velocity head $= [S^2(d - h_1)] = [0.01 \times 0.9] = 0.009$.

Table 9.12 Floor and water surface elevations for the weir crest

R_1 (m)	24	22	20	18	16	14	12	10
d (m)	2.1	2.2	2.3	2.5	2.7	2.9	3.2	3.7
h_1 (m)	1.4	1.4	1.5	1.7	1.8	2.0	2.2	2.4
Floor Elevation (m)[*]	29.8	29.7	29.6	29.4	29.2	29.0	28.7	28.2
Water Elevation (m)[#]	31.2	31.1	31.1	31.1	31.0	31.0	30.9	30.6

[*] Energy gradient elevation $- d$
[#] Floor elevation $+ h_1$

Table 9.13 Coordinates of the centre line of the freely falling jet and the thickness of the jet at the centre line

x (m)	2.0	4.0	6.0	8.0	10.0	11.0	11.5
y (m)	0.1	1.7	4.2	7.6	11.9	14.3	15.7
t (m)	1.3	1.3	1.3	1.4	1.7	2.2	2.5

Since the vertical velocity head is negligible, the assumed limiting value of $R_1 = 14.0$ m is accepted.

Thus, the width of the weir crest, $B = (R_S - R_1) = 26.4 - 14 = 12.4$ m.

Step 3: Determine the centre line and thickness of the free-falling jet.

As discussed, considering the floor of the downstream end of the weir section as the origin, the x- and y-coordinates of the centre line of the freely falling jet can be determined using Eq. (9.22). Also, the thickness of the jet at any point on the centre line can be determined using Eq. (9.23). Table 9.13 presents the values of y and t for various values of x.

The freely falling jet is plotted in Fig. 9.22 by locating the centre line and then placing the calculated thickness normal to the centre line of the jet, with one-half on each side.

As evident from Fig. 9.22, the crotch is located at $y = 17.0$ m.

Step 4: Determine the radius of the vertical shaft.

As defined earlier, the required radius R_1 of the vertical shaft below the crotch can be determined using Eq. (9.25).

At $y = 17.0$ m, neglecting the friction losses, Eq. (9.25) results in $R_1 = 6.3$ m. However, the plot of the freely falling jet in Fig. 9.22 shows that the radius of the vertical shaft should be 3.6 m. Kurtz (1925) also found discrepancies in R_1 values obtained from Eq. (9.25) and the plot.

Therefore, at the crotch, the radius of the vertical shaft, $R_1 = 3.6$ m, obtained from Fig. 9.22, is taken as the final value.

Though the vertical shaft radius may further be determined for higher y values, with R_1 decreasing with increasing y, the vertical shaft diameter obtained at the crotch is selected as its final value.

Step 5: Determine the conduit diameter, D.

Assuming the conduit diameter same as the vertical shaft diameter, $D = 7.2$ m. The throat location is at $y = 17$ m, hence, $H_a = 33 - 17 = 16.0$ m.

Velocity of flow in the conduit, $V_P = \frac{Q}{A} = \frac{750}{\left[\frac{3.14}{4} \times (7.2)^2\right]} = 18.43 \text{ m s}^{-1}$.

Velocity head in the conduit, $h_v = \frac{V_P^2}{2g} = \frac{(18.43)^2}{2 \times 9.81} = 17.31$ m.

Since conduit will flow full, $d/D = 1.0$. Hence, from Table 9.6, $(A/D^2) = 0.7854$ and $r/D = 0.25$.

$$d = D = 7.2 \text{ m}$$

Therefore,
$$r = 0.25 \times D = 0.25 \times 7.2 = 1.8 \text{ m}$$

Assuming Manning roughness coefficient, $n = 0.014$, the energy slope gradient by Manning equation,

$$s = \left(\frac{V_P n}{r^{2/3}}\right)^2 = \left(\frac{18.43 \times 0.014}{(1.8)^{2/3}}\right)^2 = 0.03$$

Hence, head loss due to friction in the conduit,

$$h_f = s \times l = 0.03 \times 125 = 3.8 \text{ m}$$

The downstream invert elevation of the conduit is determined using Eq. (9.19):

Invert elevation $=$ Throat elevation $+$ pressure head at throat

$\qquad$ $-$velocity head in the conduit$-$friction losses in the conduit

$\qquad$ $-$depth at the outlet

$\qquad$ $= 16.0 + 0.9 \times (33.0-16.0)-17.3-3.8-7.2 = 2.09 \text{ m}$

Since the calculated invert elevation is close to the given invert elevation of 2.0 m, the conduit diameter is acceptable.

Head-Discharge Computations

Table 9.14 presents the head-discharge computation. The discharge curve is plotted in Fig. 9.23.

Questions

1. Describe various types of drop-inlet structures along with their specific functions and conditions under which they are preferred.
2. Based on the information from the watersheds closer to your hometown, managed under the soil management programmes, describe various existing drop-inlet structures.

Table 9.14 Head-discharge computation Known: $R_S = 26.4$ m; $D = 7.2$ m

Head on the crest	Crest control		Pipe control	
H (m)	C	Q (m³ s⁻¹)	H_T (m)	Q (m³ s⁻¹)
(1)	(2)	(3)	(4)	(5)
0.2	1.6	24		
0.5	1.6	94		
1.0	1.6	265		
1.5	1.6	487		
2.0	1.6	750		
			26.18	732
			26.68	739
			27.18	746
			27.68	753
			28.18	759
			28.68	766

Col (1): Assumed head on the crest (maximum value limited to 1.6); Col
(3): Weir discharge using Eq. (9.2) for head in Col (1); Col (4): $H_T = 27.68$ for $H = 2.0$ (difference
between the headwater surface elevation and a point located at $0.6D$ above the downstream invert
elevation; $= 32 - 0.6D = 32 - 0.6 \times 7.2 = 27.68$), and rest of the values calculated for various
H; Col (5): Calculated using Eq. (9.9), assuming $K_e = 0.04$ and $K_c l = [h_f /(V_P^2/2\,g)$

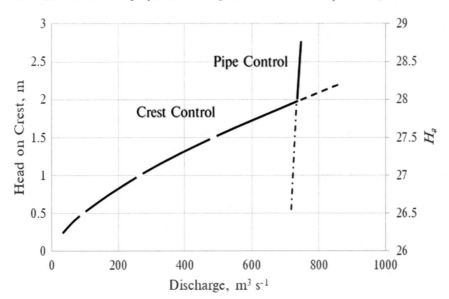

Fig. 9.23 Discharge curve of the flat-crested spillway

3. Draw a standard-crested drop-inlet spillway showing various components. Describe the function of various components.

4. A drop-inlet spillway is expected to operate under a maximum head of 2.5 m and handle a peak discharge of 8.0 m^3 s^{-1}. The length of the concrete conduit is 25.0 m. Determine the minimum uniform diameter and the slope of the conduit flowing full to carry the discharge. Assume the head loss coefficient for the entrance to the crest, $K_e = 0.25$, and the friction head loss coefficient, $K_c = 0.015$.

5. In a drop-inlet spillway, a 600 mm diameter corrugated metal pipe is used as the conduit. The conduit is 76 m long, and the elevation difference between the upstream and downstream water levels is 3 m. Determine the discharge capacity of the spillway, assuming that pipe flow controls in the conduit.

6. A drop-inlet spillway is designed for a gully having a catchment area of 75 ha. The rainfall intensity of duration equal to the time of concentration for a 25-year return period storm is 100 mm h^{-1}. The upstream water elevation 29.4 m, and the elevation of the centre of the conduit at its outlet is 24.9 m. The conduit is 46 m long, with $K_e = 0.5$ and $K_c = 0.014$. Determine the conduit diameter to discharge 50% of peak flow. Take runoff coefficient, $c = 0.40$.

7. Determine and plot the discharge curve up to 2.0 m head over crest for a standard-crested drop-inlet spillway with a crest radius of 0.80 m diameter. The elevation of the inlet crest is 28 m, and the invert elevation at the outlet end of the pipe is 24.2 m. The 1.0 m diameter concrete pipe is 75 m long, with $K_e = 0.5$ and $K_c = 0.014$. Assume that pipe flows full and controls the flow. Take the coefficient of discharge for the inlet as 0.780.

8. Design a standard-crested drop-inlet spillway for carrying a discharge of 75 m^3 s^{-1} under a maximum head of 3 m. The horizontal length of the conduit is 72 m. The crest elevation is 32 m, and outlet invert elevation is 16 m. Design should include the minimum radius of the overflow crest and the transition shaft radius. It should also include the minimum uniform conduit diameter, with the condition that the conduit must not flow more than 75 per cent full at the downstream end to prevent subatmospheric pressures in the conduit. Develop the discharge-water surface elevation relationship, considering that the crest radius must be minimised, and subatmospheric pressures along the crest may be accepted.

9. Design a standard-crested drop-inlet spillway for carrying a discharge of 60 m^3 s^{-1} under a maximum head of 2.5 m. The horizontal length of the conduit is 66 m. The crest elevation is 28 m, and outlet invert elevation is 15 m. Design should include the minimum radius of the overflow crest and the transition shaft radius. It should also include finding the minimum uniform conduit diameter, with the condition that the conduit must not flow more than 75 per cent full at the downstream end to prevent subatmospheric pressures in the conduit. Develop the discharge-water surface elevation relationships, considering that the subatmospheric pressures along the crest should be minimised, and the crest radius is flexible.

10. Design a flat-crested drop-inlet spillway to carry a discharge of 700 m³ s⁻¹ under a maximum head of 3 m. The horizontal length of the conduit is 100 m. The crest elevation is 32.0 m, and outlet invert elevation is 5.0 m. Design should include the minimum radius and width of the crest, the transition shaft radius, and the minimum uniform diameter of the conduit flowing full to carry the discharge.

Multiple Choice Questions

Answer the questions by choosing the correct option.

1. A drop-inlet spillway having the standard crest is also referred to as a

   (A) straight-drop spillway
   (B) box-inlet drop spillway
   (C) glory hole spillway
   (D) culvert spillway

2. A drop-inlet spillway is typically used for

   (A) regulating inflow to a reservoir
   (B) removing excess water from a reservoir
   (C) regulating non-uniform approach flow
   (D) minimising the downstream channel capacity

3. A standard-crested drop-inlet spillway does not have a

   (A) weir section
   (B) free-falling section
   (C) vertical shaft section
   (D) conduit

4. A flat standard-crested drop-inlet spillway does not have a

   (A) weir section
   (B) free-falling section
   (C) standard crest
   (D) conduit

5. In a drop-inlet spillway, the short-tube control occurs when

   (A) the vertical shaft is partly full
   (B) the vertical shaft is full
   (C) the conduit is full
   (D) a drawdown occurs in the vertical shaft for small heads

6. The neutral slope is defined as that slope of the conduit at which

   (A) the friction head loss is equal to the conduit slope
   (B) the friction head loss is less than the conduit slope
   (C) the friction head loss is greater than the conduit slope

(D) the friction head loss is negligible

7. In a drop-inlet spillway conduit, the cavitation may be avoided by

(A) using concrete as the construction material
(B) using a smooth pipe
(C) lowering the pipe
(D) minimising the pressure head at the entrance

8. For a given approach depth, the weir discharge coefficient of a standard-crested drop-inlet spillway

(A) increases with the head on the crest
(B) decreases with the head on the crest
(C) does not depend on the head on the crest
(D) depends on the upstream and downstream water surfaces elevations

9. A drop-inlet spillway is designed to carry a discharge of 3.0 m³ s⁻¹ under a maximum head of 3.0 m. The length of the concrete conduit is 10 m. The entrance loss coefficient, $K_e = 0.15$, and the friction head loss coefficient, $K_c = 0.04$. If the conduit is laid at a slope of 0.050, the diameter of the conduit, in m, is

(A) 0.39
(B) 0.54
(C) 0.62
(D) 0.88

10. A 500 mm diameter corrugated metal pipe is used as a conduit in a drop-inlet spillway to carry a discharge of 0.4 m³ s⁻¹. The head causing the flow is 3 m, and the friction head loss coefficient, $K_c = 0.18$. The neutral slope for the conduit is

(A) 0.001
(B) 0.01
(C) 0.77
(D) 1.00

References

Blaisdell FW (1952) Hydraulics of closed conduit spillways: Part I, Theory and its applications. US Department of Agriculture (USDA), Agricultural Research Service, University of Minnesota, St. Anthony Falls Hydraulic Laboratory. Technical Paper 12, Series B, Minneapolis, Minnesota (revised February 1958)

Camp CS, Howe JW (1939) Tests of circular weirs. Civ Eng 9:247–248

duPont RB (1937) Determination of the under nappe over a sharp Crested weir, circular in plan with radial approach. Dissertation, Case School of Applied Science, Cleveland, Ohio

Kurtz F (1925) The hydraulic design of the shaft spillway for the Davis Bridge Dam, and hydraulic tests on working models. Trans ASCE 88:1–54

Natural Resources Conservation Service (NRCS) (2011) Hydraulic design criteria for canopy, hood and drop inlet spillways. Hydraulics. Chapter 3: In: National Engineering Handbook, Part 650 Engineering Field Handbook. USDA Natural Resources Conservation Service, Washington, DC

Natural Resources Conservation Service (NRCS) (1972) Hydraulics, Chapter 3: In: Engineering field manual. USDA Natural Resources Conservation Service, Washington DC

Straub LG, Morris HM (1950) Hydraulic data comparison of concrete and corrugated metal culvert pipes. Technical Paper No. 3, Series B. University of Minnesota, St. Anthony Falls Hydraulic Laboratory. Minneapolis, Minnesota

United States Bureau of Reclamation (USBR) (1987) Design of Small Dams, 3rd edn. Govt. Printing Office, Washington, DC

Chapter 10
Chute Spillway

Abstract A chute spillway, also called a trough spillway, is designed to dispose of surplus water from upstream to downstream through a steeply sloped open channel. The chapter describes the functions of the various components of a chute spillway and presents the hydrologic, hydraulic, and structural designs of chute spillways. The hydraulic design of the entrance or approach channel, inlet or control structure, chute channel or discharge carrier and outlet or energy dissipater is presented. The structural stability is analysed considering the weight of the structure and the uplift pressure created due to the differential head between the upstream and downstream. A detailed procedure to analyse the stability of the structure against overturning, tension, and compression is demonstrated through a solved example.

10.1 Background

A chute spillway is an open channel-like structure that conveys water from the reservoir to the downstream channel at supercritical velocity (NRCS 2008). The flow enters through an inlet and flows through a steeply sloped concrete-lined channel to the outlet (Fig. 10.1). It is the simplest type of spillway in which the drop in the water surface occurs while flowing through the channel, unlike the free overfall in a drop spillway. A chute spillway is also known as a trough spillway.

10.2 Functions

The chute spillways are used for controlling and stabilising grades in a gully, having high discharge and drop of 3.0–6.0 m. These are also used for conserving water and preventing floods.

© The Author(s), under exclusive license to Springer Nature Singapore Pte Ltd. 2023   261
R. Singh, *Soil and Water Conservation Structures Design*, Water Science and Technology
Library 123, https://doi.org/10.1007/978-981-19-8665-9_10

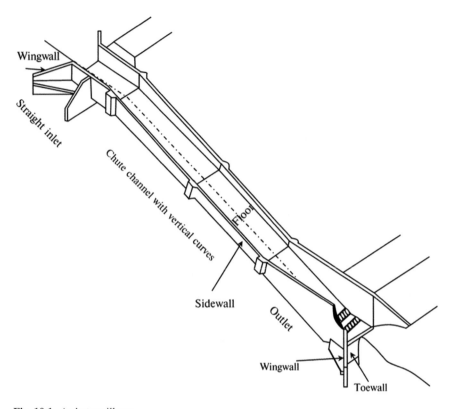

Fig. 10.1 A chute spillway

10.3   Adaptability

In general, the chute spillways are used when the total fall or drop at a site exceeds 3.0 m, i.e., inappropriate for a drop spillway, and there is no possibility of making provisions for temporary storage above the structure, i.e., inappropriate for a drop-inlet spillway.

The chute spillways may be used for a site having a total fall or drop up to 6.0 m. These spillways are often preferred with earth-fill dams.

10.4   Advantages and Limitations

The chute spillways offer the advantages of simplicity in design and construction as compared to other structures. Also, these may be adapted to almost any foundation condition. The chute spillways are economical than drop-inlet spillways for the same discharge capacity and elevation drop.

The limitations of chute spillways include the risk of damage to the structure due to rodent infestation and weakening of the foundation due to seepage in poorly drained conditions.

10.5 Chute Spillway: Components and Functions

The various components of the chute spillway are,

1. Entrance or approach channel
2. Inlet or control structure
3. Chute channel or discharge carrier
4. Outlet or energy dissipater.

The individual components are described below, along with their functions.

10.5.1 Entrance or Approach Channel

Typically, a rectangular channel or a trapezoidal channel having 1:1 side slopes is used as the entrance or approach channel. It leads the water to the inlet or the control structure. It is designed to ensure minimum head loss over the channel section and a smooth flow transition over the control structure.

10.5.2 Inlet or Control Structure

The inlet or control structure governs the discharge capacity of the structure. Two most commonly used inlets or control structures with chute spillways are,

1. Straight inlet
2. Box (rectangular) inlet.

These inlets are similar to those used in the straight or box-inlet drop spillways. As usual, a box-inlet is used for high flows that the straight inlet cannot handle for a given drop. Based on the type of inlet, auxiliary components like headwall extension and cutoff wall may protect the structure against erosion around the structure and seepage.

In addition to the straight inlet or box-inlet, ogee spillway is also used as the control structure. The ogee spillway profile follows the lower nappe shape of a free-falling jet to avoid head fluctuation, flow instability, and increased load on the structure due to negative pressure on the crest (Garg 2004).

10.5.3   Chute Channel or Discharge Carrier

An open channel of rectangular or trapezoidal shape is typically used as the chute channel or the discharge carrier. It conveys the flow passing through the inlet or the control structure to the outlet. Flows upstream of the inlet are typically subcritical, with critical flow occurring at the control structure's crest. Therefore, the flow in the chute channel or discharge carrier is usually supercritical.

The topography governs the minimum bed slope of the chute channel; however, the supercritical flow must be sustained. Consequently, the chute channel or the discharge carrier is designed with convex or concave vertical curves for a smooth convergence or divergence to avoid excessive turbulence or uneven flow distribution. The sidewalls of the chute channel confine the flow within the channel.

10.5.4   Outlet or Energy Dissipater

An outlet in the form of a stilling basin is usually provided to dissipate the kinetic energy of flow and protect the downstream channel against erosion. The hydraulic jump phenomenon is typically used for energy dissipation. Aprons are equipped with auxiliary devices like baffle blocks, chute blocks, sills to resist the supercritical flow, leading to the hydraulic jump formation. Straight aprons may be used for small structures. Saint Anthony Falls (SAF) stilling basin is also used in chute spillways.

10.6   Design of Chute Spillway

As discussed in Sect. 8.7 (Chap. 8), the design of a permanent gully control structure involves the following three phases:

1. Hydrologic design
2. Hydraulic design
3. Structural design.

As described earlier, the hydrologic design involves determining the peak flow rate. The hydraulic design includes determining the dimensions of various components of the structure for handling the design flow rate. The structural design involves analysing various forces acting on the structure and then providing the required strength and stability to the various components of the structure.

10.6.1 *Hydrologic Design*

The hydrologic design concepts and approaches discussed in Sect. 8.7.1 (Chap. 8) are also applicable to the hydrologic design of the chute spillways. A return period of 25–50 years is considered. Any of the three approaches, i.e., the rational method, frequency analysis of historical rainfall or streamflow data, or the hydrological or hydraulic modelling, may be adopted for determining the design flow rate.

10.6.2 *Hydraulic Design of Chute Spillway*

The hydraulic design of various components of a chute spillway is discussed in the following sections.

10.6.2.1 Entrance or Approach Channel

As mentioned earlier, either a rectangular channel or a trapezoidal channel having 1:1 side slopes is used as the entrance or approach channel. Also, it is designed to ensure minimum head loss over the channel section.

Manning formula is typically used to determine the friction head loss in the approach channel. The friction head loss is given as,

$$h_f = S_f L_{approach} = \left[\left(\frac{V^2 n^2}{R^{\frac{4}{3}}} \right) L_{approach} \right] \qquad (10.1)$$

where S_f = friction slope; $L_{approach}$ = length of the approach channel; V = velocity of flow; n = Manning roughness coefficient; and R = hydraulic radius.

The depth of approach, i.e., the flow depth in the approach channel, is $(H_o + P)$, with H_o being the design head over the control structure crest, including the velocity of approach head, h_a, and P being the approach height, i.e., the elevation difference between the approach channel bed and the crest level of the control structure (Fig. 10.2). It may be noted that the actual head considered for designing the structure, H_e, may differ from the design head, H_o, with both including the velocity of approach head, h_a.

The bottom width of the rectangular approach channel is usually kept equal to the crest length of the control structure. Consequently, the channel may be considered a wide channel, and the hydraulic radius may be considered equal to the depth of approach.

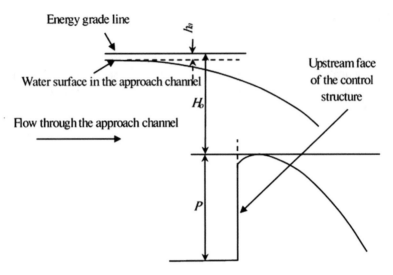

Fig. 10.2 Flow through the approach channel to the control structure

10.6.2.2 Inlet or Control Structure

As mentioned earlier, the inlet or control structure governs the discharge capacity of the structure.

The two most commonly used inlets or control structures, i.e., the straight inlet or the box-inlet, are similar to those already discussed in Sects. 8.7.2 and 8.7.3 (Chap. 8). If any of these control structures are used in a chute spillway, then the capacity formula and the design procedures described earlier will be used.

Therefore, this section will focus on the ogee spillway, which is also used as the control structure in a chute spillway. As mentioned earlier, the ogee spillway profile follows the lower nappe shape of a free-falling jet to avoid head fluctuation, flow instability, and increased load on the structure.

Crest Profile of the Ogee Spillway

The crest profile of the ogee spillway depends on the head, the slope of the upstream face of the spillway, and the approach height (that affects the velocity of the approach). The crest profile is defined with reference to the apex of the crest (Fig. 10.3).

The downstream profile is given by the following equation:

$$x_d^m = -K \ H_o^{m-1} \ y_d \qquad (10.2)$$

where x_d and y_d = downstream crest profile coordinates, with apex O as the origin; and K and m are constants. The values of K and m depend on the velocity of approach and the slope of the upstream face.

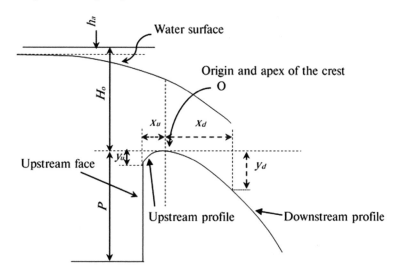

Fig. 10.3 Crest profile of the ogee spillway

Table 10.1 presents the representative values of K and m as a function of the slope of the upstream face and ratio $\left(\frac{h_a}{H_o}\right)$. After plotting most of the downstream profile, a smooth, gradual reverse curvature is provided at the end for a smooth transition of flow into the chute channel or the discharge carrier.

Table 10.2 presents the representative values of the crest profile coordinates, upstream of origin O, i.e., x_u and y_u, and the radius of the circular curves for joining the points for different slopes of the upstream face and ratios (h_a/H_o).

The centre of curvature for R_1 is located on the vertical line passing through apex O (Fig. 10.4). The centre of curvature for R_2 is located at the intersection of arc 'ab' (drawn with the tip of the vertical face as centre and R_2 as radius) and arc 'cd' (drawn with the centre of curvature for R_1 as a centre and (R_1-R_2) as radius) (Fig. 10.4).

After plotting most of the downstream profile, a smooth, gradual reverse curvature is provided at the end for a smooth transition of flow into the chute channel or the discharge carrier.

Table 10.1 K and m values (after USBR 1987)

Slope of the upstream face	$(\frac{h_a}{H_o})$					
	0.0		0.1		0.2	
	K	m	K	M	K	m
Vertical	2.00	1.870	1.96	1.835	2.14	1.836
1:3 (H:V)	2.00	1.850	1.96	1.815	2.14	1.817
2:3	1.90	1.800	1.93	1.764	2.06	1.772
3:3	1.85	1.780	1.88	1.748	2.05	1.750

Table 10.2 Crest profile coordinates upstream of origin O and the radius of the circular curves (after USBR 1987)

Slope of the upstream face	(h_a/H_o)		
	0.0	0.1	0.2
	(x_u/H_o)		
Vertical	0.284	0.232	0.165
1:3 (H:V)	0.250	0.220	0.163
2:3	0.214	0.205	0.162
3:3	0.200	0.194	0.160
	(y_u/H_o)		
Vertical	0.126	0.084	0.048
1:3 (H:V)	0.092	0.072	0.042
2:3	0.069	0.055	0.035
3:3	0.045	0.040	0.028
	(R_1/H_o)		
Vertical	0.530	0.464	0.374
1:3 (H:V)	0.530	0.550	0.384
2:3	0.445	0.492	0.350
3:3	0.450	0.462	0.422
	(R_2/H_o)		
Vertical	0.235	0.020	0.196
1:3 (H:V)	0.145	0.192	0.196
2:3	0.215	0.328	∞
3:3	∞	∞	∞

Fig. 10.4 Location of the centre of curvatures for R_1 and R_2

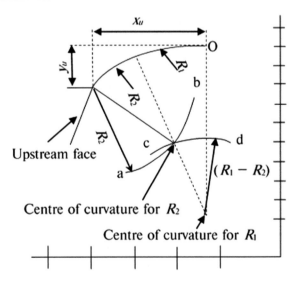

Discharge over the Ogee Crest

The following equation gives the discharge over the ogee crest:

$$Q = C \, L \, H_o^{3/2} \qquad (10.3)$$

where Q = discharge; C = discharge coefficient, which depends on several factors; and L = effective length of the crest.

It may be noted that H_e may replace H_o in Eq. (10.3) if the actual head is different from the design head.

Variation of the Discharge Coefficient, C

The discharge coefficient, C, depends on several factors, such as (1) the depth of approach, (2) the actual head different from the design head, (3) slope of the upstream face, and (4) downstream apron interference and downstream submergence. The following subsections present the variation of the discharge coefficient, C, with these factors.

Depth of Approach

The depth of approach influences the value of C through its effect on the velocity of the approach. For a high spillway, i.e., when $[P/(H_o-h_a)] > 1.33$, the depth of approach has a negligible effect on C. In such a case, C may be considered equal to the theoretical discharge coefficient, C_0, i.e., $C = C_0 = 2.2$.

However, for low spillways, i.e., when $[P/(H_o-h_a)] < 1.33$, the depth of approach affects the value of C appreciably. Thus, for $[P/(H_o - h_a)] < 1.33$, the value of C/C_o should be read from Fig. 10.5, and the value of C determined using $C_0 = 2.2$.

Actual Head different from Design Head

An actual head, different from the design head, may influence the ideal shape of the ogee crest. A broader or narrower crest shape may alter the pressure over the crest profile, thereby impacting the discharge. Therefore, the discharge coefficient will differ from the theoretical discharge coefficient, $C_0 = 2.2$. The value of C may be determined from Fig. 10.6, which presents $C's$ variation with H_e/H_o.

Slope of the Upstream Face

The theoretical discharge coefficient and subsequent corrections for the depth of approach or the actual head over the crest assume a vertical upstream face. Since the sloping upstream face may alter the discharge coefficient, the discharge coefficient should be corrected for the slope of the upstream face. Figure 10.7 presents the correction factor, as a function of P/H_o, by which the C value should be multiplied.

Downstream Apron Interference and Downstream Submergence

The vertical distance from the crest to the downstream apron and the flow depth in the downstream channel may alter the discharge coefficient. When the ogee spillway

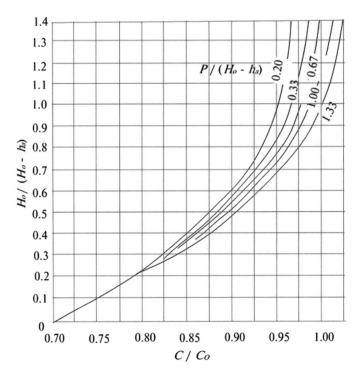

Fig. 10.5 Variation of discharge coefficient with the depth of approach (after USBR 1987)

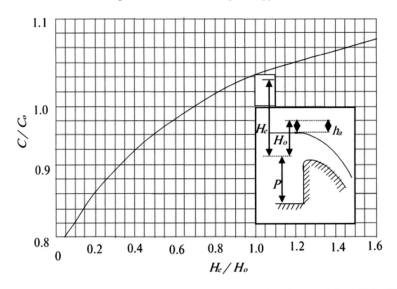

Fig. 10.6 Variation of discharge coefficient with the actual head on the crest (after USBR 1987)

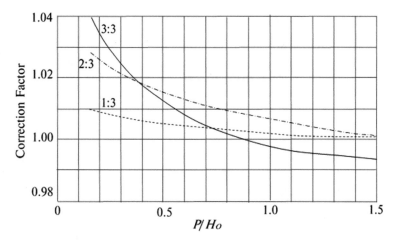

Fig. 10.7 Variation of discharge coefficient with the slope of the upstream face (after USBR 1987)

is used as a control structure in the chute spillway, a hydraulic jump occurs due to the supercritical flow in the downstream channel.

In such a case, the discharge coefficient is independent of the tailwater submergence effect and may depend only on the downstream apron position. Figure 10.8 presents the correction factor due to downstream apron as a function of $(h_d + y)/H_o$, by which the C value should be multiplied. As evident from Fig. 10.8, when $[(h_d + y)/H_0] \geq 1.7$, the downstream apron does not influence the discharge coefficient.

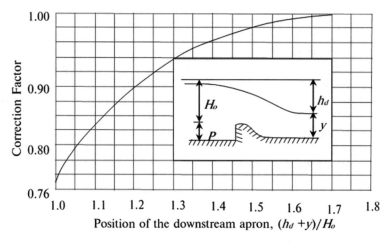

Fig. 10.8 Variation of discharge coefficient with the downstream apron position (after USBR 1987)

Effective Length of the Ogee Crest

The piers and abutments provided to impart the intermediate and end support to the multispan ogee crest may result in the end contraction of the crest. Consequently, the effective length, L, is less than the net length of the crest. The following equation may be used to determine the effective length of the ogee crest:

$$L = L' - 2(NK_p + K_a)H_o \qquad (10.4)$$

where $L' =$ net length of the crest; $N =$ number of piers; $K_p =$ pier contraction coefficient; and $K_a =$ abutment contraction coefficient.

The coefficient, K_p, is influenced by the pier shape and thickness, the design head, and the approach velocity. Similarly, the coefficient, K_a, is influenced by the abutment shape and the angle that the flow axis makes with the upstream approach wall. Tables 10.3a and 10.3b present the average K_p and K_a values, respectively, which may be used in Eq. (10.4) to determine the effective length of the ogee crest.

Example 10.1 Design an overflow ogee crest, including the crest profile, for a chute spillway discharging 55 m³ s⁻¹ at a 1.5 m head. The upstream face of the crest has a 1:1 slope, and the length of the approach channel is 30.0 m. A bridge having 0.5 m wide piers with a rounded nose is provided with the bridge spans limited to 6.0 m. The rounded abutment walls have a 1.5 m radius, and the approach walls make a 30° angle with the centreline of the spillway entrance.

Solution

Given: Discharge, $Q = 55.0$ m³ s⁻¹; Head above the crest, $H_o = 1.5$ m; Length of the approach channel, $L_{approach} = 30.0$ m; Slope of the upstream face $= 1:1$; Pier specification $= 0.5$ m wide with rounded nose; Bridge span $= 6.0$ m (maximum); Abutment specification $=$ Rounded to a radius of 1.5 m, and the approach walls making 30° with the flow axis.

To solve the problem, we begin by assuming the depth of the approach channel and the apron with reference to the crest. Let us assume that the approach channel and the downstream apron are at the same level, i.e., say 0.75 m below crest level.

Thus, $P = 0.75$ m, and the depth of approach, $(H_o + P) = 2.25$ m.

To determine the losses in the approach channel, let us begin with the theoretical discharge coefficient, i.e., $C = C_o = 2.2$.

The discharge per unit length of the crest,

$$q = \frac{Q}{L} = CH_o^{3/2} = 2.2 \times (1.5)^{3/2} = 4.04 \, \text{m}^3\text{s}^{-1}$$

Therefore, the velocity of approach, $V_a = \frac{q}{(P+H_o)} = \frac{4.04}{2.25} = 1.8 \, \text{m s}^{-1}$

Thus, the velocity head, $h_a = \frac{V_a^2}{2g} = \frac{(1.8)^2}{2 \times 9.81} = 0.17 \, \text{m}.$

Assuming that the approach channel is rectangular and wide, its hydraulic radius may be considered equal to the depth of approach.

Thus, the hydraulic radius of the approach channel, $R = (H_o + P) = 2.25$ m.

Assuming the Manning roughness coefficient, $n = 0.0225$, and substituting the values of V_a, n, R and $L_{approach}$ in Eq. (10.1), the friction head loss,

$$h_f = S_f L_{approach} = \left[\left(\frac{V^2 n^2}{R^{\frac{4}{3}}} \right) L_{approach} \right] = \left[\frac{(1.8)^2 (0.0225)^2}{(2.25)^{4/3}} \times 30 \right] = 0.02 \text{ m}$$

Thus, the effective design head, $H_o = 1.5 - 0.02 = 1.48$ m.

Correction in C due to Depth of Approach

$$\frac{H_o}{(H_o - h_a)} = \frac{1.48}{(1.48 - 0.17)} = 1.13 \text{ and } \frac{P}{(H_o - h_a)} = \frac{0.75}{(1.48 - 0.17)} = 0.6$$

From Fig. 10.5, for $\frac{H_o}{(H_o - h_a)} = 1.13$ and $\frac{P}{(H_o - h_a)} = 0.6$, $C/C_o = 0.98$.
Hence, $C = 0.98 \times 2.2 = 2.16$.

Correction in C due to Slope of the Upstream Face

Since the upstream face of the ogee crest has a slope of 1:1, the value of $C = 2.16$ should be corrected using Fig. 10.7.

For a 1:1 slope and $\frac{P}{H_o} = \frac{0.75}{1.48} = 0.50$, the correction factor from Fig. 10.7 is 1.013.

Hence, $C = 1.013 \times 2.16 = 2.19$.

Correction in C due to Downstream Apron Interference and Downstream Submergence

For correcting C due to downstream apron effects, we first need to determine $(h_d + y)/H_o$ and then use Fig. 10.8.

From the definition sketch in Fig. 10.8,

$$(h_d + y) = P + H_0 = 0.75 + 1.48 = 2.23 \text{ m}.$$

Hence, $\frac{(h_d + y)}{H_o} = \frac{2.23}{1.48} = 1.5$.

From Fig. 10.8, for $\frac{(h_d + y)}{H_o} = 1.5$., the correction factor is 0.98.

Hence, $C = 0.98 \times 2.19 = 2.15$.

Thus, the final value of C, after incorporating all corrections, is 2.15.

We now need to determine the crest length. Substituting known values of Q, C, and H_o in Eq. (10.3),

$$Q = C L H_o^{3/2} \text{ or } L = \frac{Q}{C H_o^{3/2}} = \frac{55}{2.15 \times (1.48)^{\frac{3}{2}}} = 14.2 \text{ m}$$

We may find the net length using Eq. (10.4).

Given that the bridge spans should be limited to 6.0 m let us assume that two piers will be needed for a total span of 14 m; thus, $N = 2$.

Table 10.3a Average values of the pier contraction coefficient, K_p (after USBR 1987)

Pier description	Pier contraction coefficient, K_p
Square-nosed piers with rounded corners, with a radius equal to 10% of the pier thickness	0.02
Round-nosed piers	0.01
Pointed-nose piers	0.0

From Table 10.3a, for round nose piers, pier contraction coefficient, $K_p = 0.01$.

Similarly, from Table 10.3b, for rounded abutments where $r > 0.5H$, and the headwall is placed not more than 45° to the direction of flow (given radius of 1.5 m is $> 0.5H_e$ and approach walls are at 30° with the flow axis), abutment contraction coefficient, $K_a = 0.0$.

Substituting L, N, K_p and K_a in Eq. (10.4), the net length.

$$L' = L + 2(NK_p + K_a)H_o = 14.2 + 2(2 \times 0.01 + 0) \times 1.48 = 14.3 \, \text{m}.$$

Crest Profile of the Ogee Spillway

The downstream profile of the ogee crest is given by Eq. (10.2) as,

$$x_d^m = -K \, H_o^{m-1} \, y_d$$

The values of constants K and m can be taken from Table 10.1 for known (h_a / H_o) and upstream face slope.

Here, $\frac{h_a}{H_o} = \frac{0.17}{1.48} = 0.1$ and upstream face slope $= 1:1$ (given); hence, $K = 1.88$ and $m = 1.748$.

Therefore, the equation for downstream profile,

$$x_d^{1.748} = -1.88(1.48)^{0.748} y_d \quad \text{or} \quad y_d = -\frac{x_d^{1.748}}{2.52} \tag{10.5}$$

Table 10.3b Average values of the abutment contraction coefficient, K_a (after USBR 1987)

Abutment description	Abutment contraction coefficient, K_a
Square abutments with headwall at 90° to the direction of flow	0.20
Rounded abutments with headwall at 90° to the direction of flow, when $0.5H \le r \le 0.15H$, r being the radius of the abutment rounding	0.10
Rounded abutments where $r > 0.5H$, and the headwall is placed not more than 45° to the direction of flow	0.0

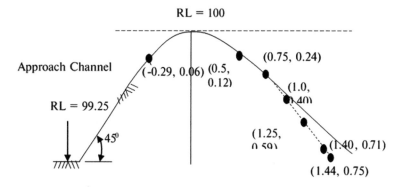

Fig. 10.9 Ogee profile (Example 10.1)

The following table presents the coordinates of the downstream crest profile determined using Eq. (10.5). Note that the maximum value of y_d is limited to the downstream apron level, i.e., 0.75 m from the spillway crest.

x_d, m	0.20	0.50	0.75	1.00	1.25	1.30	1.40	1.44
y_d, m	0.02	0.12	0.24	0.40	0.59	0.63	0.71	0.75

The representative crest profile coordinates upstream of origin O, and the radius of the circular curves can be obtained from Table 10.2.

From the table, for $\frac{h_a}{H_o} = 0.1$, $\frac{x_u}{H_o} = 0.194$; $\frac{y_u}{H_o} = 0.04$; $\frac{R_1}{H_o} = 0.462$; and $\frac{R_2}{H_o} = \infty$.
Hence, $x_u = 0.29$ m; $y_u = 0.06$ m; $R_1 = 0.68$ m; and $R_2 = \infty$.
Figure 10.9 presents the designed ogee spillway.

10.6.2.3 Chute Channel or Discharge Carrier

As already mentioned, an open channel of rectangular or trapezoidal shape is typically used as the chute channel or the discharge carrier. The critical flow usually occurs at the control structure crest, and flow enters the chute channel or discharge carrier as supercritical. The minimum bed slope of the chute channel depends on the topography, but the supercritical flow must be sustained. Consequently, the chute channel or the discharge carrier is designed with convex or concave vertical curves for a smooth convergence or divergence to avoid excessive turbulence or uneven flow distribution. The sidewalls of the chute channel confine the flow within the channel.

Design of Vertical Curves for Chute Channel or Discharge Carrier

Concave Vertical Curves

Concave vertical curves are used whenever the slope of the discharge carrier changes from steeper to milder (Fig. 10.10a). For the concave curve, the dynamic pressure exerted by the flowing water on the floor is a function of the flow energy and the radius of curvature. The radius of curvature can be obtained using the following relationships.

$$R = \frac{2q V_c}{p} \tag{10.6a}$$

$$R = \frac{2y V_c^2}{p} \tag{10.6b}$$

where R = radius of curvature, m; q = discharge per unit width, m³ s⁻¹ m⁻¹; V_c = velocity of flow in the channel, m s⁻¹; y = flow depth in the channel, m; and p = dynamic pressure exerted on the floor, N m⁻².

The value of p is typically assumed as 47,880 N m⁻² for determining the radius of curvature. However, the radius of the concave curve should always be greater than $10y$. The radius of the reverse curve at the downstream end of the ogee crest, where it joins the chute channel or the discharge carrier, should be $\geq 5y$.

Convex Vertical Curves

Convex vertical curves are used whenever the slope of the discharge carrier changes from milder to steeper (Fig. 10.10b). Convex curves should be flat enough to avoid the flow's tendency to spring away from the floor. The curvature may follow the shape defined by the equation:

$$-y_{vc} = x_{vc} \tan\theta + \frac{x_{vc}^2}{K\left[4(y + h_v)\cos^2\theta\right]} \tag{10.7}$$

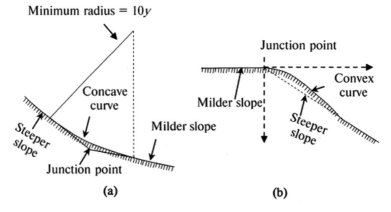

Fig. 10.10 Vertical curves for discharge carrier **a** concave, **b** convex

where $x_{yc} = y_{vc} =$ coordinates of the vertical curve; $\theta =$ slope angle of the floor upstream; $h_v =$ velocity head in the discharge carrier; and $K =$ constant, having a value equal to or greater than 1.5.

In Eq. (10.7), $(y + h_v)$ represents the specific energy at the junction point.

Convergence or Divergence of the Chute Channel or Discharge Carrier

The converging or diverging transitions may be needed when the channel cross-section is different from either the crest or the outlet. In such cases, the sidewall convergence or divergence must be gradual to ensure uniform flow across the channel. The angular variation of the sidewall may be determined using the following equation:

$$\tan \alpha = \frac{1}{3F_1} \tag{10.8}$$

where $\alpha =$ angle between the sidewall and the channel centreline; and $F_1 =$ Froude number at the beginning of the transition.

Froude number is discussed in detail in Sect. 7.3.2.2 (Chap. 7).

Freeboard for the Chute Channel or Discharge Carrier

Since the flow in the discharge carrier is supercritical, the velocity and energy of the flow govern the wave action, splash, and spray. Therefore, the required freeboard may be determined in terms of flow velocity and depth using the following equation:

$$f = 0.61 + 0.008 \, V \sqrt[3]{y} \tag{10.9}$$

where $f =$ freeboard, m.

Example 10.2 For the overflow ogee crest designed in Example 10.1, design a discharge carrier to convey the water from the ogee crest to the outlet. The vertical drop of the discharge carrier from the spillway crest to the outlet floor is 30 m. Assume additional data required to design the discharge carrier.

Solution

From Example 10.1, we know that: Discharge, $Q = 55.0 \text{ m}^3 \text{ s}^{-1}$; Effective design head above the crest, $H_o = 1.48$ m; Net length of the spillway crest, $L' = 14.3$ m.

Since, in the example, the vertical drop from the spillway crest to the outlet floor is 30 m (it was assumed as 0.75 m in example 10.1), we need to revise the coordinates of the downstream spillway profile. For this purpose, let us assume that the spillway crest is at a reduced level (RL) of 100 m (Fig. 10.11). Thus, the RLs of the approach channel bed and the apron floor are 99.25 and 70.0 m, respectively.

Now, we need to determine the position of the downstream toe of the spillway. To ensure that the spillway toe does not influence the coefficient of discharge, we may use the following relationship:

$$\frac{(h_d + y)}{H_o} \geq 1.7$$

Fig. 10.11 Reduced levels
of various components of the
chute spillway (Example
10.2)

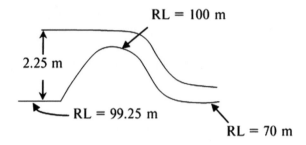

Since $H_o = 1.48$ m, $(h_d + y) \geq 1.7 \times 1.48 \geq 2.52$ m.

Let us consider, $(h_d + y) = 2.55$ m.

Hence, maximum toe elevation $= (RL$ of the spillway crest $H_o) - (h_d + y) = (100 + 1.48) - (2.55) = 98.93$ m.

Hence, the downstream spillway profile will be provided between 100.0 and 98.93 m elevation. Thus, the maximum y-coordinate of the downstream profile $= 100 - 98.93 = 1.07$ m.

Hence, we need to extend the downstream spillway profile determined in Example 10.1 using Eq. (10.5). The extended downstream profile is given in the following table.

x_d, m	0.25	0.50	0.75	1.00	1.25	1.50	1.76
y_d, m	0.04	0.12	0.24	0.40	0.59	0.81	1.07

Discharge per unit length at the downstream spillway end,

$$q = \frac{Q}{L'} = \frac{55}{14.3} = 3.85 \text{ m}^3\text{s}^{-1}\text{m}^{-1}$$

Since the flow depth is y, the velocity of flow $= \frac{3.85}{y}$ m s^{-1}.

Hence, the specific energy $= y + \frac{v^2}{2g} = y + \frac{(3.85/y)^2}{(2 \times 9.81)} = y + \frac{0.76}{y^2}$.

However, the specific energy is equal to $(h_d + y) = 2.55$ m (Referring to Fig. 10.8, $(h_d + y)$ represents the elevation difference between the apron bed and the energy grade line, and hence it represents the specific energy).

Hence, $y + \frac{0.76}{y^2} = 2.55$ or, $y^3 - 2.55y^2 + 0.76 = 0$.

Solving by trial and error, we get $y = 0.63$ m.

Design of the Discharge Carrier

$$\text{Discharge per unit length, } q = \frac{Q}{L'} = \frac{55}{14.3} = 3.85 \text{ m}^3\text{s}^1\text{m}^{-1}$$

$$\text{Critical depth, } y_c = \left(\frac{q^2}{g}\right)^{1/3} = \left(\frac{3.85^2}{9.81}\right)^{1/3} = 1.15 \text{ m}$$

Since critical depth $y_c > y$ at the spillway toe, the flow is supercritical.

The discharge carrier should now be given a milder slope for a short distance from the spillway toe, but for the supercritical flow, in no case, the slope should be less than the critical slope.

$$\text{Critical velocity of flow, } V_{\text{critical}} = \frac{q}{y_c} = \frac{3.85}{1.15} = 3.35 \text{ m s}^{-1}$$

Let us assume that the discharge carrier has a rectangular cross-section with a bottom width of 14.4 m. Thus, at critical flow,

The cross-sectional area of the flow, $A =$ Bottom width of channel × flow depth $= 14.3 \times 1.15 = 16.45 \text{ m}^2$.

Wetted perimeter, $P_r = 14.3 + 2 \times 1.15 = 16.6$ m.

Hydraulic radius, $R = \frac{A}{P_r} = \frac{16.45}{16.6} = 0.99$ m.

Manning equation now can be used to determine the critical slope, S_c. Assuming the the Manning roughness coefficient $n = 0.0225$, and substituting the values of V_{critical}, n and R in the Manning equation,

$$v_{\text{critical}} = \frac{1}{n} R^{2/3} S_c^{1/2} \text{ or } S_c = \left(\frac{n \, V_{\text{critical}}}{R^{2/3}}\right)^2 = \left(\frac{0.0225 \times 3.35}{(0.99)^{2/3}}\right)^2 = \frac{1}{174}$$

Thus, a slope of say 1/150 (i.e., > 1/174) is provided over a 30 m distance from the spillway toe.

Thus, the bed level at the end of 1/150 slope $= 98.93 - 30/150 = 98.73$ m.

The radius of the reverse curve at the toe is assumed as $5y$, and thus, its value would be equal to (5×0.63) or 3.15 m.

In between elevations 98.73 and 70.0 m, the discharge carrier may be provided with a steep slope depending on the site conditions. A slope of 8:1–4:1 may be provided initially, followed by a steeper slope of 2:1–3:1. Wherever the slope changes, vertical curves (convex or concave) may be provided.

Since the site conditions are not specified in the Example, we may use a 6:1 slope for the first 20 m fall (i.e., up to RL 78.73 m), followed by a 2:1 slope in the remaining segment.

The flow depth and velocity at the end of the 1:150 slope are assumed to be equal to the depth and velocity at the spillway toe, neglecting the possible changes over a short distance of 30 m.

The calculations of water depth, velocity, and other parameters are carried out over the entire length of the discharge carrier by dividing it into small segments. For example, we may use segments of 30 m up to RL 78.73 m and 6 m in between RL 78.73 and 70.0 m, as shown in Table 10.4.

Design of Vertical Curve No. I

(A convex vertical curve shall be provided at the junction of 150:1 and 6:1 slopes).

The convex vertical curve may be designed using Eq. (10.7),

Table 10.4 Determination of the bed elevation, flow depth, flow velocity, and Froude number along the length of the discharge carrier

S. No	Distance from the start of 6:1 slope, m	Segment length, m	Drop in bed, m	Bed level, m	Depth (y) assumed, m	Velocity, $V = q/y$, m²s⁻¹	Velocity head, $V^2/2g$, m	Sp. Energy = $y + V^2/2g$, m	Total Energy Level (calculated), m	Area, A = width × y, m²	Perimeter = width + $2y$, m	Hydraulic Radius, R, m	$S_f = (n^2 \times V^2)/R^{4/3}$	Average S_f	h_f = Average $S_f \times$ segment length, m	Total Energy Level (Actual), m	Froude Number, $F_1 = V/\sqrt{gy}$
(1)	(2)	(3)	(4)	(5)	(6)	(7)	(8)	(9)	(10)	(11)	(12)	(13)	(14)	(15)	(16)	(17)	(18)
6:1 Slope																	
1	0	0	0	98.73	0.63	6.14	1.92	2.55	101.28	8.95	15.46	0.58	0.040			100.21	2.47
2	30	30	5	93.73	0.45	8.60	3.77	4.22	97.95	6.39	15.10	0.42	0.118	0.08	2.36	97.85	4.09
3	60	30	5	88.73	0.41	9.44	4.54	4.95	93.68	5.82	15.02	0.39	0.160	0.14	4.16	93.69	4.71
4	90	30	5	83.73	0.405	9.56	4.65	5.06	88.79	5.75	15.01	0.38	0.166	0.16	4.89	88.80	4.79
5	120	30	5	78.73	0.404	9.58	4.68	5.08	83.81	5.74	15.01	0.38	0.167	0.17	5.00	83.80	4.81
2:1 Slope																	
6	126	6	3	75.73	0.346	11.18	6.38	6.72	82.45	4.91	14.89	0.33	0.278	0.22	1.34	82.46	6.07
7	132	6	3	72.73	0.319	12.13	7.50	7.82	80.55	4.53	14.84	0.31	0.362	0.32	1.92	80.54	6.86
8	137.46	5.46	2.73	70.0	0.307	12.61	8.10	8.41	78.41	4.36	14.81	0.29	0.411	0.39	2.11	78.43	7.26

Col (2): Assumed distance; Col (3): Difference in row values of Col (2); Col (4): Col (3) × slope; Col (5): Col (5)–Col (4), with the first value being RL of the toe; Col (9): Col (6) + Col (8); Col (10): Col (5) + Col (9); Col (16): Col (15) × Col (3); Col (17)–Col (16), with the first value being the sum of Col (5) + actual design head (=1.48 m, determined earlier); Calculations are done following an iterative procedure until the total energy levels in Col (10), and Col (17) match for an assumed 'y' in Col (6)

Table 10.5 Coordinates of the convex vertical curve I

x_{vc}, m	0	0.2	0.5	0.8	1.0	1.2
y_{vc}, m	0	0.004	0.02	0.05	0.07	0.1

$$-y_{vc} = x_{vc}\tan\theta + \frac{x_{vc}^2}{K\left[4(y+h_v)\cos^2\theta\right]}$$

Here, $\tan\theta = 1/150$; since θ is small, $\cos^2\theta$ will be approximately equal to 1.0.

From Table 10.4, $(y + h_v) = 2.55$ m. Taking $K = 1.5$ and substituting known values in the equation, we get,

$$-y_{vc} = x_{vc}\left(\frac{1}{150}\right) + \frac{x_{vc}^2}{1.5[4 \times 2.55 \times 1]} \quad \text{or}$$

$$-y_{vc} = \frac{x_{vc}}{150} + \frac{x_{vc}^2}{15.3} \tag{10.10}$$

Differentiating with respect to x_{vc},

$$\frac{dy_{vc}}{dx_{vc}} = -\frac{1}{150} - \frac{x}{7.65}$$

The curve would meet the downstream slope where

$$\frac{dy_{vc}}{dx_{vc}} = -\frac{1}{6} \text{ (−ve sign implies that as } x_{vc} \text{ increases, } y_{vc} \text{ decreases)}$$

Equating, we get, $\frac{1}{150} + \frac{x_{vc}}{7.5} = \frac{1}{6}$ or x_{vc} 1.2 m.

The coordinates of the curve for $x_{vc} = 0$ to $x_{vc} = 1.2$ m can be obtained using Eq. (10.10). The values are tabulated (Table 10.5).

Design of Vertical Curve No. II

A convex vertical curve shall be provided at the junction of 6:1 and 2:1 slopes. The convex vertical curve may be designed using Eq. (10.7),

$$-y_{vc} = x_{vc}\tan\theta + \frac{x_{vc}^2}{K\left[4(y+h_v)\cos^2\theta\right]}$$

Here, $\tan\theta = 1/6$; hence, $\cos^2\theta = 0.97$.

From Table 10.4, at the junction of 6:1 and 2:1, $(y + h_v) = 5.08$ m. Taking $K = 1.5$ and substituting known values in the equation, we get,

$$-y_{vc} = x_{vc}\left(\frac{1}{6}\right) + \frac{x_{vc}^2}{1.5[4 \times 5.08 \times 0.97]} \quad \text{or}$$

Table 10.6 Coordinates of the convex vertical curve II

x_{vc}, m	0	1.0	2.0	3.0	4.0	4.90
y_{vc}, m	0	0.20	0.47	0.81	1.21	1.63

$$- y_{vc} = \frac{x_{vc}}{6} + \frac{x_{vc}^2}{29.6} \tag{10.11}$$

Differentiating with respect to x_{vc},

$$\frac{dy_{vc}}{dx_{vc}} = -\frac{1}{6} - \frac{x}{14.8}$$

The curve would meet the downstream slope where

$$\frac{dy_{vc}}{dx_{vc}} = -\frac{1}{2} \ (-\text{ve sign implies that as } x_{vc} \text{ increases, } y_{vc} \text{ decreases})$$

Equating, we get,

$$\frac{1}{6} + \frac{x_{vc}}{14.8} = \frac{1}{2} \quad \text{or}$$
$$x_{vc} = 4.93 \, \text{m}$$

The coordinates of the curve for $x_{vc} = 0$ to $x_{vc} = 4.93$ m can be obtained using Eq. (10.11). The values are tabulated (Table 10.6).

Figure 10.12 shows the profile of the discharge carrier.

10.6.2.4 Outlet or Energy Dissipater

As mentioned in Sect. 10.6, an outlet in the form of a stilling basin is usually provided in the chute spillway to dissipate the flow's kinetic energy.

The hydraulic jump phenomenon is typically used for energy dissipation.

The details of the hydraulic jump phenomena and its classification are presented in Sect. 7.3.3 (Chap. 7). The section also includes the energy dissipation in the hydraulic jump phenomena and briefly mentions its application in designing stilling basins. Equation (7.21) presents the hydraulic jump phenomena as,

$$\frac{y_2}{y_1} = \frac{1}{2}\left(\sqrt{1 + 8F_1^2} - 1\right)$$

In the equation, y_1 and y_2 are the corresponding flow depths before and after the hydraulic jump, usually referred to as conjugate depth or sequent depth, and F_1 is the upstream Froude number.

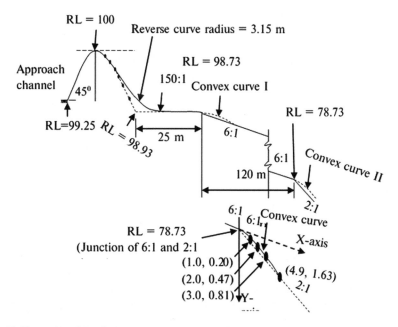

Fig. 10.12 Profile of the discharge carrier (Example 10.2)

The hydraulic jumps are classified based on F_1, and Table 7.1 (Chap. 7) presents various hydraulic jumps. Stilling basin designs vary with F_1 or the type of the hydraulic jump and are described in the following paragraphs.

Stilling Basins for Undular and Weak Jumps ($F_1 < 1.7$; $1.7 < F_1 < 2.5$)

When the upstream Froude number is less than 1.7, no particular stilling basin or dissipating devices are needed for energy dissipation. However, the channel length should be at least four times the depth of flow when the flow depth begins to change.

Even when the upstream Froude number is between 1.7 and 2.5, there is no turbulence; hence, auxiliary devices are not needed. However, the length of the stilling basin should be enough to confine the hydraulic jump. Typically, USBR Stilling Basin Type I, which has a horizontal floor, is provided (Peterka 1984).

Stilling Basins for Oscillating Jump ($2.5 < F_1 < 4.5$)

When the upstream Froude number is between 2.5 and 4.5, typically USBR stilling Basin Type IV is provided (Peterka 1984). The flow in the basin is usually in the transition stage as the hydraulic jump is not fully developed. The wave action due to the oscillating flow cannot be entirely dampened by the stilling basin alone; hence, auxiliary wave dampeners or suppressors may be provided for smooth surface flow downstream. The basin floor should be set to provide 5–10 per cent higher tailwater depth than the conjugate depth, y_2. The length of the basin varies with F_1 (Table 10.7).

Table 10.7 Length of the USBR Type IV Basin

F_1	2.5	3	3.5	4	4.5
Length of USBR Type IV Basin, L_{IV}	4.8 y_2	5.2 y_2	5.6 y_2	5.8 y_2	5.9 y_2

Stilling Basins for Steady Jump ($F_1 > 4.5$)

When the upstream Froude number is > 4.5, either aa USBR Stilling Basin Type III or USBR Stilling Basin Type II is provided (Peterka 1984). In these basins, a well-balanced steady jump is formed. Auxiliary devices like chute blocks, baffles, and sills may be provided along the basin floor to stabilise the jump. The auxiliary devices may help shorten the basin length.

Type III basins are used when the incoming flow velocity is limited to about 18 m s^{-1}. These basins are equipped with auxiliary devices to reduce the length of the jump and dissipate the energy.

On the contrary, USBR Type II basins are preferred when the incoming flow velocity exceeds 18 m s^{-1} or when impact baffle blocks are not used. Though chute blocks and dentated end sills are provided, the energy dissipation in Type II basins is primarily due to the hydraulic jump action. Therefore, the length of these is usually more than that for Type III basins.

The Type II basin floor should be set to provide 5% higher tailwater depth than the conjugate depth, y_2. However, the Type III basin floor should be set to provide tailwater depth equal to the conjugate depth, y_2. The lengths of Type II and Type III basins vary with F_1 (Table 10.8).

Bureau of Indian Standards (BIS) also has standardised stilling basins. IS: 4997 (BIS 1968) includes the standard designs of stilling basins I and II, with horizontal aprons, and stilling basins III and IV, with sloping aprons.

SAF Stilling Basin

Saint Anthony Falls (SAF) stilling basin has chute blocks, baffle blocks, and an end sill to reduce the basin size (Blaisdell 1948). The primary design quantity for the SAF stilling basin is the height of the chute blocks and floor blocks, y_1, which is equal to the depth of flow at the stilling basin entrance.

The following relationship may determine the length of the basin:

$$L_B = \frac{4.5 y_2}{F_1^{0.76}} \tag{10.12}$$

Table 10.8 Length of the USBR Type II and Type III Basins

F_1	4	5	6	8	≥ 10
Length of USBR Type II Basin, L_{II}	3.6 y_2	3.82 y_2	4.0 y_2	4.2 y_2	4.3 y_2
Length of USBR Type III Basin, L_{III}	–	2.35 y_2	2.54 y_2	2.6 y_2	2.68 y_2

where y_2 is the conjugate depth determined using Eq. (7.19). Equation (10.12) is valid for $1.7 < F_1 < 17$.

Example 10.3 Design the profile of a chute spillway to carry a discharge of 20 m³ s⁻¹. The spillway has a straight inlet with a flow depth of 1.20 m. The outlet is in the form of a SAF stilling basin. The vertical drop of the chute from the inlet crest to the SAF outlet floor is 20 m.

Solution

Given: Discharge, $Q = 20.0$ m³ s⁻¹; Head above the crest, $H = 1.2$ m; Vertical drop of the chute from the inlet crest to the SAF outlet floor, $Z_v = 20.0$ m.

The spillway has a straight inlet, and the discharge through the inlet is given by Eq. (8.9) (Chap. 8) as,

$$Q = 1.77LH^{3/2}$$

Substituting the known values of Q and H in Eq. (8.9), $20 = 1.77 \times L \times (1.2)^{3/2}$
We get, $L = 8.6$ m.

However, when a straight inlet is provided in the chute spillway, the freeboard is estimated in terms of increased discharge using the following relationship:

$$f = (0.02 + 0.0098Z_v)Q \tag{10.13}$$

where Z_v = vertical drop from the inlet crest to the outlet floor, m.
Substituting the known values of Z_v and Q in Eq. (10.13),

$$f = (0.02 + 0.0098 \times 20) \times 20 = 4.32 \ m^3 \ s^{-1}$$

Hence, the given discharge of 20 m³ s⁻¹ should be increased by 4.32 m³ s⁻¹ to account for freeboard. Thus, the increased $Q = 20 + 4.32 = 24.32$ m³ s⁻¹
Now, freeboard in terms of H can be determined by substituting increased $Q = 24.32$ m³ s⁻¹ and $L = 8.6$ m in Eq. (8.9), as,

$$24.32 = 1.77 \times 8.6 \times (H)^{3/2}$$

We get, H (including the freeboard) $= 1.4$ m. Hence, freeboard, $f = 1.4 - 1.2 = 0.2$ m.

Hence, the weir dimensions are, Crest Length $L = 8.6$ m; Depth of weir, $h = 1.4$ m.

The dimensions of various components of the straight inlet may be determined using the equations listed in Table 8.2 (Chap. 8).

Design of the Discharge Carrier

$$\text{Discharge per unit length, } q = \frac{Q}{L} = \frac{20}{8.6} = 2.33 \ m^3 s^{-1} m^{-1}$$

$$\text{Critical depth, } y_c = \left(\frac{q^2}{g}\right)^{1/3} = \left(\frac{2.33^2}{9.81}\right)^{1/3} = 0.82 \text{ m}$$

The discharge carrier should have supercritical flow; thus, its bed slope should always be greater than the critical slope.

$$\text{Critical velocity of flow, } V_{\text{critical}} = \frac{q}{y_c} = \frac{2.33}{0.82} = 2.84 \text{ m s}^{-1}$$

Let us assume that the discharge carrier has a rectangular cross-section with a bottom width of 8.6 m. Thus, at critical flow,

The cross-sectional area of the flow, $A = $ Bottom width of the channel $\times$ flow depth

$$= 8.6 \times 0.82 = 7.05 \text{ m}^2$$

$$\text{Wetted perimeter, } P_r = 8.6 + 2 \times 0.82 = 10.24 \text{ m}$$

$$\text{Hydraulic radius, } R = \frac{A}{P_r} = \frac{7.05}{10.24} = 0.69 \text{ m}$$

Manning equation now can be used to determine the critical slope, S_c. Assuming the Manning roughness coefficient, $n = 0.0225$, and substituting the values of v_{critical}, n and R in the Manning equation,

$$V_{\text{critical}} = \frac{1}{n}R^{2/3}S_c^{1/2} \quad \text{or} \quad S_c = \left(\frac{nV_{\text{critical}}}{R^{2/3}}\right)^2 = \left(\frac{0.0225 \times 2.84}{(0.69)^{2/3}}\right)^2 = \frac{1}{149}$$

Thus, the bed slope of the discharge carrier has to be greater than 1/149. The discharge carrier has a total fall of 20 m, and it may be provided with a steep slope depending on the site conditions. Wherever the slope changes, vertical curves (convex or concave) may be provided.

Since the site conditions are not specified in the Example, we may use a uniform slope of 6:1 for the entire reach of the discharge carrier.

The calculations of water depth, velocity, and other parameters are carried out over the entire length of the discharge carrier by dividing it into small segments of length 24 m, as shown in Table 10.9.

Design of Vertical Curve

(At the junction of the inlet apron and the discharge carrier, a convex vertical curve shall be provided as the slope changes from horizontal to 6:1).

The convex vertical curve may be designed using Eq. (10.7),

Table 10.9 Determination of the bed elevation, flow depth, flow velocity and Froude number along the length of the discharge carrier

S. No	Distance from the start of 6:1 slope, m	Segment length, m	Drop in bed, m	Bed level, m	Depth (y) assumed, m	Velocity, $V = q/y$, m s^{-1}	Velocity head, $V^2/2\,g$, m	Sp. Energy $= y + V^2/2\,g$, m	Total Energy Level (calculated), m	Area, A = width $\times y$, m^2	Perimeter = width $+ 2y$, m	Hydraulic Radius, R, m	$S_f = (n^2 \times v^2)/R^{4/3}$	Average S_f	$h_f =$ Average $S_f \times$ segment length, m	Total Energy Level (Actual), m	Froude Number, $F_1 = V/\sqrt{gy}$
(1)	(2)	(3)	(4)	(5)	(6)	(7)	(8)	(9)	(10)	(11)	(12)	(13)	(14)	(15)	(16)	(17)	(18)
6:1 Slope																	
1	0	0	0	100	0.8	2.91	0.43	1.23	101.23	6.88	10.20	0.67	0.007			101.23	1.04
2	24	24	4	96	0.31	7.52	2.88	3.19	99.19	2.67	9.22	0.29	0.150	0.08	1.88	99.35	4.31
3	48	24	4	92	0.3	7.77	3.07	3.37	95.37	2.58	9.20	0.28	0.166	0.16	3.79	95.56	4.53
4	72	24	4	88	0.3	7.77	3.07	3.37	91.37	2.58	9.20	0.28	0.166	0.17	3.99	91.56	4.53
5	96	24	4	84	0.3	7.77	3.07	3.37	87.37	2.58	9.20	0.28	0.166	0.17	3.99	87.57	4.53
6	120	24	4	80	0.3	7.77	3.07	3.37	83.37	2.58	9.20	0.28	0.166	0.17	3.99	83.58	4.53

Col (2): Assumed distance; Col (3): Difference in row values of Col (2); Col (4): Col (3) × slope; Col (5): Col (5)–Col (4), with the first value assumed to have RL of 100; Col (9): Col (6) + Col (8); Col (10): Col (5) + Col (9); Col (16): Col (15) × Col (3); Col (17): Col (17)–Col (16), with the first value being equal to Col (10); Calculations are done following an iterative procedure until the total energy levels in Col (10), and Col (17) match for an assumed 'y' in Col (6)

Table 10.10   Coordinates of the convex vertical curve

x_{vc}, m	0	0.1	0.2	0.3	0.5	0.62
y_{vc}, m	0	0.001	0.005	0.01	0.03	0.05

$$-y_{vc} = x_{vc} \tan \theta + \frac{x_{vc}^2}{K\left[4(y + h_v) \cos^2 \theta\right]}$$

Here, $\tan \theta = 0$; hence, $\cos^2\theta$ will be equal to 1.0.

From Table 10.9, $(y + h_v) = 1.23$ m. Taking $K = 1.5$ and substituting known values in the equation, we get,

$$-y_{vc} = x_{vc}(0) + \frac{x_{vc}^2}{1.5[4 \times 1.23 \times 1]} \quad \text{or} \quad -y_{vc} = 0 + \frac{x_{vc}^2}{7.38} \qquad (10.14)$$

Differentiating with respect to x_{vc}, $\frac{dy_{vc}}{dx_{vc}} = -\frac{x}{3.69}$

The curve would meet the downstream slope where

$$\frac{dy_{vc}}{dx_{vc}} = -\frac{1}{6}$$

($-$ve sign implies that as x_{vc} increases, y_{vc} decreases)

Equating, we get, $\frac{x_{vc}}{3.69} = \frac{1}{6}$ or $x_{vc} = 0.62$ m.

The coordinates of the curve for $x_{vc} = 0$ to $x_{vc} = 0.62$ m can be obtained using Eq. (10.14). The values are tabulated (Table 10.10).

10.6.3   Structural Design

As described earlier, the structural design involves analysing various forces acting on the structure and then providing the required strength and stability to the various components of the structure. The structural design of the straight inlet is presented in detail in Sect. 8.7.4 (Chap. 8). Here, we shall discuss the structural design of the chute channel or the discharge carrier and the stilling basin.

The thickness of the chute floor is the primary design quantity for the structural stability of the chute channel or the discharge carrier. The structural stability is analysed considering the weight of the structure and the uplift pressure created due to the differential head between the upstream and downstream. The uplift pressure is assumed to act over the entire base of the structure.

The uplift pressure is determined using a pressure diagram, drawn by assuming a pressure equal to the depth of flow at the beginning of the discharge carrier and zero at the outlet end. The pressure diagram is subsequently divided into several equal segments. The uplift pressure for the individual segments is determined by multiplying the area of the segment with the average pressure head and the density of water.

The thickness of the chute floor, which is the primary design quantity, is assumed. The weight of the structure is determined for the individual segments considered in the pressure diagram. The weight and the uplift pressure of the individual segments are then taken into account for analysing the safety of the structure against overturning, tension, and compression.

Example 10.4 Carry out the structural design of the discharge carrier designed in Example 10.3. A 2.54 m long SAF stilling basin is provided. The concrete, having a unit weight of 24 kN m^{-3}, is used as a construction material. Take the unit weight of water as 10 kN m^{-3}.

Solution

Given: Unit weight of concrete, $\gamma_{concrete} = 24$ kN m^{-3}; Unit weight of water, $\gamma_w = 10$ kN m^{-3}.

Known from Example 10.3: Head above the crest, $H = 1.2$ m; Vertical drop of the chute from the inlet crest to the outlet floor, $Z_v = 20.0$ m; Length of the stilling basin $= 2.54$ m; Slope of the discharge carrier over the entire length of 120 m (the entire fall of 20 m) $= 6{:}1$; Width of the rectangular discharge carrier $= 8.6$ m.

Let us assume that the thickness of the construction material, concrete, is 380 mm.

Figure 10.13 presents the pressure diagram. The discharge carrier is divided into three equal segments of 40.0 m each. The stilling basin represents the fourth segment, with its base equal to the basin length. The head at the beginning of the discharge carrier is assumed to be equal to the depth of flow, i.e., 1.2 m at the beginning of segment 1. The head at the outlet end, i.e., the end of segment 4, is assumed to be zero.

A trapezium represents the area of water in various segments. Therefore, the area of water in segment 1, for example, can be determined as follows:

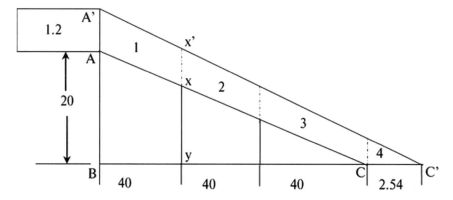

Fig. 10.13 Pressure diagram for the discharge carrier (Example 10.4)

The pressure head at the ends of segments 1–3 can be calculated by considering two triangles ABC and A′BC′. Considering angle A′C′B to be θ' and angle ACB to be θ,

$$\tan \theta' = \left(\frac{21.2}{122.54}\right) \quad \text{and} \quad \tan \theta = \left(\frac{20}{120}\right)$$

Then, the pressure head at the end of segment 1, $x'x$, can be determined as,

$$x'x = x'y - xy$$

From triangle ABC, $\tan \theta = \left(\frac{xy}{80}\right)$, hence $\frac{xy}{80} = \frac{20}{120}$ or $xy = \frac{20 \times 80}{120} = 13.33$.
Similarly, from triangle A′BC,

$$\tan \theta' = \left(\frac{x'y}{82.54}\right), \text{ hence, } \frac{x'y}{82.54} = \frac{21.2}{122.54} \text{ or } x'y = \frac{21.2 \times 82.54}{122.54} = 14.28 \text{ m}$$

Hence, $x'x = x'y - xy = 14.28 - 13.33 = 0.95$ m
Similarly, the pressure heads at the end of segments 2 and 3 are determined as 0.70 m and 0.39 m, respectively.

$$\text{Area of segment } 1 = \frac{1}{2} \times (1.2 + 0.95) \times 40 = 43 \text{ m}^2$$

Therefore, the uplift force in segment 1

$$= \text{area of the segment} \times \text{discharge carrier width} \times \gamma_w$$
$$= 43 \times 8.6 \times 10 = 3698 \text{ kN}$$

The length of the discharge carrier (length over segments 1–3) $= \sqrt{20^2 + 120^2} = 121.66$ m
Hence, length of individual segments $= 121.66/3 = 40.55$ m
The weight of the individual segment may be determined as follows:

$$\text{Weight} = \text{Cross} - \text{sectional area of the segment} \times \text{thickness of the concrete} \times \gamma_{\text{concrete}}$$

For example, the weight of segment 1

$$= (40.55 \times 8.6) \times 0.38 \times 24 = 3180.4 \text{ kN}$$

However, the discharge carrier segments would also have sidewalls, whose weight needs to be considered and added to the weight of the floor portion calculated above.
The height of the sidewalls may be assumed as 1.42 m.
Hence, the weight of the sidewalls for segment 1

$= 2$ (length of the segment $\times$ height of the sidewall $\times$ thickness of concrete $\times \gamma_{concrete}$)

$= 2 \,(40.55 \times 1.42 \times 0.38 \times 24)$

$= 1050.3 \, \text{kN}$

Hence, the total weight of segment $1 = 3180.4 + 1050.3 = 4230.7$ kN

Table 10.11 presents the uplift force and weight calculations for different segments of the discharge carrier.

The weight of the SAF outlet may be determined as follows:

Weight of the stilling basin base $= L_B \times$ bed width $\times$ thickness of concrete $\times \gamma_{concrete} = 2.54 \times 8.6 \times 0.38 \times 24 = 199.2$ kN

Note that the stilling basin bed width is considered equal to the discharge carrier bed width.

Weight of the side walls $= 2 \times (2.54 \times 1.42 \times 0.38 \times 24) = 65.8$ kN

Total weight of the stilling basin $= 199.2 + 65.8 = 265.0$ kN

Table 10.11 includes the uplift force and weight calculations for the SAF outlet.

Safety of the Structure against Overturning

We need to check the structure for safety against overturning about the toe of the structure. The uplift force acting on the structure produces the turning moment (TM), which must be resisted by the resisting moment (RM), produced by the weight of the structure.

The factor of safety against overturning (FS_O) is defined in Sect. 8.7.4.1 (Chap. 8). It is expressed as the ratio of the total resisting moment to the total turning moment about the toe given by Eq. (8.31) as,

$$FS_O = \frac{\sum RM}{\sum TM}$$

For the structure to be safe against overturning, FS_o should be more than 1.5.

The uplift force and weight of individual segments (Table 10.11) would be used to determine the turning moment (TM) and the resisting moment (RM).

Since uniform segments are considered for determining the uplift force and weight, the distances from the toe would also be the same.

Hence, the sum of weights and uplift forces may be considered representative of $\sum RM$ and $\sum TM$, respectively. Thus, in this example,

$$FS_O = \frac{\sum RM}{\sum TM} = \frac{\sum \text{ Cumulative weight of the structure}}{\sum \text{ Cumulative uplift force}} = \frac{12957.1}{8453.4} = 1.53$$

Since calculated $FS_o > 1.5$, the structure is safe against overturning.

Table 10.11 Uplift force and weight calculations for the discharge carrier segments and the SAF outlet

Segment	Upstream pressure head, m	Downstream Pressure head, m	Horizontal width of the water segment, m	Area, m²	Uplift force, kN	Cumulative uplift force, kN	Weight of the segment, including the weights of sidewalls, kN	Cumulative weight of the structure, kN	Remarks
1	2	3	4	5	6	7	8	9	10
1	1.2	0.95	40	43	3698.0	3698.0	4230.7	4230.7	Weight > Uplift force
2	0.95	0.7	40	33	2838.0	6536.0	4230.7	8461.4	Weight > Uplift force
3	0.7	0.39	40	21.8	1874.8	8410.8	4230.7	12,692.1	Weight > Uplift force
4	0.39	0	2.54	0.49	42.6	8453.4	265.0	12,957.1	Weight > Uplift force

Table 10.12 Resisting and turning moments for the discharge carrier segments and the SAF outlet

Segment	Uplift force, kN	Distance of the line of action from the toe, m	Moment, kN−m	Weight of the segment, including the weights of sidewalls, kN	Distance of the line of action from the toe, m	Moment, kN-m
1	2	3	4	5	6	7
1	3698.0	102.54	379,192.9	4230.7	102.54	433,815.5
2	2838.0	62.54	177,488.5	4230.7	62.54	264,587.7
3	1874.8	22.54	42,258.0	4230.7	22.54	95,359.9
4	42.6	1.27	54.1	265.0	1.27	336.6
$\sum$TM			598,993.5	$\sum$RM		794,099.6

Safety of the Structure against Tension

We need to check the structure for safety against tension that may develop due to irregular forces. As described in Sect. 8.7.4.4 (Chap. 8), the resultant force acting on the structure must pass through the middle-third of the structure's base to ensure the safety of the structure against tension.

The position of the resultant force is determined using Eq. (8.45) as:

$$Z = \frac{\sum M}{\sum V}$$

where $\sum M = \left(\sum RM - \sum TM \right)$ and $\sum V =$ Cumulative weight of the structure − Cumulative uplift force.

Table 10.12 presents the resisting moments and turning moments.

From Table 10.12, $\left(\sum RM - \sum TM \right) = 794099.6 - 598993.5 = 195106.1$

It may be noted that the distance of the line of action from the toe in Col (3) is the distance of the mid-point of the segment from C′.

For example, for segment 1, the distance is $(20 + 40 + 40 + 2.54 = 102.54$ m).

Hence, $Z = \frac{\sum M}{\sum V} = \frac{195106.1}{4503.7} = 43.3$ m.

Base length $d = 122.54$, thus, $Z < (d/2)$; hence, structure is safe against tension.

Also, using Eq. (8.46), eccentricity, $e = \frac{d}{2} - Z = \frac{122.54}{2} - 43.3 = 17.95$ m

Safety of the Structure against Compression or Contact Pressure

We need to check the structure for safety against compression or contact pressure. As described in Sect. 8.7.4.5 (Chap. 8), the contact pressure may be determined using Eq. (8.47) as:

$$P = \frac{\sum V}{A} \left(1 \pm \frac{6e}{d} \right)$$

The base area, $A = (8.6 \times 122.54) = 1053.8$ m^2

Substituting $\sum V = 4503.7$, $e = 17.9$, $d = 122.54$ and A in Eq. (8.47),

$$P = \frac{\sum V}{A}\left(1 \pm \frac{6e}{d}\right) = \frac{4503.7}{1053.8} \times \left(1 \pm \frac{6 \times 17.9}{122.54}\right) = 8.0 \text{ or } 0.5\text{kN} \ \text{m}^{-2}$$

Since both P values are positive, the structure is safe against compression or contact pressure.

Since the structure is safe against various failures, the design dimensions of various components are acceptable.

Questions

1. Draw a chute spillway showing its structural components. Describe the function of various components.
2. Based on the information from the watersheds closer to your home town, managed under the soil management programmes, describe various existing chute spillways.
3. Describe the type of control structure and outlet included in the spillways. Describe the stilling basins that are usually adopted as an energy dissipater in a chute spillway.
4. Describe the factors that affect the coefficient of discharge in the flow equation for an ogee spillway. How is the discharge computation accomplished?
5. Design an overflow ogee crest, including its crest profile, for a chute spillway discharging 25 m^3 s^{-1} at a 2.0 m head. The crest has a vertical upstream face, and the length of the approach channel is 50.0 m. A bridge having square-nosed piers with rounded corners (with a radius equal to 10% of the pier thickness) is provided with the bridge spans limited to 6.0 m. The square abutments have a headwall at 90° to the direction of flow.
6. For the overflow ogee crest designed in Question 5, design a discharge carrier to convey the water from the ogee crest to the outlet. The vertical drop of the discharge carrier from the spillway crest to the outlet floor is 20 m. Assume additional data required to design the discharge carrier.
7. For the overflow ogee crest and the discharge carrier designed in Questions 5 and 6, design a stilling basin to convey the water safely to the downstream channel. Assume additional data required to design the stilling basin.
8. Design the profile of a chute spillway to carry a discharge of 30 m^3 s^{-1}. The spillway has a straight inlet with a flow depth of 1.25 m. The outlet is in the form of a standard USBR stilling basin. The vertical drop of the chute from the inlet crest to the basin floor is 30 m.
9. Design the profile of a chute spillway to carry a discharge of 30 m^3 s^{-1}. The spillway has an ogee spillway as the inlet with a head of 1.5 m and a stilling basin. The vertical drop of the chute from the inlet crest to the outlet floor is 25 m.

10. Carry out the structural design of the discharge carrier and the stilling basin designed in Question 8. The concrete, having a unit weight of 24 kN m^{-3}, is used as a construction material. Take the unit weight of water as 9.8 kN m^{-3}.

Multiple Choice Questions

Answer the questions by choosing the correct option.

1. When drop exceeds 4.0 m, and there is no possibility of upstream flow accumulation, then the most suitable permanent gully control structure, i.e., spillway, is

   (A) straight drop   (B) drop-inlet
   (C) chute   (D) box-inlet drop

2. When the actual head causing flow through an ogee spillway is higher than the design head

   (A) the pressure on the crest will be zero
   (B) the pressure on the crest will be negative
   (C) the pressure on the crest will be positive
   (D) the discharge coefficient of the spillway will be reduced

3. A standard USBR stilling basin II is useful for energy dissipation in a chute spillway when the value of the upstream Froude number is

   (A) < 1.7   (B) 1.7–2.5
   (C) 2.5–4.5   (D) > 4.5

4. In a SAF stilling basin, the height of the end sill is equal to

   (A) 1.7 × (conjugate depth, y_2)   (B) 0.7 × (conjugate depth, y_2)
   (C) 0.07 ≥ (conjugate depth, y_2)   (D) 0.007 × (conjugate depth, y_2)

5. In a chute spillway, the hydraulic jump is formed in the

   (A) approach channel (B) inlet
   (C) chute channel   (D) outlet

6. The downstream profile of the ogee spillway is best represented by the equation (H being the design head, and x and y being the coordinates of the crest profile measured from the apex of the crest):

   (A) $x^{1.85} = -2H^{0.85}y$ (B) $x = -2H^{1.85}y^{0.85}$
   (C) $x^{0.85} = -2H^{1.85}y$ (D) $x = -2H^{0.85}y^{1.85}$

7. While carrying out the structural analysis of a discharge carrier, the cumulative weight of the segments considered for the analysis is 12500 kN. If the cumulative uplift force for the segments considered is 8400 kN, then for the discharge carrier, the factor of safety against overturning is

(A) 0.7   (B) 0.9
(C) 1.5   (D) 4,100

8. The flow depth and the flow velocity at the junction of the discharge carrier and the SAF outlet are 0.25 m and 7.5 m s^{-1}, respectively. The length of the SAF stilling basin, in m, is

(A) 0.47   (B) 1.57
(C) 2.15   (D) 4.79

9. When the bed slope of the discharge carrier changes from milder to steeper, we need to provide a

(A) concave vertical curve   (B) convex vertical curve
(C) reverse curve   (D) converging transition

10. Water emerging from an ogee spillway has a velocity of 13.5 m s^{-1} and a depth of 1.0 m at its toe. The tailwater depth required to form a hydraulic jump at the toe, in m, is

(A) 10.3   (B) 5.6
(C) 2.6   (D) 1.3

References

Bureau of Indian Standards (BIS) (1968) IS 4997: Criteria for design of hydraulic jump type stilling basins with horizontal and sloping apron. Water Resources, Dams and Spillways (WRD 9), New Delhi

Blaisdell FW (1948) Development and hydraulic design, Saint Anthony Falls stilling basin. Trans ASCE 113:483–561

Garg SK (2004) Irrigation engineering and hydraulic structures, 18th edn. Khanna Publishers, Delhi

Natural Resources Conservation Service (NRCS) (2008) National engineering handbook. Section 14: Chute spillways. USDA Natural Resources Conservation Service, Washington DC

Peterka AJ (1984) Hydraulic design of spillways and energy dissipaters. Engineering Monograph 25, Water Resources Technical Publication, US Bureau of Reclamation, Colorado

United States Bureau of Reclamation (USBR) (1987) Design of small dams, 3rd edn. Govt. Printing Office, Washington DC

Chapter 11
Wind Erosion

Abstract Wind erosion is a serious environmental hazard, which causes land degradation and air pollution and adversely affects human health. Dust emission generated by wind erosion is the most prominent aerosol source that directly or indirectly influences the global radiation balance. The chapter presents the factors influencing wind erosion and describes the mechanics of soil particle movement in wind erosion. The Wind Erosion Equation (WEQ), the first empirical wind erosion model for estimating the annual soil loss, and its revised version, the Revised WEQ (RWEQ), are discussed. A few popular process-based wind erosion models are introduced. The basic principles adopted for controlling wind erosion are presented. The chapter also describes the benefits of windbreaks and shelterbelts, two popular mechanical measures of wind erosion control. The design of the windbreaks and shelterbelts is discussed in terms of their height, length, continuity, density, orientation, and number of rows and plant species.

11.1 Background

Wind erosion is the detachment, transportation, and deposition of soil particles due to wind. Strong winds remove the loose, fine soil particles, and organic matter from the soil surface; thus altering the soil structure and damaging the crop productivity. Wind erosion is a serious environmental hazard, which causes air pollution and adversely affects human health. It also affects the dust particle concentration in the atmosphere and influences the climate by altering the global radiation budget.

Wind erosion is a widespread phenomenon causing severe land degradation and desertification over about 28% of the global land area. The arid and semi-arid regions worldwide, with little or no vegetation cover and inadequate precipitation, are susceptible to wind erosion. The major areas of agricultural land subject to wind erosion are in Asia, North Africa, Australia, and parts of North and South America (UNEP WMO UNCCD 2016). In India, Rajasthan, Gujrat, Punjab, and Haryana suffer from wind erosion.

Free (1911) was probably the first person to study wind erosion. He first used the term *saltation* to define the movement of soil particles by bouncing and jumping.

Bagnold (1941) may be credited to initiate basic research on eolian sediment transport. He introduced the term *static threshold* to define the wind speed at which sand particles move due to the direct shear of the wind. He also defined *dynamic threshold* as the wind speed required to sustain an initial movement of sand particles.

11.2 Factors Affecting Wind Erosion

Wind erosion factors include atmospheric conditions, soil characteristics, land surface characteristics, and land use practice.

The atmospheric conditions that affect wind erosion are precipitation, temperature, humidity, and wind speed and direction. Usually, wind speed and soil moisture determine the severity of wind erosion. Since dry soil particles are highly susceptible to movement by wind, wind erosion is more likely to occur in areas having light precipitation and high temperature. Wind velocity at the ground level is a significant factor for initiating soil movement.

Besides, soil moisture, soil texture, structure, and organic matter content affect wind erosion. Usually, fine- or medium-textured loam, like clay loam and silt loam, is more resistant to wind erosion than coarse-textured loams like sandy loams and loamy sands and sands. Soil aggregates, formed due to the combination of individual soil particles, are more resistant to the wind than individual soil particles. Since organic matter holds the aggregates together, soils with higher organic matter content are more resistant to wind erosion.

Land-surface characteristics, including topography, vegetative cover, and surface roughness, also govern wind erosion. Level land is usually more susceptible to wind erosion than rolling land as it offers a slight resistance to the wind. A good vegetative cover, i.e., a growing crop, standing stubble, or mulch, enhances the surface roughness and resists the erosive wind force; thus reducing wind erosion. A 30% ground cover is usually needed to prevent wind erosion.

Land use practices, like farming, grazing, and mining, also impact wind erosion. Tillage practices that maintain a rough land surface, besides maintaining ridges and depression, help reduce the wind speed, thus, the wind erosion. On the contrary, grazing and mining activities expose the land surface and enhance the wind erosion hazard.

11.3 Mechanics of Movement

The soil particles movement in wind erosion has the following three phases:

1. Initiation of movement
2. Transportation
3. Deposition

These phases are discussed in the following sections.

11.3.1 Initiation of Movement

Soil particle movement is initiated due to forces exerted by the wind against the ground surface. The shear forces exerted by the wind induce the soil particle movement by superseding the threshold friction velocity or the gravitational force of the soil particles. Though soil particles or other substances on the soil surface resist the force exerted by the wind, the lighter or loose soil particles are lifted in the initiation process. Depending on the nature of the soil, a certain critical wind velocity is required to initiate the soil particle movement. This critical wind velocity at a given reference height is referred to as the *threshold wind velocity*.

11.3.2 Transportation

After the soil particle movement is initiated, particles are transported from one location to another, based on their size, and wind velocity and direction. The quantity of soil transported depends on the actual and threshold wind velocities and the mean soil particle diameter (Bagnold 1941). The relationship can be expressed as

$$\text{Quantity of soil transported} \propto (V - V_{th})^3 \sqrt{\frac{D}{D_{250}}} \qquad (11.1)$$

where V = actual wind velocity; V_{th} = threshold wind velocity; D = mean soil particle diameter; and D_{250} = particle diameter of a standard 0.25 mm sand.

Soil particles move by saltation, suspension, or surface creep (Fig. 11.1).

11.3.2.1 Saltation

Saltation is the process of jumping and bouncing soil particles over the ground surface and is responsible for the bulk of the soil mass movement. It carries almost 50–75% of the total transported soil mass. In saltation, soil particles detached during the initiation process rise vertically until their kinetic energy is converted to potential energy. They travel a horizontal distance 10–15 times their height of the vertical rise and return to the ground surface. Usually, fine soil particles, 0.10–0.50 mm in diameter, move in saltation, with the vertical rise varying with their size, density, and wind velocity. After returning to the ground surface, the soil particles either rebound or collide with other particles and blow them into the air.

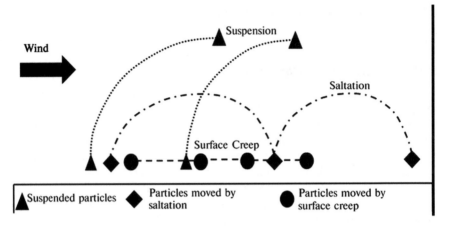

Fig. 11.1 Soil particle movement during wind erosion

11.3.2.2 Suspension

If the soil particles detached during the initiation process are 0.02–0.1 mm in diameter, then wind draws and carries such particles to considerable heights and long distances. This soil movement process is called suspension. Suspension represents the floating of fine soil particles in the air stream and accounts for about 3–10% of the total transported soil mass.

Suspension is often initiated by the impact of soil particles in saltation. It is the most noticeable form of soil movement as it results in blowing dust and dust storms. The suspended soil may travel hundreds of kilometres over an extended period.

11.3.2.3 Surface Creep

The large soil particles, 0.5–2.0 mm in diameter, cannot be lifted into the air due to their weight. Such particles roll or slide along the soil surface, mainly due to the impact of the saltating soil particles. This soil movement process is called surface creep. It accounts for about 5–25% of the total transported soil mass.

11.3.3 Deposition

Soil particles are deposited on the surface when the gravitational forces exceed the forces keeping the soil particle in the air. The deposition is caused primarily due to a reduction in the wind velocity owing to vegetative or other physical barriers. Raindrops may also take out soil particles from the air and lead to deposition.

11.4 Estimation of Soil Loss Due to Wind Erosion

Chepil (1959) proposed a generalised wind erosion equation, similar to the Universal Soil Loss Equation (USLE), for estimating wind erosion. The equation is expressed as

$$E = IRKFBWD \tag{11.2}$$

where E = soil loss due to wind erosion; I = soil cloddiness factor; R = vegetative material factor; K = ridge roughness equivalent factor; F = soil abradability factor; B = wind barrier factor; W = field width factor; and D = wind direction factor.

The generalised wind erosion equation formed the basis for developing the Wind Erosion Equation (WEQ), the first empirical wind erosion model for estimating the annual soil loss (Woodruff and Siddoway 1965). The WEQ and its revised version, the Revised Wind Erosion Equation (RWEQ) (Fryrear et al. 1998), are discussed in the following sections.

11.4.1 Wind Erosion Equation (WEQ)

WEQ predicts the potential annual soil loss from an agricultural field (Woodruff and Siddoway 1965). It considers different factors and their interactions to predict the average wind erosion in mass per unit area per year. WEQ, the most widely used wind erosion model, is represented as

$$E = f(I, K, C, L, V) \tag{11.3}$$

where E = average annual soil loss, Mg ha^{-1} yr^{-1}; I = soil erodibility index, Mg ha^{-1} yr^{-1}; K = soil ridge roughness factor; C = climate factor; L = unsheltered length, m; and V = vegetative cover factor.

The term f shows that WEQ depends on the factors included in the equation. Initially, simple tables and graphs were provided for obtaining the values of different factors (Woodruff and Siddoway 1965). Subsequently, equations were developed for determining the factors.

11.4.1.1 Soil Erodibility Index, I

Soil erodibility index, I, is the potential annual soil loss from a wide, unsheltered field having a bare, smooth, and un-crusted surface. It is a function of the non-erodible soil aggregates (AGG) having > 0.84 mm diameter and is given as

$$I = 525 \times (2.718)^{(-0.04AGG)} \tag{11.4}$$

In Eq. (11.4), I is in Mg ha^{-1} yr^{-1}; and AGG is in percentage.

11.4.1.2 Soil Ridge Roughness Factor, K

The soil ridge roughness factor, K, represents the soil surface roughness due to ridges or slight undulations caused by the primary and secondary tillage implements. Ridges and undulations increase the surface roughness by absorbing wind energy, deflecting the wind and trapping the soil particles being transported. Therefore, the ridge roughness governs the soil ridge roughness factor, K.

The ridge roughness, K_r, is a function of the ridge height and the ridge spacing and can be determined using the following equation (Zingg and Woodruff 1951):

$$K_r = \frac{4h^2}{s} \tag{11.5}$$

where K_r = ridge roughness, mm; h = ridge height, mm; and s = ridge spacing, mm.

Using ridge roughness, K_r, the soil ridge roughness factor, K, can be determined using the following relationship (Woodruff and Siddoway 1965; Schwab et al. 1993):

$$K = 0.34 + \frac{12}{(K_r + 18)} + 6.2 \times 10^{-6} K_r^2 \tag{11.6}$$

For a bare, smooth field, the K_r is 0; thus, the soil ridge roughness factor, K, is 1.0.

Climate Factor, C

The climate factor, C, includes the effect of wind velocity, taken as the mean monthly wind velocity, and soil moisture, considered in terms of the Thornthwaite precipitation effectiveness index, PE (Thornthwaite 1931) on wind erosion.

The climatic factor, C, can be determined using the following relationship (FAO 1979; Skidmore 1986):

$$C = \frac{1}{100} \sum_{i=1}^{i=12} U^3 \left(\frac{ETP_i - P_i}{ETP_i} \right) d \tag{11.7}$$

where U = mean monthly wind velocity at 2-m height, m s^{-1}; P_i = monthly precipitation, mm; ETP_i = monthly evaporation, mm; and d = number of days in the month.

Equation (11.7) is the modified form of the original climate factor (Chepil et al. 1962), which used the Thornthwaite precipitation effectiveness index, PE, as a function of mean monthly precipitation and temperature. However, the original form had the problem of the climatic factor approaching significantly large values in arid conditions. Therefore, FAO (1979) modified the equation and replaced the temperature with monthly evaporation.

11.4.1.3   Unsheltered Length, L

The unsheltered length, L, represents the unsheltered distance along the prevailing wind direction. The unsheltered distance is measured from the sheltered edge of a field to the end of the unsheltered field. Typically, erosion is zero at the upwind field edge for an unsheltered field and increases with distance across the field downwind. If the field is long enough, the soil erosion rate reaches a maximum for the given wind. Alternatively, L is taken as the total distance across the field, measured along the prevailing wind direction, and adjusted to account for the shelter effect of the barriers present.

In a practical implementation of WEQ, L is considered for determining the field length factor, WF. WF can be determined using the following relationship (Skidmore and Williams 1991):

$$WF = E_2 \times \left[1.0 - 0.122 \left(\frac{L}{L_0} \right)^{-0.383} \right] e^{L/L_0} \tag{11.8}$$

where WF = field length factor, Mg ha^{-1} yr^{-1}; and L = unsheltered length, m.

E_2 in Eq. (11.8) represents wind erosion considering soil and surface properties, determined as,

$$E_2 = I K \tag{11.9}$$

where I = soil erodibility index, determined from Eq. (11.4); and K = soil ridge roughness factor, determined from Eq. (11.6).

L_0 in Eq. (11.8) is the maximum field length in m, determined as,

$$L_0 = 1.56 \times 10^6 (E_2)^{-1.26} exp(-0.00156 E_2) \tag{11.10}$$

11.4.1.4   Vegetative Cover Factor, V

The vegetation cover factor, V, represents the combined effect of the type, quantity, and orientation of the vegetative material, expressed as the small grain equivalent (Skidmore 1994). The following equation expresses it:

$$SG = a R_w^b \tag{11.11}$$

where SG = small grain equivalent, kg ha^{-1}; a, b = crop residue coefficients (Lyles and Allison 1981; Skidmore 1994); and R_w = quantity of residue converted to small grain equivalent, kg ha^{-1}.

Typical values of a and b for corn silage are 0.229 and 1.135, whilst for cotton, the recommended a and b are 0.188 and 1.145.

Skidmore (1994) recommended using a weighted small grain equivalent for fields in which more than one crop exists. If two crops are involved, then the weighted small grain equivalent can be determined using the following relationship:

$$SG = a_1^{p_1} a_2^{p_2} (R_{w1} + R_{w2})^{b_1 p_1 + b_2 p_2} \tag{11.12}$$

where p_1 and p_2 are the fractions of individual crop residues.

Using small grain equivalent, SG, the vegetative cover factor, V, Mg ha^{-1}, can be determined using the following relationship:

$$V = 2.533 \times 10^{-4} (SG)^{1.363} \tag{11.13}$$

11.4.1.5   Application of WEQ for Estimating Wind Erosion

Once the factors I, K, C, L, and V of Eq. (11.3) are determined, the average annual soil loss due to wind erosion can be estimated using the following steps (Skidmore 1994; NRCS 2002):

- The initial estimate of wind erosion E_1, in Mg ha^{-1} yr^{-1}, is:

$$E_1 = I \tag{11.14}$$

  where I = soil erodibility index, determined from Eq. (11.4).
- Determine E_2 by including soil ridge roughness factor, K, determined from Eq. (11.6):

$$E_2 = I \, K \tag{11.15}$$

- Determine E_3 by including climate factor, C, determined from Eq. (11.7):

$$E_3 = I \, K \, C \tag{11.16}$$

- Determine the field length factor, WF, using Eq. (11.8).
- Determine the interaction of soil, surface, climate, and field length effects using the following relationship:

$$E_4 = \left(W F^{0.348} + E_3^{0.348} - E_2^{0.348} \right)^{2.87} \tag{11.17}$$

- Determine the vegetation factors, V_1 and V_2, which represent the vegetation effect on erosion using the following relationships:

$$V_1 = e^{-0.759V - 4.74 \times 10^{-2} V^2 + 2.95 \times 10^{-4} V^3}$$ (11.18)

$$V_2 = 1.0 + 8.93 \times 10^{-2} V + 8.51 \times 10^{-3} V^2 - 1.5 \times 10^{-5} V^3$$ (11.19)

Equations (11.18) and (11.19) include the vegetation cover factor, V, determined using Eq. (11.13).

- Determine the final value of soil loss due to wind erosion by incorporating the vegetation factors using the following relationship:

$$E_5 = V_1 E_4^{V_2}$$ (11.20)

where $E_5 =$ average annual soil loss due to wind erosion, Mg ha^{-1} yr^{-1}.

Example 11.1 Estimate the annual soil loss due to wind erosion on a 600 m long, recently tilled field. The field has several 100 mm high ridges, and the spacing between ridges is 750 mm. The field has 20% non-erodible clods (>0.84 mm) and 1.5 Mg ha^{-1} of corn silage residue. The C factor for the region is 0.80.

Solution

Given: Non-erodible clods or soil aggregates (>0.84 mm), $AGG = 20\%$; ridge height, $h = 100$ mm; ridge spacing, $s = 750$ mm; unsheltered field length, $L = 600$ m; quantity of residue, $R_w = 1500$ kg ha^{-1}; climate factor, $C = 0.8$.

The steps in Sect. 11.4.1.6 are followed to estimate the annual soil loss due to wind erosion.

1. Soil erodibility index using Eq. (11.4),

$$I = 525 \times (2.718)^{(-0.04 AGG)} = 525 \times (2.718)^{(-0.04 \times 20)} = 235.9 \, \text{Mg ha}^{-1} \, \text{yr}^{-1}$$

Thus, the initial estimate of wind erosion,

$$E_1 = 235.9 \, \text{Mg ha}^{-1} \, \text{yr}^{-1}$$

The ridge roughness using Eq. (11.5),

$$K_r = \frac{4h^2}{s} = \frac{4 \times (100)^2}{750} = 53.3 \, \text{mm}$$

The soil ridge roughness factor using Eq. (11.6),

$$K = 0.34 + \frac{12}{(K_r + 18)} + 6.2 \times 10^{-6} K_r^2$$

$$= 0.34 + \frac{12}{(53.3 + 12)} + 6.2 \times 10^{-6} \times (53.3)^2 = 0.53$$

Thus, $E_2 = I\,K = 235.9 \times 0.53 = 125 \text{ Mg ha}^{-1} \text{ yr}^{-1}$

2. Substituting climate factor, $C = 0.8$, and earlier determined I and K in Eq. (11.16),

$$E_3 = I\,K\,C = 235.8 \times 0.53 \times 0.8 = 100 \text{ Mg ha}^{-1} \text{ yr}^{-1}$$

Thus, the maximum field length using Eq. (11.10),

$$
\begin{aligned}
L_0 &= 1.56 \times 10^6 (E_2)^{-1.26} exp(-0.00156 E_2) \\
&= 1.56 \times 10^6 (125)^{-1.26} exp(-0.00156 \times 125) \\
&= 2926 \text{ m}
\end{aligned}
$$

Thus, the field length factor, WF, using Eq. (11.8),

$$
\begin{aligned}
WF &= E_2 \times \left[1.0 - 0.122 \left(\frac{L}{L_0} \right)^{-0.383} \right] e^{L/L_0} \\
&= 125 \times \left[1.0 - 0.122 \left(\frac{600}{2926} \right)^{-0.383} \right] e^{600/2926} = 119
\end{aligned}
$$

3. Substituting WF, E_3 and E_2 in Eq. (11.17),

$$
\begin{aligned}
E_4 &= \left(WF^{0.348} + E_3^{0.348} - E_2^{0.348} \right)^{2.87} \\
&= \left((119)^{0.348} + (100)^{0.348} - (125)^{0.348} \right)^{2.87} \\
&= 94 \text{ Mg ha}^{-1} \text{ yr}^{-1}
\end{aligned}
$$

4. The crop residue coefficients for corn silage are $a = 0.229$ and $b = 1.135$. Thus, small grain equivalent using Eq. (11.11),

$$SG = a R_w^b = 0.229 \times (1500)^{1.135} = 922 \text{ kg ha}^{-1}$$

Hence, vegetative cover factor using Eq. (11.13),

$$V = 2.533 \times 10^{-4} (SG)^{1.363} = 2.533 \times 10^{-4} (922)^{1.363} = 2.78$$

Thus, the vegetation factors using Eqs. (11.18) and (11.19),

$$
\begin{aligned}
V_1 &= e^{-0.759V - 4.74 \times 10^{-2} V^2 + 2.95 \times 10^{-4} V^3} \\
&= e^{-0.759(2.78) - 4.74 \times 10^{-2} (2.78)^2 + 2.95 \times 10^{-4} (2.78)^3}
\end{aligned}
$$

$$= 0.08$$

$$V_2 = 1.0 + 8.93 \times 10^{-2} V + 8.51 \times 10^{-3} V^2 - 1.5 \times 10^{-5} V^3$$
$$= 1.0 + 8.93 \times 10^{-2}(2.78) + 8.51 \times 10^{-3}(2.78)^2 - 1.5 \times 10^{-5}(2.78)^3$$
$$= 1.31$$

5. The annual soil loss using (11.20),

$$E_5 = V_1 E_4^{V_2} = 0.08 \times (94)^{1.31} = 30.8 \, \text{Mg ha}^{-1} \, \text{yr}^{-1}$$

Answer: The annual soil loss due to wind erosion is 30.8 Mg ha^{-1} yr^{-1}.

11.4.1.6 Limitations of WEQ

Despite being empirical, WEQ provides valuable wind erosion estimates. However, it has several limitations, like it cannot predict soil erosion rates for an isolated storm or at a daily time step. WEQ only considers average wind and precipitation conditions and cannot handle variations from the average. Also, it performs poorly for locations with extreme rainfall patterns and extremely large or narrow fields. Therefore, WEQ was revised and refined continuously, which led to the development of the RWEQ (Fryrear et al. 1998). The details of the RWEQ are presented in the following section.

11.4.2 Revised WEQ (RWEQ)

RWEQ model describes wind erosion processes using a combination of process-based and empirical components. RWEQ predicts soil loss as a function of the wind capacity to transport soil, independent of the soil type. It estimates horizontal mass transport by wind within the first 2-m height for a specified period in a single event. The following relationship expresses the model in the sigmoidal form (Fryrear et al. 1998; Jarrah et al. 2020):

$$Q_x = Q_{\max} \left[1 - e^{\left(\frac{x}{s}\right)^2} \right] \tag{11.21}$$

where Q_x = maximum amount of soil transported at downwind field length x, kg m^{-1}; $Q_{\max}$ = maximum transport capacity, kg m^{-1}; x = total field length, m; and s = critical field length at which 63% of $Q_{\max}$ is reached.

The $Q_{\max}$ and s are estimated as

$$Q_{\max} = 109.85(wf \times K \times EF \times SCF \times COG) \tag{11.22}$$

$$s = 150.71((wf \times K \times EF \times SCF \times COG))^{-0.3711} \tag{11.23}$$

where wf = weather factor; K = soil roughness factor; EF = erodible fraction of the soil; SCF = soil crust factor; and COG = crops on ground factor.

The weather factor, wf, kg m^{-1}, can be estimated as

$$wf = W_f \frac{\rho}{g}(SW)SD \tag{11.24}$$

where W_f = wind factor; ρ = air density, kg m^{-3}; SW = soil wetness factor; and SD = snow cover factor.

W_f and SW are estimated as

$$W_f = \frac{W}{N}N_d \tag{11.25}$$

$$SW = \frac{ET_p - (R + I_r)\frac{R_d}{N_d}}{ET_p} \tag{11.26}$$

where W = wind value, m^3 s^{-3}; N = number of wind speeds used (minimum of 500); N_d = number of days in the study period; ET_p = potential relative evapotranspiration, mm day^{-1}; R = amount of rainfall, mm; I_r = cumulative irrigation, mm; and R_d = the number of rainfall and irrigation days.

W and ET_p are estimated as

$$W = \sum_{i=1}^{i-n} U_2(U_2 - U_t)^2 \tag{11.27}$$

$$ET_p = 0.0162 \left(\frac{SR}{58.5}\right)(DT + 17.8) \tag{11.28}$$

where U_2 = wind velocity at 2 m, m s^{-1}; U_t = threshold wind velocity at 2 m equal to 5 m s^{-1}; SR = total solar radiation for the study period, J m^{-2} day^{-1}; and DT = average temperature, °C.

Soil roughness factor, K, is estimated as

$$K = e^{\left(1.86K_r R_c - 2.41 K_r R_c^{0.934} - 0.124 C_{rr}\right)} \tag{11.29}$$

where R_c = wind direction correction factor; and C_{rr} = random roughness.

R_c and C_{rr} are estimated as

$$R_c = 1 - 3.2 \times 10^{-4}(A) - 3.49 \times 10^{-4}(A^2) + 2.58 \times 10^{-6}(A^3) \tag{11.30}$$

$$C_{rr} = 17.46RR^{0.738} \tag{11.31}$$

where A = wind angle (taken as $0°$ for perpendicular and $90°$ for parallel) and RR = random roughness index.

Erodible fraction of the soil, EF, is estimated as

$$EF = \frac{29.09 + 0.31S_a + 0.17S_i + 0.33\frac{S_a}{Cl} - 2.59SOM - 0.95CaCO_3}{100} \tag{11.32}$$

where S_a = soil sand content, %; S_i = soil silt content, %); S_a/Cl = ratio of sand to clay contents; SOM = soil organic matter content, %; and $CaCO_3$ = calcium carbonate content, %.

Surface crust factor, SCF, is estimated as

$$SCF = \frac{1}{\left[1 + 0.0066(Cl)^2 + 0.21(SOM)^2\right]} \tag{11.33}$$

The vegetation/residue crops on ground factor, COG, represent the effect of crop residue, plant orientation, and crop canopy and is estimated as

$$COG = e^{(-0.0438(SC))} + e^{(-0.0344(SA))} + e^{\left(-5.614\left(CC^{0.7366}\right)\right)} \tag{11.34}$$

where SC = land surface covered by crop residue, %; SA = standing stem area index, $m^2\ m^{-2}$; and CC = land surface covered by crop canopy, %.

The average soil loss, SL, $kg\ m^{-2}$, is estimated as

$$SL = \frac{Q_{max}}{x} \tag{11.35}$$

RWEQ shows much improvement in soil loss prediction compared to WEQ, but it still has many limitations. For example, it cannot consider changes in weather and management to simulate variations in soil roughness and freezing/thawing on a daily basis. Also, it cannot consider within-field variations, fields with eroding boundaries, and transport of suspended soil particles. Therefore, several process-based models have been developed in the recent past. The following sections present a brief description of a few process-based models.

11.4.3   Process-Based Models for Wind Erosion

11.4.3.1   Wind Erosion Prediction System (WEPS)

The wind erosion prediction system (WEPS) model was developed in the 1980s to overcome the limitations of the WEQ and RWEQ (Hagen 1991; Tatarko et al. 2016).

It is a process-based, continuous, and daily time-step model, which can simulate total wind erosion, including saltation, suspension, and surface creep processes and estimate air pollution with dust emissions (PM-10, i.e., particulate matter <10 μm in diameter). The model includes relationships between surface conditions and erosion, developed based on field and laboratory experiments.

The model helps plan soil conservation systems for controlling wind erosion from agricultural fields. The limitation of WEPS lies in the detailed input data requirement. Also, WEPS under-predicts wind erosion for both small and large storms.

11.4.3.2   Wind Erosion Stochastic Simulator (WESS)

The wind erosion stochastic simulator (WESS) is an event-based model for simulating wind erosion. It is the wind erosion module of erosion productivity impact calculator (EPIC) (Potter et al. 1998; Van Pelt et al. 2004). It uses soil aggregate size distribution, soil roughness, soil texture, soil water content, the quantity of crop residue, 10-min average wind speeds, and field size as input.

11.4.3.3   Texas Erosion Analysis Model (TEAM)

The Texas erosion analysis model (TEAM) is also a process-based model that can simulate the suspension and saltation movement of soil particles (Gregory et al. 2004). It determines the soil detachment and transport using soil roughness, particle size distribution, relative humidity, and wind velocity and distribution. The outputs can be used to design wind erosion controls, especially for stabilising moving dunes in desert regions. However, TEAM cannot adjust soil erodibility due to rainfall and irrigation.

11.4.3.4   Dust Production Model (DPM)

The dust production model (DPM) combines a sandblasting model with a saltation model to determine aerosols released under the prevailing soil and wind conditions (Alfaro and Gomes 2001). It determines the release of <20 μm dust particles (PM20), along with their particle size distribution. The model is primarily developed for agricultural lands in Spain and Niger. The wind friction velocity, soil roughness length, and dry size distribution of the aggregates constituting the soil erodible fraction are the main input parameters.

11.4.3.5   Wind Erosion on European Light Soils (WEELS)

Wind erosion on European Light Soils (WEELS) is an integrated model jointly developed by researchers from Germany, the Netherlands, Sweden, and the United

Kingdom (Böhner et al. 2003). WEELS can determine erosion rates at different spatial and temporal scales. The model comprises two groups of modules. The first group accounts for temporal variations of climatic erosivity using wind, wind erosivity, and soil moisture modules. In contrast, the second group accounts for erodibility through soil erodibility, surface roughness, and land use models. The WEELS has limitations of simulating only the saltation process for fine-textured soils. Therefore, it cannot evaluate air quality as particles in suspension are not simulated.

11.4.3.6 Aeolian Erosion (AERO) Model

The aeolian erosion (AERO) model is a wind erosion decision-support tool that simulates wind erosion and dust emissions (Edwards et al. 2018). The model uses meteorological, soil, and vegetation data to simulate horizontal and vertical mass flux on a plot scale. The generalised formulation of AERO makes it applicable across different land cover settings.

11.5 Wind Erosion Control

For effective wind erosion control, the following basic principles may be adopted (Tibke 1988):

1. Reduce the field width along the prevailing wind direction
2. Establish and maintain vegetative cover on the surface
3. Maintain stable aggregates or clods on the surface
4. Roughen the land surface.

11.5.1 Reduce the Field Width Along the Prevailing Wind Direction

The field width along the prevailing wind direction may be reduced by planting or erecting barriers, such as windbreaks and shelterbelts, grass barriers, artificial barriers, or adopting agronomical measures, such as strip cropping. The barriers lead to wind erosion control by reducing wind speed and field length along the wind direction. The impact of barriers on erosion control depends on the wind characteristics like speed and direction, the threshold wind speed needed to initiate the soil movement, and the barrier characteristics like height, width, shape, and porosity.

Windbreaks and Shelterbelts

Windbreaks and shelterbelts are popular mechanical measures of wind erosion control. The details of windbreaks and shelterbelts are presented in Sect. 11.6.

Grass Barriers

Annual and perennial grass barriers help control wind erosion. Perennial barriers are particularly recommended for sandy soils where mulching and strip cropping are inadequate in wind erosion control (Tibke 1988). These barriers also trap snow and reduce evaporation. A low cost and ease of establishment are the advantages of such barriers.

Artificial Barriers

Artificial barriers such as earthen embankments, stone walls, and bamboo fences are also adopted to a limited extent for wind erosion control. These barriers are used only if natural barriers cannot be established, say in dune areas. However, since the cost of material and construction is high, these are used only for high-value crops.

Strip Cropping

As discussed in Sect. 2.5.2.1 (Chap. 2), strip cropping consists of dividing a large field into narrow strips and growing crops of varying root and shoot characteristics in alternate strips. Strip cropping controls erosion effectively by reducing the surface exposure to winds. Strips are usually oriented perpendicular to the prevailing wind direction. The design of strip cropping systems involves deciding the strip widths and crops to be grown. The design depends on the prevailing wind speed and direction, soil texture, farm equipment width, and other control practices used in conjunction with the strip cropping.

11.5.2 Establish and Maintain Vegetative Cover on the Surface

Keeping the soil surface covered using crop residues is an effective means of wind erosion control. Therefore, farming practices like zero-till or stubble-mulch that preserve surface residues are practical approaches to control wind erosion.

No-till, also known as zero-till, is a cultivation practice in which a crop is planted directly into a seedbed without any tillage operation. A unique planter or drill is used for planting the seed into the soil with minimum disturbance. No additional tillage occurs, and herbicides are used for weed control. Crop residues are distributed evenly and left on the soil surface. No-till practice reduces erosion by about 90%.

Stubble mulch tillage is a cultivation practice in which plant residues are maintained on the soil surface throughout the year. Various agricultural operations like seedbed preparation, planting, cultivating, and harvesting are performed to leave previous crop residues in the field. Residues alter the wind velocity close to the soil surface and absorb wind energy. Stubble mulch tillage practice reduces erosion by about 60–99%.

11.5.3 Maintain Stable Aggregates or Clods on the Surface

Consolidation of soil particles, both as aggregates or clods on the surface, holds them together and prevents erosion. A soil surface with about 70% clods >0.84 mm in diameter considerably reduces erosion. The soil consolidation and stability depend on clay content, organic matter content, soil moisture, and microorganism activities. Loams, silt loams, and clay loams are best suited for developing and maintaining stable aggregates. Tillage operations may alter clods at the surface. Therefore, care should be taken whilst selecting the tillage implement and the speed of operation for ensuring the maintenance of clods on the surface.

11.5.4 Roughen the Land Surface

Roughening the soil surface by tillage operations that create ridges and depressions help reduce wind erosion. The soil ridges with clods reduce the wind velocity close to the soil surface, whilst the depressions trap soil particles in saltation. However, since the soil ridges protrude from the soil surface, they are subjected to larger wind forces. Therefore, clods must be maintained on top of the ridge; else the ridges will erode and lose their beneficial effects.

11.6 Windbreaks and Shelterbelts

Windbreaks and shelterbelts are popular mechanical measures of wind erosion control. Windbreaks and shelterbelts act as environmental buffers and include strips of trees, shrubs, or grasses planted to protect agricultural fields, buildings, farmsteads, and other areas from the wind.

Whilst windbreaks may include stone walls, plank, and brush fences or cloth screens, in addition to the vegetative measures, shelterbelts typically comprise living trees and shrubs. Windbreaks are constructed by one row or maximum up to two rows of protective fences across the prevailing wind direction. Shelterbelts are usually longer barriers than windbreaks and may have more than two rows of trees and shrubs. Shelterbelts are preferred for protecting agricultural fields.

Windbreaks and shelterbelts block and redirect wind and reduce the mechanical action of wind on soil, crops, and pasture. Consequently, windbreaks and shelterbelts reduce wind erosion, protect growing crops, manage snow, and regulate climate conditions. Windbreaks may also be used to protect structures and livestock, improve aesthetics, and as a living fence to demarcate the boundaries of a property.

The first reported application of windbreaks in agriculture was in the mid-1400s in Scotland (Droze 1977). Since then, windbreaks and shelterbelts have been used extensively in semi-arid tropics and semi-arid temperate regions of Africa, Asia,

Australia, Europe, and North America for protecting soils and crops against wind and wind erosion. Shelterbelts have also been traditionally used along coasts, for example, along the Bay of Bengal coast of India and Bangladesh.

11.6.1   Benefits of Windbreaks and Shelterbelts

Windbreaks and shelterbelts are valuable conservation control measures and serve multiple purposes. Their notable benefits are as follows (van Eimern et al. 1964; Kort 1988):

Reduced Wind Erosion

Windbreaks and shelterbelts typically act as a barrier to wind movement as these are designed and established to reduce wind velocity and turbulence. The decrease in the wind velocity reduces erosion significantly. For example, a 50% reduction in wind speed may reduce erosion by about 80% as erosion is directly proportional to the cube of wind speed (Eq. (11.1)). Wind erosion reduces productivity due to the removal of top fertile soil and leads to desertification. Field shelterbelts have been found effective in controlling erosion, even to the extent of 50%, leading to long-term soil retention.

Improved Microclimatic Conditions

Windbreaks and shelterbelts modify the microclimatic conditions creating favourable cropping conditions. Besides reducing wind speed, windbreaks and shelterbelts result in lower evapotranspiration, higher soil moisture, higher daytime temperatures, lower night-time temperatures, and higher night-time carbon dioxide levels. These positive impacts in sheltered areas lead to increased crop yield. Also, the reduced sand and soil abrasion may improve crop quality.

Snow Retention

In regions where snowfall contributes significantly to precipitation, snow entrapment and retention by windbreaks and shelterbelts enhance the crop yield. The trapped snow also acts as an insulating layer and reduces crop damage by winds. Field shelterbelts also reduce lodging due to snow and rain and prevent a reduction in crop yields. Windbreaks may also serve as snow fences and control snow drifts near roads, buildings, or livestock farms.

Reduced Wind Damages

Wind may cause damage to field crops in several ways, such as seed removal from soils, uprooting of emerging crops, and sandblast injury, i.e., abrasion of plant tissues by soil particles, to plants. Windbreaks and shelterbelts reduce wind damages by reducing wind velocities and redirecting winds. They also trap wind-borne soil to help restore marginal and degraded lands.

Energy Conservation

Windbreaks lead to energy savings in residential areas and farmsteads. They check the cold air entry into buildings and reduce winter heating costs by 20–40%. Similarly, in summer, windbreaks provide a cool surrounding, thus reducing the cost of air-conditioning.

Though windbreaks and shelterbelts offer multiple benefits, they also have a few drawbacks. For example, they occupy land and reduce the cropped area. Also, shelterbelts being live barriers, compete with crops for soil moisture and nutrients.

11.6.2   Design of Windbreaks and Shelterbelts

The design of windbreaks and shelterbelt includes critical factors like height, length, continuity, density, orientation, and number of rows and species selection. The following sections include a description of the critical factor.

11.6.2.1   Height

The height of the windbreak or shelterbelt, H, is the most vital factor governing the distance downwind protected by them. The height varies from one wind-break/shelterbelt to the other and increases as the trees included in the measure mature. In multiple-row windbreaks/shelterbelts, the average height of the tallest tree row determines the height.

Windbreaks/shelterbelts reduce wind speed over a distance of two to five times their height upwind and 10 to 20 times their height downwind (van Eimern et al. 1964) (Fig. 11.2). For example, if the height of a field shelterbelt is 5 m, then the wind speeds would reduce over 10 m–25 m on the windward side and 50–100 m on the leeward side.

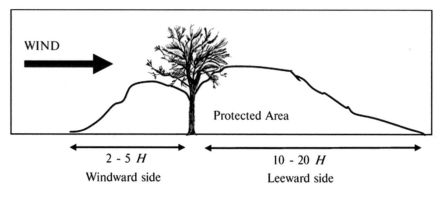

Fig. 11.2 Distance of wind speed reduction due to a windbreak or shelterbelt

The distance of full protection due to a windbreak or shelterbelt is given by the following relationship (Woodruff and Zingg 1952):

$$d = 17H\left(\frac{v_m}{v}\right)\cos\theta \qquad (11.36)$$

where d = distance of full protection, m; H = height of the barrier, m; v_m = minimum wind velocity at the height of 15 m required to move the most erodible soil fraction, km h^{-1}; v = actual wind velocity at 15 m height above the ground surface, km h^{-1}; and θ = angle of deviation of the prevailing wind from a line perpendicular to the windbreak or shelterbelt.

As evident from Eq. (11.36), the maximum (theoretical) value of d would be 17 H, when v equals v_m, and θ is zero. However, due to variable wind direction and poor maintenance of the trees, the actual protection seldom achieves the theoretical level of 17 H. Therefore, the realistic distance of full protection is maybe 10 or 12 times the height of the windbreak or shelterbelt. The distance of full protection from Eq. (11.36) contradicts the upper limit (20 H) of protection downwind (Fig. 11.2). For a dry, smooth, bare surface, v_m may be taken as 34 km h^{-1}.

Example 11.2 Determine the distance of full protection for 15 m high windbreaks. The 5-year return period wind velocity at 15 m height from the ground surface is 52 km h^{-1}. The angle of deviation of the prevailing wind from a line perpendicular to the windbreak is 15°. Assume that the soil surface in the field is smooth, dry, and bare.

Solution

Given: Height of the windbreak, H = 15 m; wind velocity at 15 m height from the ground surface, v = 52 km h^{-1}; angle of deviation of the prevailing wind from a line perpendicular to the windbreak, θ = 15°.

Assuming v_m = 34 km h^{-1}, and substituting known values H, v, v_m, and θ in Eq. (11.36),

$$d = 17H\left(\frac{v_m}{v}\right)\cos\theta = 17 \times 15 \times \left(\frac{34}{52}\right) \times \cos 15 = 161 \text{ m}$$

Answer: The distance of full protection for the given windbreak is 161 m.

11.6.2.2   Length

The length of the windbreak or shelterbelt governs the overall area under protection. The length of the windbreaks or shelterbelts should usually be more than the length of the field and at least ten times the height of the windbreaks or shelterbelts. The length-height ratio of >10:1 is recommended to overcome the *end effect*. Since wind tends to flow around the end of the barrier, it causes the end effect, which results in a 10–30% increase in the wind speeds around the end of the barrier due to higher turbulence.

11.6.2.3 Continuity

The continuity of windbreaks or shelterbelts influences the efficiency. Gaps in the windbreaks or shelterbelts reduce their effectiveness as gaps introduce a wind tunnel effect, due to which the wind speeds exceed the open field wind speeds. Lanes or field access through the windbreak or shelterbelt should be avoided and should ideally be placed at the end of the barrier.

11.6.2.4 Density

The density of windbreaks or shelterbelts is the ratio of the solid portion to the total area. It governs the amount of air passing through the barrier. Windbreaks or shelterbelts, having <20% density, are ineffective in reducing the wind speed, whilst those having >80% density face the problems of reduced effectiveness. In dense windbreaks or shelterbelts, the wind is deflected over the structure, but low pressure on the downwind side draws the wind back down and reduces the protected area. Therefore, 20–80% density is recommended based on the function. For example, a barrier density between 60 and 80% is favourable for protecting farmsteads, residential sites or feedlots, whilst density between 40 and 60% is beneficial for large areas like agricultural fields.

The windbreak or shelterbelt density depends on the number of rows, spacing between rows, tree species, and distance between trees in a row.

11.6.2.5 Orientation

Windbreaks or shelterbelts are most effective when the orientation is perpendicular to the prevailing wind direction. Since wind direction varies considerably, windbreak or shelterbelt may be single-leg (protecting against the wind from a single direction) or multiple-leg (protecting against the wind from multiple directions). The design objective of an individual barrier also governs the orientation. For example, field crops usually need to be protected against dry and hot summer winds; thus, windbreaks or shelterbelts should be planted on the south or west side. On the contrary, for the farmsteads and feedlots that need to be protected against the cold, winter winds, multiple-leg windbreaks, or shelterbelts should be planted on the north and west sides.

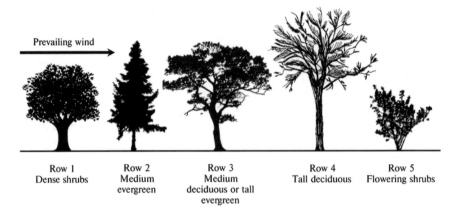

Fig. 11.3 Multiple-row shelterbelt having diverse species of trees and shrubs

11.6.2.6 Number of Rows and Plant Species

The number of rows in windbreaks or shelterbelts depends on the space availability and design objectives. When space is limited, a single-row windbreak or shelter-belt consisting of coniferous trees for year-round protection, and densely-branched deciduous trees or shrubs may be used. Alternatively, a twin-row barrier having two coniferous rows with a tree in one row filling a gap in the next row may be used.

When space is not a constraint, typical multiple-row shelterbelts are preferred. These may include dense coniferous trees to block and reduce the wind speed, tall broadleaf coniferous, or deciduous trees for the longer distance of protection and low shrubs to catch snow and provide wildlife habitat (Fig. 11.3). As evident, a combination of different species of trees and shrubs is used in multiple-row shelter-belts. Besides serving usual purposes, species diversity reduces the risk of diseases or insects and promotes bio-diversity. However, the choice of trees and shrubs is location-specific and depends on the local climate and soil conditions.

Questions

1. Describe the mechanics of soil particle movement during the wind erosion process.
2. Describe the factors affecting wind erosion.
3. Describe various measures that could be adopted for controlling wind erosion.
4. Describe the multiple benefits derived from windbreaks and shelterbelts.
5. Describe the Wind Erosion Equation, WEQ. What are the shortcomings of the model?
6. Describe the Revised Wind Erosion Equation, RWEQ. What are its advantages over the WEQ model?

7. 10 m high windbreaks have been established to provide full protection over a distance of 75 m. The minimum wind velocity at the height of 15 m required to move the most erodible soil fraction is 9.6 m s^{-1}. Determine the maximum wind velocity at 15 m height if the angle of deviation of wind direction from the perpendicular to the windbreak is 20°.

8. In a field with a dry, smooth, bare soil surface, 10 m high shelterbelts are established. The minimum wind velocity at the height of 15 m required to move the most erodible soil fraction is 10 m s^{-1}. The angle of deviation of the prevailing wind from a line perpendicular to the proposed shelterbelt is 15°. The wind velocity at 15 m height from the ground surface is 50 km h^{-1}. Determine the spacing between the shelterbelts.

9. Determine the soil ridge roughness factor if the ridge height is 70 mm, and the spacing between ridges is 200 mm.

10. Estimate the annual soil loss due to wind erosion on an 800 m long, recently tilled field. The field has several 80 mm high ridges, and the spacing between ridges is 500 mm. The field has 25% non-erodible clods (>0.84 mm) and 1.0 Mg ha^{-1} of sunflower residue. Assume C factor as 0.80.

Multiple Choice Questions

Answer the questions by choosing the correct option.

1. In wind erosion, the process in which the soil particles less than 0.1 mm in diameter move is called

   (A) suspension
   (B) saltation
   (C) deflation
   (D) surface creep

2. In wind erosion, the amount of soil transported varies

   (A) directly as the cube of the threshold wind velocity and inversely as the square of the mean soil particle diameter
   (B) directly as the cube of the actual wind velocity and inversely as the square of the mean soil particle diameter
   (C) directly as the cube of the difference in actual and threshold wind velocity and the square root of the mean soil particle diameter
   (D) directly as the cube of the difference in actual and threshold wind velocity and the square of the mean soil particle diameter

3. In wind erosion, soil erodibility is a function of the non-erodible soil aggregates (*AGG*) having a diameter greater than

   (A) 0.10 mm
   (B) 0.64 mm
   (C) 0.84 mm
   (D) 1.0 mm

4. The most noticeable form of soil movement in wind erosion is

    (A) suspension
    (B) saltation
    (C) surface creep
    (D) rolling and sliding

5. During wind erosion, the process in which the soil particles move by jumping and bouncing is called

    (A) suspension
    (B) saltation
    (C) deflation
    (D) surface creep

6. In WEQ, the most widely used wind erosion model, the average annual soil loss is considered as a function of

    (A) soil erodibility, soil ridge roughness, rainfall erosivity, unsheltered length, and vegetative cover
    (B) soil erosivity, soil ridge roughness, rainfall erodibility, unsheltered length, and vegetative cover
    (C) soil erodibility, soil ridge roughness, climate factor, sheltered length, and vegetative cover
    (D) soil erodibility, soil ridge roughness, climate factor, unsheltered length, and vegetative cover

7. Out of the following wind erosion prediction models, the one which is not process-based is

    (A) DPM
    (B) RWEQ
    (C) WEELS
    (D) WEPS

8. The ridge roughness, K_r, a function of the ridge height, h, and the ridge spacing, s, is expressed as

    (A) $\frac{h}{s}$
    (B) $\frac{h^2}{s}$
    (C) $\frac{(2h)^2}{s}$
    (D) $\frac{(2h)^2}{s^2}$

9. Shelterbelts are established by planting trees and shrubs in

    (A) one row
    (B) two rows
    (C) more than two rows
    (D) a row of trees and shrubs each

10. In a field having a dry, smooth, bare soil surface, 10 m high windbreaks are established. The minimum wind velocity at the height of 15 m required to move the most erodible soil fraction is 9.6 m s^{-1}. The direction of the prevailing wind is perpendicular to the windbreak, and its velocity at 15 m height from the ground surface is 52 km h^{-1}. The distance of full protection by the windbreaks, in m, is

(A) 7
(B) 31
(C) 113
(D) 256

References

Alfaro SC, Gomes L (2001) Modeling mineral aerosol production by wind erosion: emission intensities and aerosol size distributions in source areas. J Geophys Res 106:18075–18084

Bagnold RA (1941) The physics of blown sand and desert dunes. Methuen and Co., Ltd., London

Böhner J, Schäfer W, Conrad O, Gross J, Ringeler A (2003) The WEELS model: methods, results and limitations. CATENA 52:289–308

Chepil W (1959) Wind erodibility of farm fields. J Soil Water Conserv 14:214–219

Chepil W, Siddoway FH, Armbrust DV (1962) Climatic factor for estimating wind erodibility of farm fields. J Soil Water Conserv 17:162–165

Droze WH (1977) Trees, prairies, and people: a history of tree planting in the plains states. USDA Forest Service and Texas Women's University Press, Denton, TX, p 313

Edwards BL, Webb NP, McCord SE (2018) AERO: a wind erosion modeling framework with applications to monitoring data. In: 73rd Soil and Water Conservation Society Annual Conference, Albuquerque, USA

FAO (1979) A provisional methodology for soil degradation assessment. Food and Agriculture Organization of the United Nations, Rome

Free EE (1911) The movement of soil material by the wind. US Department of Agriculture, Bureau of Soils, Bulletin no. 68

Fryrear D, Saleh A, Bilbro JD, Schomberg H, Stout JE, Zobeck TM (1998) Revised Wind Erosion Equation (RWEQ). Technical Bulletin 1. Wind Erosion and Water Conservation Research Unit, USDA-ARS, Southern Plains Area Cropping Systems Research Laboratory, Texas

Gregory JM, Wilson GR, Singh UB, Darwish MM (2004) TEAM: integrated, process-based wind-erosion model. Environ Model Softw 19:205–215

Hagen L (1991) A wind erosion prediction system to meet user needs. J Soil Water Conserv 46:106–111

Jarrah M, Mayel S, Tatarko J, Funk R, Kuka K (2020) A review of wind erosion models: data requirements, processes, and validity. CATENA 187:104388

Kort J (1988) Benefits of windbreaks to field and forage crops. Agric Ecosyst Environ 22(23):165–190

Lyles L, Allison BE (1981) Equivalent wind-erosion protection from selected crop residues. Trans ASAE 24:405–408

NRCS (Natural Resources Conservation Service) (2002) National agronomy manual. US Department of Agriculture, Washington DC

Potter K, Williams J, Larney F, Bullock M (1998) Evaluation of EPIC's wind erosion submodel using data from southern Alberta. Canadian J Soil Sci 78:485–492

Schwab GO, Frevert RK, Edminster TW, Barnes KK (1993) Soil and water conservation engineering, 4th edn. Wiley, New York

Skidmore EL (1986) Wind erosion climatic erosivity. Clim Change 9:195–208

Skidmore EL, Williams J (1991) Modified EPIC wind erosion model. In: Hanks J, Ritchie (eds) Modeling plant and soil systems. https://doi.org/10.2134/agronmonogr31.c19

Skidmore EL (1994) Wind erosion. In: Lal R (ed) Soil erosion research methods soil and water conservation society, Ankeny, Iowa

Tatarko J, van Donk S, Ascough J, Walker D (2016) Application of the WEPS and SWEEP models to non-agricultural disturbed lands. Heliyon 2:e00215

Thornthwaite CW (1931) The climates of North America according to a new classification. Geography Rev 21:633–655

Tibke G (1988) Basic principles of wind erosion control. Agric Ecosyst Environ 22(23):103–122

UNEP WMO UNCCD (2016) Global assessment of sand and dust storms. United Nations Environment Programme, Nairobi

van Eimern J, Karschon R, Razumava LA, Robertson GW (1964) Windbreaks and Shelterbelts. Technical Note No. 59 (WMO-No.147.TP.70), Geneva, Switzerland

Van Pelt RS, Zobeck TM, Potter KN, Stout JE, Popham TW (2004) Validation of the wind erosion stochastic simulator (WESS) and the revised wind erosion equation (RWEQ) for single events. Environl Model Softw 19:191–198

Woodruff NP, Zingg AW (1952) Wind-tunnel studies of fundamental problems related to wind-breaks. SCS-TP-112, Soil Conservation Service, US Department of Agriculture, Washington DC

Woodruff NP, Siddoway F (1965) A wind erosion equation. Soil Sci Soc Am J 29:602–608

Zingg AW, Woodruff NP (1951) Calibration of a portable wind tunnel for the simple determination of roughness and drag on field surfaces. Agronomy J 43:191–193

Chapter 12
Earthen Embankments and Farm Ponds

Abstract Earthen embankments are used extensively for impounding and diverting water for irrigation. Similarly, farm ponds are prevalent water harvesting structures that store excess runoff from the catchment to serve multiple purposes, like supplemental irrigation, fish culture, duck farming, and livestock consumption. Section 12.1 of this chapter presents various types of earthen embankments. The design of earthen embankments is discussed in terms of their height, side slopes, top width, freeboard, and settlement allowance. The chapter includes the procedures to determine the seepage through the body of the earthen embankments and ways to control it. The possible causes of the failure of the earthen embankments are also discussed. Section 12.2 of the chapter presents a broad classification of farm ponds into two types: embankment type and excavated farm ponds. The design of farm ponds includes the capacity, shape, and dimension of the pond, besides the inlet channel and spillway and the outlet or waste weir. The chapter also contains farm ponds construction, operation, and maintenance. Sealing of farm ponds for reducing seepage and percolation losses is discussed. The chapter includes solved examples to demonstrate the design procedures for both earthen embankments and farm ponds.

12.1 Earthen Embankments

Earthen embankments are water impounding and diverting structures that are widely used as levees, dikes, and dams. These are usually constructed by compacting the locally available soil and, therefore, are susceptible to water erosion. When constructed across streams, the earthen embankments are called earthen dams, the most common form of dams found worldwide. Though earthen embankments rely on their mass to resist sliding and overturning, unlike earthen dams with well-designed foundations, these two terms are often used interchangeably.

Earthen embankments have been used since the pre-Christian era to impound and divert water for irrigation. The oldest known embankment dam was built in Ceylon (Sri Lanka) in 504 BCE (Schuyler 1905). The Kallanai dam, popularly known as the Grand Anicut, is another ancient dam built across the Cauvery River in 100 BCE. It is one of the oldest irrigation systems globally that is still functioning (Agoramoorthy et al. 2008). Until the 1930s, the earthen embankments were designed empirically based on experience; consequently, several of these structures failed worldwide due to overtopping, internal erosion and piping (an internal erosion process in which water seeping through the body or the foundation of the embankment transports soil particles and creates small channels resembling pipes). However, since then, the design of the earthen embankments has significantly improved due to advances in soil mechanics and geotechnical engineering.

12.1.1 Functions

Earthen embankments mainly function either as storage or detention structures. Storage embankments impound water during surplus supply periods, such as monsoon or spring, and meet the irrigation and other water demands during the dry summer season. In contrast, detention embankments control floods by confining the surface runoff and allowing groundwater recharge. The earthen embankments are also used for providing additional storage in the dugout ponds.

12.1.2 Advantages and Limitations

The advantages of earthen embankments include the use of excavated or locally available materials. These embankments involve relatively simple design and construction procedures. Also, their topographical requirements are less stringent, and they have minimal foundation requirements.

The principal disadvantage of earthen embankments is that they may easily be damaged or destroyed by the overflowing water or the erosive action of water against them. Therefore, their upstream section should be amply protected. Also, the compaction during the construction must be adequate to avoid seepage through them. Earthen embankments may require regular maintenance to prevent erosion, plant growth, animal and insect damages, and seepage.

12.1.3 Types of Earthen Embankments

Earthen embankments may be of three types, namely, homogenous embankments, diaphragm embankments, and zoned embankments. The selection and design of earthen embankments primarily depend on the site conditions, including the foundation and the availability of appropriate construction materials. This chapter includes a discussion on only the rolled-fill type of earthen embankments. In rolled-filled embankments, a considerable portion of the embankment is constructed by placing earthen materials in successive layers and compacting them at optimum water content.

12.1.3.1 Homogeneous Embankments

Homogeneous embankments are constructed using a single material; however, a different material may be used for slope protection. The homogenous embankments are used when only a single type of material is economically or locally available. However, the construction material must be impervious enough to enable water storage upstream of the embankment. Also, the material should have sufficient shear strength to support moderately flat side slopes for stability.

When homogeneous embankments are constructed using a relatively impervious soil material on an impervious foundation, the material is typically keyed into the foundation (Fig. 12.1a). However, if the soil material or the foundation is pervious, a blanket of relatively impervious material is usually placed on the upstream face (Fig. 12.1b). These embankments are susceptible to the seepage of the downstream face. Therefore, they typically have large cross-sections to ensure stability and safety against seepage failure.

The design of homogeneous embankments has been modified in the recent past by incorporating an internal drainage system to control seepage. The internal drainage system may consist of a horizontal filter (Fig. 12.2a) or a rock toe (Fig. 12.2b). The drainage filter or the rock toe contains the top seepage line, i.e., the phreatic line, well within the body of the embankment. Consequently, much steeper side slopes may be used in the modified design.

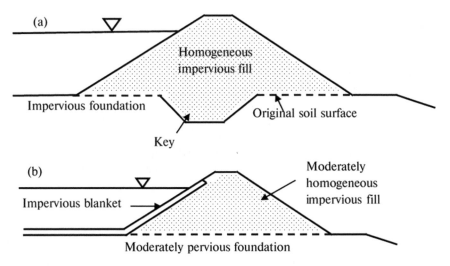

Fig. 12.1 Homogeneous earthen embankments with **a** impervious material keyed to the foundation **b** impervious blanket on the upstream face

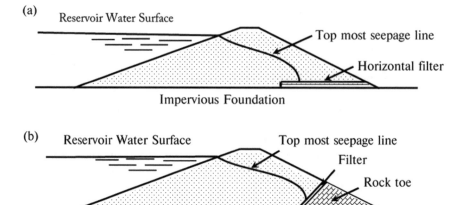

Fig. 12.2 Modified homogeneous earthen embankments with internal drainage system **a** horizontal filter, **b** rock toe

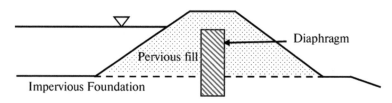

Fig. 12.3 Diaphragm earthen embankment

12.1.3.2 Diaphragm Embankments

In diaphragm embankments, a thin diaphragm of impermeable material is provided within the embankment section, constructed using pervious materials like sand, gravel, or rock. The diaphragms, usually located at the centre of the embankment, act as a water barrier and control seepage (Fig. 12.3).

The diaphragms may be constructed using asphalt concrete, reinforced concrete, metals, compacted earth fill, or geomembranes. When compacted earth fill is used as the construction material, the diaphragms are typically called a thin core. A significant disadvantage of diaphragms lies in the inability to inspect or repair them in case of rupture due to settlement of the embankment or its foundation.

12.1.3.3 Zoned Embankments

Zoned embankments typically have a central impervious core flanked by a comparatively pervious transition zone and a much more pervious outer zone called a shell (Fig. 12.4). The central impervious core checks the seepage through the embankment.

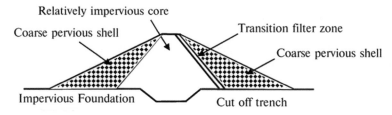

Fig. 12.4 Zoned earthen embankment

The transition zone protects the central core against erosion during rapid drawdown and piping due to cracks in the core. The outer zone or shell distributes the load over a large foundation area and provides stability to the central core.

The impervious core is usually constructed using an appropriate mixture of the less pervious fine-grained soils like clays, silts, silty clays, sandy clays, sandy silts and gravelly clays. In contrast, the outer zone or the shell is constructed using pervious materials like sands, gravels, rocks, or a mixture. The transition zones are included only when the permeability of the materials used to construct the central core and the shell are significantly different. In such cases, transition zones of an appropriate soil mix, having permeability between the central core and the shell, are provided. When the transition zone is absent, the thickness of the central impervious core or diaphragm differentiates the zoned and the diaphragm embankments. In zoned embankments, the thickness of the central impervious core at any elevation exceeds 10 m or the embankment height.

Zoned embankments are widely preferred because of their superior stability and seepage performance. However, their cost of construction is high because of the complex earthwork.

12.1.4 Design of Earthen Embankments

12.1.4.1 Preliminary Investigations

Once the site(s) for constructing the embankment is identified, a field visit is essential for assessing the topography, foundation condition, and the available fill material. The visit may also include a field survey using a theodolite or a Global Positioning System (GPS) to collect accurate geographical features and appropriate field tests to ascertain the nature of the underlying soil and geologic conditions. The information may be used to analyse the economic implications of constructing the embankment at the chosen site(s).

12.1.4.2 Foundation and Earth-Fill Conditions

Since foundations reportedly account for about 10–20% of embankment failures (Foster et al. 2000), special efforts should be made to characterise and understand the foundation properties. USBR (1987) has classified the possible foundation conditions into the following categories:

1. **Rock Foundations**: Homogeneous and competent rock foundations are usually the most favoured foundations as they are resistant to erosion and percolation. Moreover, these possess relatively high sheer strength. However, rock foundations may have faults, fractures, or soluble zones that are susceptible to seepage. If not adequately controlled, the excessive seepage may lead to instability due to excessive internal erosion or uplift pressure.
2. **Sand and Gravel Foundations**: Usually, sand and gravel foundations have adequate strength to withstand the loads exerted by the embankment and reservoir. However, these foundations are pervious and may face the problems of under-seepage and internal erosion. Therefore, adequate arrangements should be made to safeguard the embankment against failure due to internal erosion or instability caused by seepage and uplift pressures.
3. **Silt and Clay Foundations**: These foundations are usually impermeable and, thus, theoretically free from under-seepage and internal erosion problems. However, these foundations pose the problem of instability due to non-uniform settlement and soil collapse upon saturation. Therefore, proper foundation treatment should be done based on the soil type, the location of the water table, and the density of the soil.
4. **Non-uniform Foundations**: It may not always be possible to find the foundations described above at every site. In such cases, special efforts should be made to collect relevant data to understand the foundations and devise treatment plans to avoid instability or failure.

Earthen embankments are designed economically using locally available materials. The quality of fill material near or at the site, along with the foundation characteristics, helps determine the type of earthen embankment. The fill materials should be well-graded and capable of being well compacted. The fill materials containing 20% gravel, 20–50% sand, less than 30% silt, and 15–25% clay are best suited for compaction to maximum density.

Homogeneous embankments are selected where the foundation is impervious, and quality fill materials are available. Zoned embankments are preferred when the depth to the impervious foundation is low and where materials are available to construct a central impervious core. However, if the depth to the impervious foundation is high and quality core materials are not available, diaphragm embankments are chosen.

In levee construction, the quality of material may change along the length of the levee, and hence, the design may vary based on the availability of the material.

Table 12.1 Side slopes for earthen embankments (after Sharda et al. 2007)

Type of material	Upstream slope	Downstream slope
Clayey gravels, silty gravels, gravel sand clay mixtures, and gravel sand silt mixtures	2.5:1	2:1
Sandy clay, silty clay, clay, inorganic silt, and clay	3:1	2.5:1
Inorganic clays of high plasticity and inorganic silts	3.5:1	2.5:1

12.1.4.3 Catchment Yield

Once the site is selected, the catchment area needs to be delineated, and the expected peak rate of runoff and annual runoff estimated. This information is vital for determining the embankment cross-section. The runoff volume and the peak rate of runoff can be determined using the *SCS curve number method* (Sect. 3.2.3, Chap. 3) and the *rational method* (Sect. 3.2.4, Chap. 3), respectively.

12.1.4.4 Height of the Embankment

The height of the earthen embankment must be sufficient to prevent overtopping during the passage of the design flow. However, the height must compensate for settlement and include the freeboard required for wave action. The overall height usually is limited to 15 m (Stephens 2010).

12.1.4.5 Side Slopes

The side slopes depend on the embankment height, the nature of the construction material and the duration of inundation. Table 12.1 presents the side slopes usually adopted for the earthen embankments (Sharda et al. 2007).

As evident from the table, the upstream slope may vary between 2.5:1 and 3.5:1. Flat upstream slopes are preferred at times to minimise the cost of slope protection. Similarly, the downstream slope may vary between 2:1 and 2.5:1. Earthen embankments used for storing water may require flat slopes if subject to rapid drawdown. In contrast, levees constructed along major rivers may be subject to prolonged flood peaks and require side slopes as flat as 7:1. However, conservation levees that protect banks on the upper tributaries may have 2:1 side slopes as they only face flood peaks over short-duration.

12.1.4.6 Top Width

The top width of the earthen embankments depends on the embankment's height and the intended use of the surface. For embankments having a height up to 5 m, a minimum top width of 2.5 m is provided. For embankments having a height greater than 12.5 m, the following equation may be used to determine the minimum top width (NRCS 2005):

$$W = 0.2H + 2.1 \tag{12.1}$$

where W = top width, m; and H = maximum height of the embankment, m.

The minimum top width determined by Eq. (12.1) may be increased if the embankment top is used as a public road. In such a case, the minimum width should be 5 m. The minimum width may also be increased to provide structural stability.

12.1.4.7   Freeboard

The freeboard is the vertical distance between the reservoir water surface elevation and the top of the embankment after settlement. It prevents overtopping of the embankment by wave action and accounts for uncertainties in estimating the flow and the water surface elevation. The reservoir water surface elevation could be either *normal* or *maximum*.

The Normal Reservoir Water Surface (NRWS) elevation is the maximum elevation corresponding to the live reservoir capacity, i.e., the typical storage used in daily reservoir operations (excluding the flood operations). When the reservoir level is at NRWS, a normal freeboard is specified to protect the embankment against the wave action caused by the maximum wind velocity at the site.

The wave height for moderate-sized water bodies may be determined using Hawksley's formula (Wood and Richardson 1975):

$$h_{\text{wave}} = 0.014(D_f)^{1/2} \tag{12.2}$$

where h_{wave} = wave height for maximum wind velocity, m; and D_f = fetch or exposure, m.

The normal freeboard includes the wave height determined by Eq. (12.2). It may also include the provisions for the frost action if needed.

The Maximum Reservoir Water Surface (MRWS) elevation is the reservoir water level attained when the flow corresponding to a long return period is routed through the reservoir. When the reservoir level is at MRWS, a minimum freeboard is specified to prevent the embankment overtopping due to wind loads. The minimum freeboard should be greater than 1 m or the sum of the run-up and setup caused by the wind during the flood event. Wind run-up is the vertical height that the water reaches above still water level when the water moves towards the embankment on the breaking of a wave. The wind setup is the vertical rise in the still water level downstream of the fetch, i.e., on the upstream slope of the embankment. The procedure to determine the wind run-up and setup can be referred to from USBR Design Standards No. 13 (USBR 2021).

12.1.4.8 Settlement Allowance

The settlement represents the consolidation of the fill and foundation materials due to self-weight and an increase in the moisture content of the fill material due to stored water. It depends on the method of construction and the foundation material.

The settlement allowance accounts for the settlement of the earthen embankments after construction. It is added to the height of the embankment over and above the calculated freeboard.

Rolled-fill embankments constructed on firm foundations and compacted at optimum water content may settle about 1% of the total height. However, on soft foundations, settlement maybe about 6%. Since farm ponds embankments are not compacted thoroughly, the settlement allowance of about 5–10% may be considered.

Example 12.1 Design a homogeneous earthen embankment with the following data:

Level of the Highest Flood Level (HFL) of the reservoir $= 208$ m; level of the deepest stream bed $= 198$ m. The length of the embankment is 300 m. Gravel sand silt mixture is available for the fill, and the foundation is impervious. Assume a fetch of 1.6 km.

Solution Given: Level of the Highest Flood Level (HFL) of the reservoir $= 208$ m; level of the deepest stream bed $= 198$ m; length of the embankment $= 300$ m; fetch, $D_f = 1600$ m; fill material $=$ gravel sand silt mixture.
Head of water upstream, $h = 208 - 198 = 10$ m.

Using Eq. (12.2), the wave height,

$$h_{wave} = 0.014(D_f)^{1/2} = 0.014 \times (1600)^{1/2} = 0.56\,m$$

Therefore, freeboard $= 0.56$ m.
Thus, the height of the embankment $= 10 + 0.56 = 10.56$ m.
Since the foundation is impervious, a settlement allowance of 1% of the fill height may be provided.
Hence, the settlement allowance $= 0.01 \times 10.56 = 0.11$ m.
Thus, the maximum height of the embankment, $H = 10.56 + 0.11 = 10.67$ m.
Equation (12.1) may be used to determine the top width of the embankment as,

$$W = 0.2H + 2.1 = 0.2 \times 10.67 + 2.1 = 4.23 \approx 4.25\,m$$

Referring to Table 12.1, for gravel sand silt mixture fill, upstream and downstream slopes are 2.5:1 and 2:1.
Hence, the bottom width of the embankment,

$$B = 2.5 \times 10.67 + 4.25 + 2 \times 10.67 = 52.3 \approx 52.5\,m$$

Answers: Height of the embankment, $H = 10.67$ m; top width of the embankment, $W = 4.25$ m; bottom width of the embankment, $B = 52.5$ m; upstream side slope = 2.5:1; downstream side slope = 2:1.

12.1.5 Seepage Analysis

Earthen embankments are porous and usually face seepage that may saturate a part of the embankment. The embankment may face instability due to sloughing (the progressive removal of soil from the downstream face due to saturation) and piping if its downstream face gets saturated. Therefore, seepage analysis is essential to keep the downstream face free from saturation.

The flow through an earthen embankment is two-dimensional, i.e., the horizontal and vertical velocity components vary in the flow domain. Laplace's equation governs the two-dimensional flow through a soil mass. Laplace's equation has the following assumptions:

1. Water and soil are incompressible.
2. The flow is steady.
3. The flow is laminar, i.e., Darcy's law is valid.
4. The saturated soil mass is homogeneous and isotropic.
5. The flow is two-dimensional. Considering the soil element in Fig. 12.5 for illustration, it has a dimension, dy, which gives the element the volume, and no flow takes place perpendicular to the plane of the figure.

In the figure, v_x and v_z represent the flow velocity in the X- and Z-directions, respectively, and $\left(v_x + \frac{\partial}{\partial x} v_x d_x\right)$ and $\left(v_z + \frac{\partial}{\partial z} v_z d_z\right)$ represent the flow velocities at the exit in the X- and Z-directions, respectively.

Laplace's equation representing the two-dimensional flow through the soil element for isotropic conditions is expressed as:

Fig. 12.5 Flow through a soil element

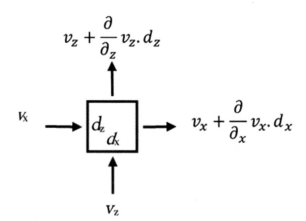

$$v_z + \frac{\partial}{\partial z} v_z . d_z$$

$$v_x + \frac{\partial}{\partial x} v_x . d_x$$

$$K \frac{\partial^2 h}{\partial x^2} + K \frac{\partial^2 h}{\partial z^2} = 0$$

Or,

$$\frac{\partial^2 h}{\partial x^2} + \frac{\partial^2 h}{\partial z^2} = 0 \qquad (12.3)$$

where K = coefficient of permeability; and h = head causing the flow.

Laplace's equation, thus, states that the sum of changes of gradients in X- and Z-directions is zero.

12.1.5.1 Graphical Solution of Laplace's Equation

The graphical solution of Laplace's equation results in a flow net having two sets of curves, viz., streamlines (or flow lines) and equipotential lines (Fig. 12.6).

The flow net has the following characteristics:

1. Streamlines or flow lines represent the flow path of water particles.
2. Equipotential lines represent the lines joining points having an equal head.
3. Flow lines and equipotential lines intersect at 90°, i.e., they are orthogonal to each other.
4. The passage between two flow lines is called a flow channel.
5. A constant flow takes place through a flow channel, say, Δq.
6. A flow line cannot intersect another flow line, and flow cannot occur across flow lines.
7. The velocity of flow is normal to the equipotential line.
8. The difference in head between two equipotential lines is called the potential drop or head loss (Δh).

A flow net can be used to determine the seepage through the earthen embankment. From Darcy's law, the seepage through a flow channel is equal to,

Fig. 12.6 Typical flow net

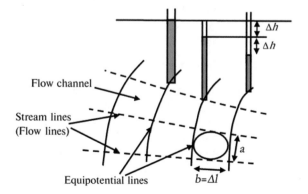

$$\Delta q = K i A$$

Considering unit thickness of the field,

$$\Delta q = K \frac{\Delta h}{l}(b \times 1) \tag{12.4}$$

where l = length of the field, and b = width of the field.

If N_d is the total number of potential drops in the flow net, then

$$\Delta h = \frac{h}{N_d} \tag{12.5}$$

Substituting Δh from Eq. (12.5) in (12.4),

$$\Delta q = K \frac{h}{N_d} \frac{b}{l} \tag{12.6}$$

If N_f is the total number of flow channels in the flow net, then the total discharge passing through the flow net is,

$$q = \Delta q \times N_f$$

Or,

$$q = K h \frac{N_f}{N_d} \frac{b}{l} \tag{12.7}$$

Since the field is square, hence, $b = l$. Thus, Eq. (12.7) reduces to,

$$q = K h \frac{N_f}{N_d} \tag{12.8}$$

Thus, the seepage discharge passing through the embankment can be determined if a flow net is drawn for an earthen embankment. Figure 12.7 presents a typical flow net for a homogeneous earthen embankment (without any internal drainage system to control seepage).

Phreatic Line in an Earthen Embankment

The phreatic line, also called the top seepage line or saturation line, is the top flow line that separates the unsaturated and saturated zones in an earthen embankment. A positive hydrostatic pressure exists below the phreatic line, while the atmospheric pressure acts on the phreatic line. The seepage flow below the phreatic line reduces the effective weight of the soil and leads to the reduced shear strength of the soil.

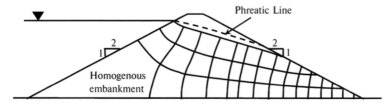

Fig. 12.7 Typical flow net for a homogeneous embankment (without any internal drainage system)

Locating Phreatic Line

The location of the phreatic line is essential for drawing the flow net for an earthen embankment. The phreatic line may be drawn using the Casagrande method (Casagrande 1937). Figure 12.8a shows the actual phreatic line MNOP. The phreatic line coincides with the parabola M'NOP'Q', but with some deviations at the upstream and downstream faces of the embankment. The parabola has point Q as its focus. The phreatic line MNOP begins at point M on the dam's upstream face by making a 90°angle with the face. Also, $MM' = 0.3\Delta$, Δ being the horizontal distance between the upstream toe and the point of interaction of the upstream water surface and the upstream slope (Fig. 12.8a).

The parabola M'NOP'Q' can be drawn as follows:

1. Let the distance QQ' be p. Consider a point A, having coordinates of (x, z), on the parabola with focus Q as the origin (Fig. 12.8b). Now, based on the property of the parabola, AQ = AD.

From Fig. 12.8b, $AQ = \sqrt{x^2 + z^2}$ and $AD = 2p + x$. Thus,

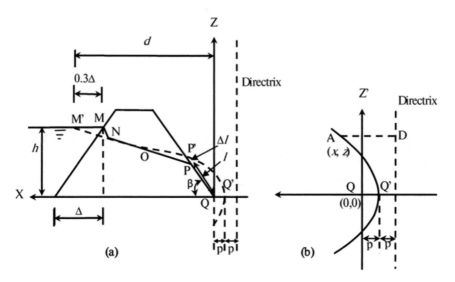

Fig. 12.8 Locating phreatic line in an earthen embankment

$$\sqrt{x^2 + z^2} = 2p + x \qquad (12.9)$$

From Fig. 12.8a, at $x = d$, $z = h$. Putting these values of x and z in Eq. (12.9) and rearranging, we get

$$p = \frac{1}{2}\left(\sqrt{d^2 + h^2} - d\right) \qquad (12.10)$$

Since d and h are known, the value of p can be determined.

2. From Eq. (12.9),

$$x^2 + z^2 = 4p^2 + x^2 + 4px$$

Or,

$$x = \frac{z^2 - 4p^2}{4p} \qquad (12.11)$$

With p determined in step (1), the values of x can be obtained for different values of z, and parabola constructed.

3. For completing the phreatic line, the portion MN has to be approximated and drawn by hand.
4. In Fig. 12.8a, PQ is known as seepage face. Let the distance $PQ = l$ and $PP' = \Delta l$.

When $\beta < 30°$, the value of l can be approximated by the following equation (Fukuchi 2020):

$$l = \frac{d}{\cos \beta} - \sqrt{\frac{d^2}{\cos^2 \beta} - \frac{h^2}{\sin^2 \beta}} \qquad (12.12)$$

When $\beta > 30°$, the value of l can be referred from Casagrande (1937).

Once point P is located, curve OP can be approximately drawn.

Example 12.2 Plot the phreatic line for the homogeneous earthen embankment shown in Fig. 12.9. Assume that the fill material is isotropic.

Solution Given: Head of water on the upstream, $h = 14$ m; freeboard, $f = 2$ m; top width, $T = 3$ m; upstream side slope, $Z = 1$; downstream side slope, $Z = 2$.

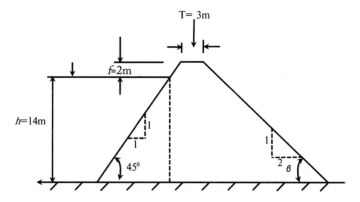

Fig. 12.9 Homogeneous earthen embankment (Example 12.2)

From the given data in Fig. 12.9 and referring to Fig. 12.8a,

$$\Delta = h \cot 45° = 14 \times 1 = 14\,\text{m}$$
$$\beta = \tan^{-1}(1/2) = 26.5°$$
$$MM' = 0.3\Delta = 0.3 \times 14 = 4.2\,\text{m}$$
$$d = MM' + f \cot 45° + T + (h + f) \cot 26.5°$$
$$= 4.2 + 2 \cot 45° + 3 + 16 \cot 26.5° = 41.2\,\text{m}$$

From Eq. (12.10),

$$p = \frac{1}{2}\left(\sqrt{d^2 + h^2} - d\right) = \frac{1}{2}\left(\sqrt{41.2^2 + 14^2} - 41.2\right) = 1.16\,\text{m}$$

Using Eq. (12.11), i.e., $x = \frac{z^2 - 4p^2}{4p}$, we can determine the coordinates of the basic parabola by assuming different values of z, as shown in the following table.

z, m	14	12	10	8	6	4	3
x, m	41.1	29.9	20.4	12.6	6.6	2.3	0.8

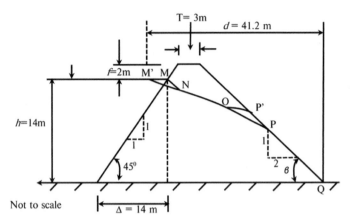

Fig. 12.10 Phreatic line drawn for the homogeneous earthen embankment in Example 12.2

Since $\beta < 30°$, the value of l can be approximated by Eq. (12.12). Hence,

$$l = \frac{d}{\cos \beta} - \sqrt{\frac{d^2}{\cos^2 \beta} - \frac{h^2}{\sin^2 \beta}} = \frac{41.2}{\cos(26.5)} - \sqrt{\frac{41.2^2}{\cos^2(26.5)} - \frac{14^2}{\sin^2(26.5)}} = 12.3 \, \text{m}$$

The parabola is drawn (Fig. 12.10) using the (x, z) coordinates calculated and subsequently approximating portions MN and OP.

12.1.5.2 Analytical Solution of Laplace's Equation

An approximate solution of Laplace's equation is discussed here (Justin et al. 1944). The method can be used for homogeneous or zoned embankments and yields reasonably accurate results for small earthen embankments.

Referring to the case of a homogenous embankment with an impervious foundation (Fig. 12.11), the method assumes that,

$$e = h/3 \tag{12.13}$$

where e = the vertical distance from the ground level to the point where seepage line meets the downstream face, and h = head of water upstream.

Applying Darcy's law, the discharge per unit length of the embankment is given as,

$$q = -K \frac{dh}{dL} \times \text{Area}$$

Or,

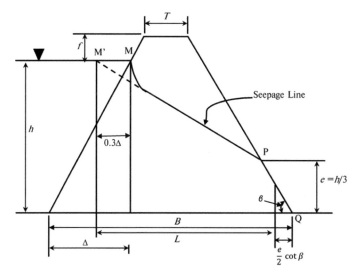

Fig. 12.11 Approximate method of determining seepage through a homogenous earthen embankment

$$q = -K\left(\frac{e-h}{L}\right) \times \left(\frac{e+h}{2} \times 1\right)$$

Or,

$$q = K\left(\frac{h-e}{L}\right) \times \left(\frac{e+h}{2}\right)$$

Or,

$$q = \frac{K}{2}\left(\frac{h^2-e^2}{L}\right) \qquad (12.14)$$

where L = mean length of the seepage line, i.e., the distance between the starting point of the seepage line, M, and the mid-point of the seepage face PQ.

Substituting e from Eq. (12.13) in Eq. (12.14), we get

$$q = \frac{K}{2}\left(\frac{h^2-(h/3)^2}{L}\right) = \frac{4Kh^2}{9L} \qquad (12.15)$$

In Fig. 12.11,

$$L = B - \left(\Delta + \frac{e}{2}\cot\beta\right) + MM' \qquad (12.16)$$

where B = bottom width of the embankment.

Example 12.3 Using the analytical solution, determine the discharge per unit length of the homogeneous earthen embankment in Example 12.2. The coefficient of permeability of the fill material is 2.5×10^{-5} m s^{-1}.

Solution Given: Head of water on the upstream, $h = 14$ m; freeboard, $f = 2$ m; top width, $T = 3$ m; upstream side slope, $Z = 1$; downstream side slope, $Z = 2$; coefficient of permeability of the fill material, $K = 2.5 \times 10^{-5}$ m s^{-1}.

Using Eq. (12.13),

$$e = \frac{h}{3} = \frac{14}{3} = 4.67 \, \text{m}$$

Also, from Example 12.2, we know that

$$\Delta = h \cot 45° = 14 \times 1 = 14 \, \text{m}$$
$$\beta = \tan^{-1}(1/2) = 26.5°$$
$$MM' = 0.3\Delta = 0.3 \times 14 = 4.2 \, \text{m}$$

The bottom width of the embankment,

$$B = (h + f) \cot 45° + T + (h + f) \cot 26.5°$$
$$= (14 + 2) \times 1 + 3 + (14 + 2) \times 2 = 51 \, \text{m}$$

Using Eq. (12.16),

$$L = B - \left(\Delta + \frac{e}{2} \cot \beta\right) + MM' = 51 - \left[14 + \left(\frac{4.67}{2} \times 2\right)\right] + 4.2 = 36.53 \, \text{m}$$

Putting the known values of K, h and L in Eq. (12.15), we get

$$q = \frac{4Kh^2}{9L} = \frac{4 \times 2.5 \times 10^{-5} \times 14^2}{9 \times 36.53} = 6 \times 10^{-5} \, \text{m}^3\text{s}^{-1}$$

Answer: The discharge per metre length of the embankment is 6.0×10^{-5} m^3 s^{-1}.

12.1.5.3   Flow Net for Anisotropic Soil

An anisotropic soil usually has greater horizontal permeability than vertical. Thus, the flow net, which is based on the assumption of isotropic porous media, would be different for anisotropic soil. For taking the anisotropy of the soil into account, the horizontal dimensions of the embankment are drawn using a transformed scale, i.e., by multiplying them with a factor equal to $\sqrt{K_v/K_h}$, K_v and K_h representing the vertical and horizontal permeability. For example, if the top width of the dam is 5 m, it would be changed to $(5\sqrt{K_v/K_h})$. Subsequently, the flow net is drawn on the transformed embankment section, and the number of flow channels (N_f) and equipotential drops (N_d) is counted.

The seepage discharge through the embankment may be determined using the following equation:

$$q = \sqrt{K_h K_v} h \frac{N_f}{N_d}$$ (12.17)

12.1.6 Internal Drainage System

An internal drainage system is provided to contain the phreatic line within the body of the embankment such that it does not intersect the downstream face of the embankment. The internal drainage system may vary, but a horizontal filter or a rock toe is usually preferred.

12.1.6.1 Horizontal Filter

A horizontal filter extending from the downstream toe of the embankment towards the centre is usually provided in a homogenous embankment (Fig. 12.2a). The filter length may vary from 25 to 100% of the distance between the centreline of the embankment and the downstream toe. The filter may have a minimum thickness of one metre, and it should be sufficiently pervious to drain off the seepage water. It may be provided with a slope of 1 in 100 towards the downstream toe to facilitate rapid drainage.

12.1.6.2 Rock Toe

The rock toe also referred to as a toe filter is provided at the downstream toe of the embankment (Fig. 12.2b). It consists of stones of sizes varying from 150 to 200 mm, usually provided in layers. The height of the toe filter is kept in between 25 and 35% of the upstream head in the reservoir.

12.1.6.3 Filter Material Characteristics

The material chosen for the internal drainage system should meet the following conditions:

Table 12.2 Maximum filter criteria as per Indian Standard Code (after BIS 1999)

Base material category	Base material description and percent finer than 75 micron	Filter criteria
1	Fine silt and clay, more than 85% finer	$D_{15}(F) \leq 9D_{85}(B) \geq 0.2$ mm
2	Sand, silt, clay, and silty and clayey sand, 40–85% finer	$D_{15}(F) \leq 0.7$ mm
3	Silty and clayey sands and gravel, 15–39% finer	$D_{15}(F) \leq \{(40\text{-}A)/25\}(4D_{15}(B) - 0.7$ mm$) \geq 0.7$ mm
4	Sand and gravel, less than 15% finer	$D_{15}(F) \leq 4D_{85}(B)$

'F' denotes the filter material and 'B' the base material. The subscripts 10, 15, 85, or 90 refer to the percentage finer by weight
'A' is the percent passing the 75-micron sieve

1. The permeability of the filter material should be higher than those of the base or fill material.
2. The smaller grained fill material should not block the pores of the larger filter material.

To satisfy the permeability requirement of filter material, the minimum D15(F) shall be greater than or equal to 5 times D15(B) but not less than 0.1 mm, i.e., D15(F) $\geq$ 5D15(B) > 0.1 mm. Maximum D15(F) shall be determined from Table 12.2, but it should not be smaller than 0.2 mm (BIS 1999).

Filter design criteria may also be referred from USBR (1987).

12.1.6.4 Seepage Discharge Through an Earthen Embankment with a Horizontal Filter

As discussed in Sect. 12.1.6, the horizontal filter contains the phreatic line within the body of the embankment such that it does not intersect the downstream face of the embankment. Figure 12.12 shows an earthen embankment with a horizontal filter. As evident, the actual phreatic line MNQ′ is pushed down to Q′ on the filter. As described in Sect. 12.1.5.1, the phreatic line coincides with the parabola M′NQ′, with some deviation at the upstream face of the embankment. The parabola has point Q as its focus. The phreatic line MNQ′ begins at point M on the dam's upstream face by making a 90° angle with the face. Also, $MM' = 0.3\Delta$.

Let the distance QQ′ be (s/2), s being the focal distance of the parabola, i.e., the distance between the origin and the directrix. Consider a point A, having coordinates of (x, z), on the parabola with focus Q as the origin (Fig. 12.12). Now, based on the property of the parabola, AQ = AD.

Thus, $AQ = \sqrt{x^2 + z^2}$ and $AD = s + x$,

Or,

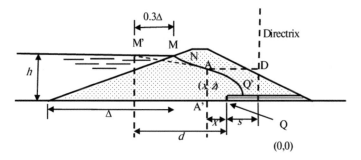

Fig. 12.12 Flow through an earthen embankment with a horizontal filter

$$\sqrt{x^2 + z^2} = s + x \qquad (12.18)$$

Or,

$$s = \sqrt{x^2 + z^2} - x \qquad (12.19)$$

From Figure, at M', at $x = d, z = h$.
Thus,

$$s = \sqrt{d^2 + h^2} - d \qquad (12.20)$$

Since d and h are known, the value of s can be determined using Eq. (12.20).
From Eq. (12.18),

$$x^2 + z^2 = s^2 + x^2 + 2xs$$

Or,

$$z = \left(s^2 + 2xs\right)^{1/2} \qquad (12.21)$$

Let q = discharge per unit length of the embankment.
Thus, considering section AA', from Darcy's law,

$$q = KiA$$

Considering unit thickness of the field,

$$q = K\frac{dz}{dx}(z \times 1) \qquad (12.22)$$

Differentiating Eq. (12.21) with respect to x,

$$\frac{dz}{dx} = \frac{s}{\left(s^2 + 2xs\right)^{1/2}} \tag{12.23}$$

Substituting z and dz/dx from Eqs. (12.21) and (12.23) into (12.22),

$$q = K\frac{dz}{dx}(z \times 1) = K\frac{s}{\left(s^2 + 2xs\right)^{1/2}}\left(s^2 + 2xs\right)^{1/2}$$

Or,

$$q = Ks \tag{12.24}$$

Equation (12.24) is the simple equation to determine the discharge q in terms of the focal distance.

12.1.7   Causes of Failure of Earthen Embankments

Earthen embankments usually fail due to defective design, inappropriate construction and improper upkeep. The causes of failures may be categorised as,

1. Hydraulic failure
2. Seepage failure
3. Structural failure.

12.1.7.1   Hydraulic Failure

Hydraulic failures account for more than 40% of earthen embankment failures. The prominent reasons for the hydraulic failure are as follows:

1. **Overtopping**: Water may pass over the earthen embankments if the design flood is underestimated, if the freeboard or spillway capacity is insufficient, or if the spillway gates are not properly operated. A sufficient freeboard should be provided to avoid overtopping.
2. **Erosion of upstream face**: The waves developed near the top water surface due to wind action may erode the upstream face. The upstream face should be protected with stone pitching or riprap to avoid erosion.
3. **Erosion of downstream toe**: The toe may be eroded due to tailwater. The downstream face should be protected with stone pitching or riprap up to the normal tailwater depth.
4. **Erosion of downstream face**: The downstream face may get eroded due to rainwater flowing over the face. It may lead to gully formation and collapse. The downstream face should be protected with grass plantation. Also, regular maintenance should be done to fill the cuts or cracks.

12.1.7.2   Seepage Failure

Seepage continuously occurs through the body and foundation of an earthen embankment. However, excessive seepage may erode soil particles and lead to embankment failure. Seepage failures account for more than 30% of the earthen embankment failures. Seepage failures may occur in the following ways:

1. **Piping through the body or the foundation**: Water seeping through the body or the foundation of the embankment may cause progressive erosion and removal of the soil particles. Subsequently, small channels may be formed to transport the fill and foundation materials downstream. The channels may grow bigger leading to embankment washout.
2. **Sloughing of the downstream face**: Sloughing is the progressive removal of soil from the wet downstream face. It is caused due to saturation of the downstream face of the embankment. The soil mass gets softened and leads to the slump and slide of the downstream face.

12.1.7.3   Structural Failure

Structural failures account for about 25% of earthen embankment failures, mainly due to shear failure. The structural failure may occur due to one of the following reasons:

1. **Sliding due to weak foundation**: Due to the presence of fine silt, clay, or similar soft soils, the foundation may not withstand the pressure of the embankment. In such failures, the lower slope moves outwards along with a part of the foundation, and the top of the embankment subsides, forming large mud waves near the toe.
2. **Sliding of the upstream face due to sudden drawdown**: The earthen embankment develops pore water pressure within the embankment's body due to standing water upstream. The pore pressure is not released immediately when the reservoir is suddenly emptied, maybe to accommodate an incoming flood, causing a slump in the upstream face.
3. **Sliding of the downstream face due to steep slope**: When the downstream slope of the earthen embankment is too steep to provide shear strength of the shoulder material, it may cause instability. The instability may lead to the sliding of the downstream face.
4. **Damage caused by burrowing animals or water-soluble materials**: Earthen embankments may get damaged due to animal burrows, which may cause the seepage water to flow quickly, carrying fine soil materials along. The phenomena may lead to piping failure within the body of the embankment. Similarly, water-soluble materials within the body of the embankment may get washed along with the seepage flow, causing piping and consequent failure.

12.1.8   Construction

The construction site is first cleared by removing all trees, stumps, and major roots. Once the site is cleared, the core trench is excavated. Subsequently, the earthen materials are placed in successive layers and compacted at optimum water content. Optimum water content is essential as it permits the fill material to mix with the original surface and eliminates the separation of the fill and the foundation. The ideal moisture content can be determined using the Proctor density test. For earthen embankments, the degree of compaction, defined as the ratio of the densities of the compacted fill to the original fill, should ideally be 85–100% (BIS 1995; Das 2020).

The nature of the fill material also governs the compaction. Soils with high clay content are usually compacted when their water content is lower than the plastic limit to avoid creating voids. On the contrary, the pervious materials, like sands and gravels, are compacted when their moisture content is near saturation.

Compaction is achieved by placing 200–250 mm thick layers of the fill material and rolling them with either pneumatic tyre or sheep foot rollers at optimum moisture content. Pneumatic tyre rollers are preferred for the fill materials of low plasticity. The roller size is chosen based on the construction needs.

12.1.9   Maintenance and Protection

A properly designed and constructed earthen embankment does not require regular maintenance. However, a regular inspection is essential to detect damages due to animal or human activities, erosion, seepage of the downstream face, and any other indication of piping. If such conditions are found, they need to be tendered immediately to avoid severe damages.

For protecting the embankment, the area may be fenced. An extensive catchment management program should be initiated to minimise the erosion and sediment inflow to the reservoir. Close-growing vegetation should be maintained on the reservoir banks to minimise the entry of pollutants into the water body.

12.2   Farm Ponds

Farm ponds are the common water harvesting structures used to store surface runoff generated from the catchment. In semi-arid regions having more than 500 mm of annual rainfall, farm ponds are adopted to serve multiple purposes, like supplemental irrigation, fish culture, duck farming, and livestock consumption. These also help in flood mitigation by reducing the peak flow in the catchment. Farm ponds also conserve soil and nutrients.

Farm ponds provide food and financial security in dryland agriculture areas by augmenting irrigation to boost crop productivity, besides generating additional income through fish and duck farming. Therefore, the design and construction of farm ponds are vital for sustained agriculture and allied activities in semi-arid regions.

12.2.1 Types of Farm Ponds

Farm ponds are broadly classified into the following two categories:

1. Embankment type
2. Excavated type.

12.2.1.1 Embankment Type Farm Ponds

The embankment type farm ponds are created by constructing an earthen dam across a stream or watercourse. These ponds are preferred in areas having gentle to moderately steep slopes. Also, the stream valley should permit storing a sufficient volume of water.

Since the design of embankment type farm ponds is similar to that of earthen embankments, only the design of excavated ponds is included here.

12.2.1.2 Excavated Farm Ponds

The excavated farm ponds are constructed by excavating soil to create a dugout. These ponds are usually designed to serve individual farms because their capacity is limited as it depends entirely on the excavation.

Different sources can feed the excavated farm ponds; thus, based on the source of water and location, these are grouped into the following four categories:

1. Dugout ponds
2. Surface ponds
3. Spring or creek fed ponds
4. Off-stream ponds.

Dugout Ponds

The dugout ponds are constructed by excavating soil and creating a dugout. These ponds are preferred in low-lying areas to utilise the shallow water table as the water source. However, these ponds may also be fed by the surface runoff. Besides storing water, these ponds also lower the water table in low-lying areas with poorly drained soils. The stored water is lifted using the indigenous water-lifting devices or pumps for various purposes.

Surface Ponds

Surface ponds, the most common types of farm ponds, are constructed partly by excavating the ground and partly by utilising the excavated soil to construct embankments around the pond. Thus, the desired capacity of these ponds is achieved by both excavated and embanked portions of the pond. Therefore, these are also referred to as excavated-cum-embankment type ponds. These ponds may also be classified as embankment type ponds if the water depth impounded against the embankment exceeds one metre.

Surface ponds are fed mainly by surface runoff. Thus, these usually have no supply during the dry season. Their outlet is designed so that the water flows out by gravity till the water reaches the ground level. Subsequently, the stored water is pumped out. An existing local depression or the lowest portion of the farm is preferred for locating these ponds to achieve maximum storage with minimum excavation.

Spring or Creek Fed Ponds

Spring or creek fed ponds receive the water from a spring or a creek. Therefore, these are usually constructed near a natural spring or creek. The foothills of a hilly catchment are the preferred location as the subsurface flow of the catchment typically emerges as a spring at the foothills.

Off-stream Ponds

Off-stream ponds are constructed by the side of the seasonal streams and collect water from the stream using diversion ditches or pipes. These ponds aim at storing water from the seasonal streams for the lean season.

12.2.2 Site Selection for Farm Ponds

The site selection for farm ponds is vital as it governs the cost of construction and the usefulness of the pond. It depends on multiple factors, such as topography, soil characteristics, and catchment characteristics, including rainfall characteristics. As a thumb rule, the site should be such that ample storage can be obtained with the least amount of excavation, i.e., the site having a high water storage-earth work ratio. Also, ponds should be located as near as possible to the intended place of their use. While selecting the site, care should also be taken such that drainage from farmsteads, feeding lots, sewage lines, and similar areas would not reach and pollute the pond. The factors governing the site selection are discussed below.

12.2.2.1 Topography

The topographic features of the farm, such as slope and drainage flow pattern, govern the site selection to a large extent. For achieving high water storage-earth work ratio, the embankment ponds are usually preferred in deep, narrow valleys. Such sites also minimise the area of shallow water, thus, reducing the evaporation losses.

The dugout ponds are usually located in areas having flat topography and a shallow groundwater table. On the other hand, surface ponds are usually located in a natural depression or the lowest portion of the farm so that the excavation is minimum. Thus, a natural drainage line in a rolling topography could be preferred for surface ponds.

12.2.2.2 Soil Characteristics

For the excavated ponds, soils having stones and boulders should be avoided to minimise the construction cost. Also, the soil profile must be thoroughly investigated, and sites having low pH and EC and deep soils should be selected.

For surface ponds, sites having relatively impervious soils are preferred to minimise seepage losses. The black soils having good water retention potential are ideal. Clays and silty clays extending below the pond depth are also desirable. Sites having porous soils, underlain by sands or gravel, should be avoided. Also, deep soils, especially those having more than 1.0 m depth, are preferred as these help store water for a longer duration.

For dugout ponds, aquifer properties are more vital. Therefore, a thorough subsurface investigation is needed to analyse the features of the water-bearing strata. Test holes may be bored to determine the volume of excavation needed to get the desired pond capacity. The rate at which the water rises in the test holes would govern the pond refilling process after extensive use.

12.2.2.3 Catchment Characteristics

The catchment characteristics are vital, especially for surface ponds, as surface runoff is the primary source of their water supply. The catchment area should be sufficiently large to yield enough surface runoff to replenish the pond. The catchment characteristics, such as land slope, soil infiltration, land use, and land cover, affect the pond design. The rainfall characteristics such as amount, duration and intensity significantly impact the runoff generation in the catchment. The land use and land cover help reduce soil erosion and keep the inflow to the pond free from sediment, thus, keeping the pond capacity intact.

12.2.3 Design of Excavated Ponds

The primary objective of the farm pond design is to provide enough water to meet the water requirement for various purposes the pond is intended to serve. The design of an excavated pond includes the determination of its capacity, shape, dimensions, and inlet and outlet structures to control the inflow and outflow. For designing a pond, input data like catchment rainfall–runoff relationship, soil infiltration characteristics, evaporation losses, and water requirement for various uses should be available.

12.2.3.1 Estimation of Catchment Runoff

The peak rate of runoff and the runoff volume from the pond catchment are essential inputs for designing farm ponds. To accurately estimate the catchment runoff, we need to conduct a probability analysis of the long-term rainfall to estimate the design rainfall. The probability analysis of the rainfall data may be carried out using Weibull's formula described in Sect. 8.7.1.2 (Chap. 8). Once the design rainfall is known, the SCS curve number method (Sect. 3.2.3, Chap. 3) and rational method (Sect. 3.2.4, Chap. 3) could determine the runoff volume and peak rate of runoff from the catchment, respectively.

12.2.3.2 Determination of Pond Capacity

The capacity of the farm ponds depends on the water required to satisfy the intended uses and the runoff volume expected from the catchment. The pond capacity could be 30–40% of annual runoff volume from the catchment so that pond could be filled more than once a year (Reddy et al. 2012). However, while deciding the pond capacity, the area under the excavated pond should be kept within 10–12% of the farm area (Reddy et al. 2012).

A farm pond is typically designed to meet the water requirements for various purposes like supplemental irrigation, livestock consumption, and domestic use. However, the storage capacity of the pond is also influenced by the seepage and evaporation losses. Typically, an allowance of about 10–20% is made in the pond capacity to handle the seepage and evaporation losses. Also, a provision of about 5–10% of the storage capacity should be made to handle siltation during the life span of the pond. However, the silt deposit reduces the seepage losses. The pond capacity could be expressed as follows (Reddy et al. 2012):

$$
\begin{aligned}
\text{Pond capacity} = \ & (\text{Irrigation} + \text{Livestock} + \text{Domestic}) \text{ requirements} \\
& + 20\% \text{ of}(\text{Irrigation} + \text{Livestock} + \text{Domestic}) \\
& \text{requirements towards seepage,} \\
& \text{evaporation, and siltation losses}
\end{aligned}
\tag{12.25}
$$

12.2.3.3 Irrigation Water Requirement

Irrigation water requirement depends on the crops cultivated and the local climate. CropWat model (Clarke et al. 2001) can estimate the irrigation water requirement based on the crop and climate data.

12.2.3.4 Livestock and Domestic Water Requirement

The livestock water requirement is determined using the number and types of livestock maintained on the farm and the average daily water consumption by various kinds of livestock. However, the condition of the livestock, such as age, weight, lactation stage, besides weather, fodder, and feed, governs the water consumption. Thus, the livestock water requirement needs to be determined based on the recommendations made by the local animal husbandry authorities.

The domestic water requirement is also determined using the number of persons living on the farm and the recommended average per capita daily water consumption. In India, the domestic water requirement is recommended by the Ministry of Housing and Urban Affairs (PIB 2020).

Equation (12.25) can be used to determine the pond capacity once the irrigation, livestock and domestic water requirements are known. If the annual runoff volume from the catchment is sufficient to fulfil the estimated pond capacity, the capacity is finalised. However, if the annual runoff volume is insufficient to satisfy the estimated pond capacity, the required pond capacity is revised by reducing the cultivated area. The concept of catchment and cultivated area ratio given by Critchley et al. (1991) may be used in a modified form as follows:

The calculation of the catchment: cultivated area ratio is based on the concept that the design must comply with the rule:

$$\text{Water Harvested} = \text{Extra Water Required} \tag{12.26}$$

The water harvested from the catchment area may be determined as follows:

$$\text{Water Harvested} = C \times V_Q \tag{12.27}$$

where C = catchment area, ha; and V_Q = runoff volume estimated using SCS CN method, mm.

The water required for irrigation can be determined as follows:

$$\text{Water Required for Irrigation} = CA \times IRR \tag{12.28}$$

where CA = cultivated area, ha; and IRR = Irrigation water requirement estimated by the CropWat, mm.

The extra water required will be 1.2 times the sum required for irrigation, livestock, and domestic consumption (WR$_{LD}$). Therefore, Eq. (12.26) can be written as follows:

$$C \times V_Q = 1.2[(CA \times IRR) + (WR_{LD})] \qquad (12.29)$$

Equation (12.29) can determine the cultivated area for the chosen crops that the pond can sustain. It may be noted that WR_{LD} is usually expressed in litres, and it has to be converted into ha-mm for use in Eq. (12.29).

12.2.3.5 Shape and Dimensions of Pond

After determining the capacity of the pond, the next step is to decide the shape and dimensions of the pond. The excavated farm ponds may have a rectangular, square or inverted cone shape. Though the inverted cone-shaped ponds may have higher storage volume per unit surface area, these are difficult to construct and maintain. Therefore, either rectangular or square shapes are preferred. However, the rectangular shape is usually favoured due to the ease of construction.

Next, the pond dimensions, i.e., length, width and depth, are determined to provide the desired capacity. Of the three dimensions of the pond, selecting an appropriate depth is most vital considering the ease of excavation. Also, it governs water retention as choosing the maximum possible depth for a given capacity may minimise the seepage and evaporation losses. In dugout ponds, i.e., the ponds fed by groundwater, the depth is governed by the water table depth. However, the depth is usually kept within 2.5–3.5 m for ease of construction and operation.

The side slopes of the pond sidewalls depend on the soil type. The side slopes are usually kept flatter than the angle of repose of the soil to prevent caving in of the walls. Typically, 1:1 or flatter side slopes are provided. The side slopes may vary between 1:1 and 2:1 for clay, 1.5:1 and 2:1 for clay loam, and 2:1 and 2.5:1 for sandy loam. For sandy soil, the recommended side slope is 3:1 (Stephens 2010).

Once the volume, depth and side slopes are known (Fig. 12.13), the length and width of the pond can be determined using the following prismoidal formula:

$$V = \frac{(A + 4B + C)}{6} \times D \qquad (12.30)$$

where V = volume of excavation, m^3; A = area of the excavation at the ground surface, m^2; B = area of the excavation at the mid-depth (D/2) point, m^2; C = area

Fig. 12.13 Definition sketch of a farm pond (after Venkateswarlu et al. 2013)

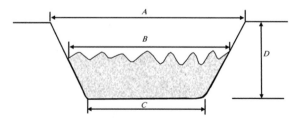

of the excavation at the bottom of the pond, m^2; and D = average depth of the pond, m.

Considering the plan and section views of a rectangular pond (Fig. 12.14), the bottom dimensions can be derived from Eq. (12.30) as follows:

Let us consider that L and B are two sides of the pond, and Z is the side slope (Z:1, Z horizontal to 1 vertical).

Thus, the area of the excavation at the bottom of the pond,

$$C = LB \tag{12.31}$$

Area of the excavation at the ground surface,

$$A = (L + 2ZD)(B + 2ZD) \tag{12.32}$$

Area of the excavation at the mid-depth (D/2) point,

$$B = (L + ZD)(B + ZD) \tag{12.33}$$

Substituting the values of A, B and C in Eq. (12.30), we get,

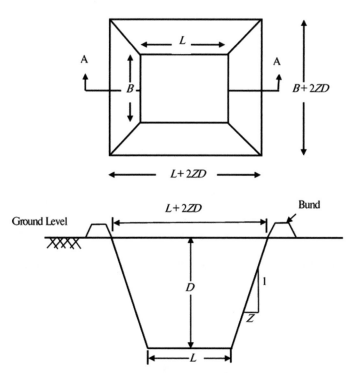

Fig. 12.14 Plan and section views of a rectangular pond (after Venkateswarlu et al. 2013)

$$V = \frac{(\{(L + 2ZD)(B + 2ZD)\} + \{4(L + ZD)(B + ZD)\} + LB)}{6} \times D \quad (12.34)$$

Assuming length–width ratio, $M = L/B$; thus, substituting $L = BM$ in Eq. (12.34), and simplifying,

$$\frac{6V}{D} = (BM + 2ZD)(B + 2ZD) + 4(BM + ZD)(B + ZD) + B^2M$$

Or,

$$\frac{6V}{D} = 6B^2M + 6ZMBD + 6ZBD + 8Z^2D^2$$

Or,

$$\frac{V}{D} = B^2M + ZBD(1 + M) + \frac{4}{3}Z^2D^2$$

Or,

$$B^2M + ZD(1 + M)B + \left(\frac{4}{3}Z^2D^2 - \frac{V}{D}\right) = 0 \quad (12.35)$$

Equation (12.35) is a quadratic equation, the solution of which would yield,

$$B = \frac{0.5}{M}\left[\sqrt{\{ZD(1 + M)\}^2 - 4M\left(\frac{4}{3}Z^2D^2 - \frac{V}{D}\right)} - ZD(1 + M)\right] \quad (12.36)$$

Thus, Eq. (12.36) can be used to determine the two sides of a rectangular pond. For a square pond, $L = B$, i.e., $M = 1$. Thus, Eq. (12.36) is simplified as follows:

$$B = \sqrt{\frac{V}{D} - \frac{Z^2D^2}{3}} - ZD \quad (12.37)$$

Let us now consider the plan and section views of an inverted cone pond (Fig. 12.15).

Let us consider that D_1 and D_3 are diameters at the base and the top of the pond, D is the depth of the pond, and Z is the side slope.

Thus, area of the excavation at the bottom of the pond,

$$C = \frac{\pi}{4}D_1^2 \quad (12.38)$$

Area of the excavation at the ground surface,

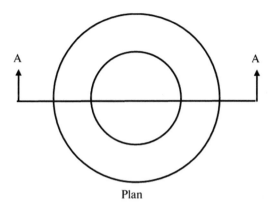

Plan

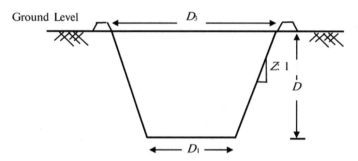

Fig. 12.15 Plan and section views of an inverted cone pond (after Venkateswarlu et al. 2013)

$$A = \frac{\pi}{4}D_3^2 = \frac{\pi}{4}(D_1 + 2ZD)^2 \qquad (12.39)$$

Area of the excavation at the mid-depth ($D/2$) point,

$$B = \frac{\pi}{4}(D_1 + ZD)^2 \qquad (12.40)$$

Substituting the values of A, B and C in Eq. (12.30), we get,

$$V = \frac{\left(\left\{\frac{\pi}{4}(D_1 + 2ZD)^2\right\} + \left\{4 \times \frac{\pi}{4}(D_1 + ZD)^2\right\} + \frac{\pi}{4}D_1^2\right)}{6} \times D \qquad (12.41)$$

Or,

$$\frac{24V}{\pi D} = 6D_1^2 + 12ZDD_1 + 8Z^2D^2$$

Or,

$$\frac{4V}{\pi D} = D_1^2 + 2ZDD_1 + \frac{4}{3}Z^2D^2$$

Or,

$$D_1^2 + 2ZDD_1 + \left(\frac{4}{3}Z^2D^2 - \frac{4V}{\pi D}\right) = 0 \qquad (12.42)$$

Equation (12.42) is a quadratic equation, the solution of which would yield,

$$D_1 = \sqrt{\left(\frac{4V}{\pi D} - \frac{1}{3}Z^2D^2\right)} - ZD \qquad (12.43)$$

Thus, Eq. (12.43) can be used to determine the pond diameter at the base. Subsequently, the pond diameter at the top can also be determined.

Example 12.4 Design a pond in an arid region to supply water to a farm having ten milk cows, four calves, four bulls, and a family of six. The pond also supplies irrigation to fruits and vegetables grown over 0.5 ha. Assume that seepage and evaporation losses of about 40% in the pond. The average annual rainfall is 400 mm, and the runoff curve number for the farm area is 70.

Solution Given: Average annual rainfall, $P = 400$ mm; runoff curve number for the farm area, $CN = 70$; number of persons on the farm $= 06$; number of livestock on the farm: milk cows $= 10$, calves $= 4$, bulls $= 4$; area under fruits and vegetables cultivation $= 0.5$ ha.

Determination of the annual water requirement and pond capacity

The recommended water requirement for farm cattle is as follows (Phansalkar 2006):
  Milk cows—70 L per day; bulls—30 L per day; calves—20 L per day.
  Therefore, the annual water requirement for livestock consumption

$$= [\{(70 \times 10) + (30 \times 4) + (20 \times 4)\} \times 365] = 328.5 \,\mathrm{m}^3$$

The recommended water consumption for the rural population is 55 L per capita per day (PIB 2020).
  Therefore, the annual water requirement for domestic consumption

$$= [(55 \times 6) \times 365] = 120.45 \,\mathrm{m}^3$$

Let us assume that seasonal vegetables are grown over 0.4 ha and bananas over 0.1 ha. Also, potatoes and tomatoes are grown over four months in summer, and peppers and peas are grown over four months in winter. The water requirements of potato, tomato, pepper, peas, and banana are 1130, 907, 801, 816, and 3369 mm (Elnashar et al. 2021).

Therefore, the annual water requirement for irrigation

$$= [\{(1130 + 907 + 801 + 816) \times 10^{-3} \times 0.4 \times 10^4\}$$
$$+ \{(3369 \times 10^{-3}) \times 0.1 \times 10^4\}] = 17{,}981 \text{ m}^3$$

Substituting the annual irrigation, livestock, and domestic requirements and 40% losses towards seepage and percolation losses in Eq. (12.18),

$$\text{Pond capacity} = (17{,}981 + 328.5 + 120.45) + 0.4$$
$$\times (17{,}981 + 328.5 + 120.45) = 25{,}800 \text{ m}^3$$

Therefore, the pond must have a capacity of 25,800 m³ to meet the annual water requirement at the farm.

Determination of the pond dimensions

Let us assume that the pond is rectangular, having 3.0 m depth, 2:1 side slope and a length–width ratio of 2.0.

Therefore, putting $V = 25{,}800 \text{ m}^3$, $D = 3.0 \text{ m}$, $M = 2.0$, and $Z = 2$ in Eq. (12.36),

$$B = \frac{0.5}{M}\left[\sqrt{\{ZD(1+M)\}^2 - 4M\left(\frac{4}{3}Z^2D^2 - \frac{V}{D}\right)} - ZD(1+M)\right]$$

$$= \frac{0.5}{2}\left[\sqrt{\{2 \times 3(1+2)\}^2 - (4 \times 2)\left(\frac{4}{3}(2)^2(3)^2 - \frac{25{,}800}{3}\right)} - (2 \times 3)(1+2)\right]$$

$$= 61 \text{ m}$$

Hence, length of the pond, $L = B \times M = 61 \times 2 = 122 \text{ m}$.

Hence, the dimensions of the pond are: Length = 122.0 m; width = 61.0 m; depth = 3.0 m; and side slope = 2:1.

Determination of the catchment area

The catchment runoff can be estimated using the SCS curve number method, described in Sect. 3.2.3 (Chap. 3).

Since CN is known, the potential maximum retention can be determined using Eq. (3.18):

$$S = \frac{25{,}400}{CN} - 254 = \frac{25{,}400}{70} - 254 = 108.9 \text{ mm}$$

Substituting S in Eq. (3.17), we get

$$V_Q = \frac{(P - 0.2S)^2}{(P + 0.8S)} = \frac{(400 - 0.2 \times 108.9)^2}{(400 + 0.8 \times 108.9)} = 293.7 \text{ mm}$$

Therefore, catchment area required,

$$A = \frac{\text{Volume of water required}}{\text{Runoff depth}} = \frac{25{,}800}{0.2937} = 87{,}845\,\text{m}^2 = 8.8\,\text{ha}$$

Answers: Pond capacity $= 25{,}800\,\text{m}^3$; pond dimensions: length $= 122.0$ m; width $= 61.0$ m; depth $= 3.0$ m; and side slope $= 2{:}1$; catchment area $= 8.8$ ha.

12.2.3.6 Inlet Channels and Spillways

The excavated farm ponds are provided with an inlet channel of sufficient capacity to receive the entire surface runoff generated from the catchment. The inlet channel is typically provided on one side of the farm without affecting the cultivated area. A grassed waterway or a stone-pitched earthen channel having non-erosive flow velocity may be used as the inlet channel to minimise erosion.

The flow from the inlet channel to the pond bottom may act like a waterfall and cause serious erosion problems. Consequently, the pond capacity may reduce over time. Therefore, the ponds are provided with inlets designed as a chute spillway for carrying the runoff smoothly to the pond bottom.

In the case of off-stream ponds, a pipeline of appropriate size should be used to divert sufficient flow from the stream to the pond. Alternatively, pumping arrangements may be made to fill the ponds during storm flows.

12.2.3.7 Outlet/Waste Weir

The excavated farm ponds may also be provided with an outlet or a waste weir to remove the excess runoff. The outlet is usually provided at the lowest end of the pond and designed to ensure a low, uniform flow to minimise erosion. Also, the outlet should be protected using either vegetation or stone pitching. The outlet elevation should be slightly lower than the inlet, and its discharge capacity should be about half of the inlet capacity.

12.2.4 Construction of Farm Ponds

Once the pond site and dimensions are finalised, the site is cleaned by removing woody vegetation and stones. The stakes are placed to mark the area to be excavated. The area where the excavated soil would be dumped should also be appropriately marked. The depth of cut from the ground surface to the pond bottom is marked on the stakes for proper guidance during digging.

Tractor-pulled wheeled scrapers, tracked excavators or diggers, and dragline excavators are commonly used for excavating ponds. The dragline excavators are used exclusively at sites having a shallow groundwater table.

The construction is usually initiated by digging at the centre to the designed depth and moving slowly towards the outer edged. Once the cut soil is entirely removed from the pond bottom, the desired slope is provided towards one corner of the pond by cutting the soil. Subsequently, the sides are shaped by cutting or scraping the soil to provide the desired slopes. After the excavation, berms are provided, and the pond bottom and sides are compacted thoroughly to get the final shape.

12.2.5 Operation and Maintenance of Farm Ponds

Proper operation and maintenance of the farm ponds are essential for beneficial use over their designed life. However, the practice requires periodic inspection and maintenance of the facilities. The farm ponds must be inspected for excessive erosion, settlement or cracks after every heavy rainfall, and damages, if any, should be immediately repaired. The damages may appear small in the beginning, but if neglected, these may become serious. The inlets and outlets should be checked for leakages and blockages. The pond should be kept free from rodents or burrowing animals, and any damage caused by them or due to trampling by livestock should be immediately repaired. For ensuring the pond water quality for irrigation and livestock drinking, efforts should be made to keep it clean and free from pollution. The pond may also be desilted annually to maintain its design capacity.

12.2.6 Sealing of Farm Ponds

Since seepage and percolation losses are significant in farm ponds, sealing is recommended for reducing the losses and ensuring the availability of water over a longer period. Sealing of farm ponds helps maintain water storage at design capacity for meeting the design commitments. Seepage losses depend on the soil permeability, with light textured soils having significantly high losses. Table 12.3 presents the seepage losses in different soils.

The seepage losses can be minimised by sealing the soil surface and reducing the soil permeability. There are several popular methods for sealing the soil surface (Kumar and Singh 2010). A few of these methods are discussed in the following sections.

Table 12.3 Seepage losses in different soils (after TNAU Agritech Portal 2021)

Soil type	Seepage loss (cumec per million m^3 of wetted surface)
Heavy clay loam	1.21
Medium clay loam	1.96
Sandy clay loam	2.86
Sandy loam	5.12
Loose sandy soil	6.03
Porous gravelly soil	10.54

12.2.6.1 Sealing by Compaction

Ponds constructed in areas having well-graded, coarse-grained soils can be sealed by compaction. In ponds having up to 3.0 m depth, a 200-mm-thick compacted seal could prove effective. However, if the pond depth exceeds 3.0 m, two or more layers of the compacted seal of 200 mm thickness each should be provided. This method is the most economical.

12.2.6.2 Sealing by Clay Lining

Ponds constructed in areas having coarse-grained soils can be sealed by clay lining or clay blanketing. A well-graded soil containing about 20% by weight of clay particles is used as the sealing material. Soil cement, a mixture of Portland cement and natural soil, is sometimes used as a sealing material. In ponds having up to 3.0 m depth, a 300-mm-thick compacted seal should be provided over the impounding area. Since cracks may develop in the clay lining or clay blankets, a 300–400-mm-thick gravel cover may be placed over the lining. This method is inexpensive and cost-effective.

12.2.6.3 Sealing with Bentonite

Bentonite, a natural colloidal clay, having the characteristics of absorbing water and swelling 8–15 times its dry size, is an efficient and effective sealing material. It is preferred in ponds constructed in well-graded coarse-grained soils. It is usually mixed with the coarse-grained material and spread as a 25–50-mm-thick membrane over a thoroughly compacted subgrade. Subsequently, it is mixed with the soil to a 150–200 mm depth using tillage equipment and compacted. The high cost of Bentonite makes this method expensive.

12.2.6.4 Sealing with Plastic Films and Synthetic Rubber

A variety of plastic films and synthetic rubbers are available for pond lining. The popular plastic films include polyvinyl chloride (PVC), high-density polyethylene (HDPE), low-density polyethylene (LDPE), reinforced polyethylene (RPE), and flexible polypropylene (fPP). In contrast, the popular synthetic rubbers include ethylene propylene diene monomer (EPDM) and butyl (synthetic) rubber. Besides cost, the choice of a lining material depends on its flexibility, weight, thickness, and UV and ozone resistance.

Polyvinyl chloride (PVC) was probably the first plastic material used for lining. PVC offers the advantages of low cost and flexibility. The flexibility makes it easier to handle and install. PVC can resist tears and punctures and withstand chemicals. However, PVC includes toxic plasticisers that may pollute the pond water. Also, these films could prematurely degrade as they have low UV and ozone resistance. PVC is also prone to cracks and splits at locations having very high or low temperatures. Therefore, PVC films should be used only after carefully weighing drawbacks against their low cost and ease of installation.

High-density polyethylene (HDPE) is historically one of the most popular plastic films worldwide as it is strong, safe against length-wise tears, and resistant to various chemicals. HDPE is very durable and could last for more than 36 years if properly maintained. However, HDPE is quite stiff and heavy; thus, their transportation and installation costs are high. Though their unit price is low, the excessive transportation and installation costs make it cost-effective only for large ponds. Also, HDPE being less flexible; they tend to expand and become loose at high temperatures.

Low-density polyethylene (LDPE) is made from the same plastic as HDPE. Therefore, like HDPE, LDPE is non-toxic, inexpensive, and functions well in the typical temperature range of pond water. Since LDPE has a low density, they are softer and more flexible than HDPE. The softness and flexibility of LDPE make it easier to install and mould around corners and less prone to cracking. Therefore, LDPE has replaced HDPE as one of the most favoured ponds lining materials, especially for small (capacity: < 200 m^3; depth: < 2.0 m) or medium-sized ponds (capacity: 200–1000 m^3; depth: 3–3.5 m). However, the softness of LDPE makes it susceptible to length-wise tears. Also, they have lower resistance to UV rays, chemicals, and oxidation than HDPE, and they are less durable.

Reinforced polyethylene (RPE) is made from the same plastic as HDPE and LDPE. Therefore, like HDPE and LDPE, RPE is non-toxic and functions well in the typical temperature range of pond water. However, being reinforced, it is stronger and more puncture resistant than either LDPE or HDPE, besides being moderately resistant to chemicals and UV rays. Also, RPE weighs only one-third of HDPE and LDPE, and thus, easy to transport and install. Though RPE is stiffer than LDPE, it is still easy to install and mould around corners. However, RPE films are more expensive than HDPE and LDPE films. Also, RPE films are reasonably new, and therefore, insufficient information is available concerning their expected lifespan.

Ethylene propylene diene monomer (EPDM) is a synthetic rubber and one of the most popular materials used for lining ponds. EPDM rubber is non-toxic, soft and durable, and offers a unique combination of flexibility and toughness. Besides being resistant to puncture, UV rays, ozone, and weathering, EPDM can withstand temperature variations. They have a life span of 27 years if properly maintained. However, EPDM geomembranes are comparatively expensive and relatively heavy, which results in high transportation costs.

Butyl (synthetic) rubber is also a synthetic rubber-like EPDM. It is highly flexible, easy to mould, durable, and resistant to UV radiation, ozone, and weathering. Like EPDM, it is non-toxic and withstands temperature variations. However, it is pretty expensive. Also, a significant drawback of butyl is its weakness, making it susceptible to rupture and puncture.

Though no plastic film or synthetic rubber discussed above is perfect, each possesses qualities fit for application as a lining material for farm ponds. However, as mentioned earlier, LDPE is popularly used as lining material for small and medium-sized farm ponds. LDPE film of 200-micron thickness covered with 300 mm soil is recommended as the most efficient and economical pond lining material (Kumar and Singh 2010). However, HDPE films of 500 microns are preferred for the lining of large farm ponds having 3–4 m depth of water storage (Reddy et al. 2012). In India, the Bureau of Indian Standards has recommended polyethylene films as per code number IS 15828 to design and construct lined farm ponds (BIS 2009).

For lining the farm ponds, the bed and the sides of the pond are levelled and compacted thoroughly. It is ensured that the bedsides are free from rocks, large stones or other projected articles that might damage the film. Once the bed and sides are ready, weedicides recommended for agricultural use are sprayed to suppress the growth of weeds to avoid damage to the film. In arid and semi-arid regions, insecticides may also be applied to make the soil free from termites. Subsequently, a cushion layer of fine sand is spread over the soil surface to protect the film against damage. The film is then spread over the cushion layer uniformly by fixing one end of the film. If multiple film rolls are used, then adjacent layers should be joined properly. After laying, the film is typically covered with a soil layer to avoid puncturing or tearing the film.

Questions

1. Describe various types of earthen embankments along with the specific conditions under which they are preferred.
2. Describe various types of farm ponds along with the specific conditions under which they are preferred.
3. Based on the information from the watersheds closer to your home town, managed under the soil management programmes, describe various existing earthen embankments and farm ponds.
4. Describe various methods of sealing farm ponds for seepage control. Discuss the advantages and disadvantages of each method, along with the specific conditions under which they are preferred.
5. Design a homogeneous earthen embankment with the following data:

Level of the Highest Flood Level (HFL) of the reservoir = 200 m; Level of the deepest river bed = 190 m; Length of the embankment = 350 m. Homogenous well-graded material is available for the fill, and the foundation is impervious. Assume a fetch of 1.4 km.

6. Derive an expression for determining the seepage discharge through an earthen embankment constructed using isotropic soils. Also, describe the correction needed in the expression if the construction material is anisotropic.

7. Calculate the seepage discharge through a homogenous earthen embankment having an upstream water level of 12.0 m. The top width of the embankment is 3.0 m, and its upstream and downstream side slopes are 2:1 and 2.5:1, respectively. A freeboard of 2.0 m is provided. The coefficient of permeability of the fill material is 2.5×10^{-5} m s^{-1}. What would be the seepage discharge if a 20 m long horizontal filter is placed inward from the downstream toe?

8. Plot the phreatic line for the homogeneous earthen embankment described in Question 7. Assume that the fill material is isotropic.

9. Determine the dimensions of an inverted cone dugout pond for storing 400 m^3 of water. The site permits a maximum depth of 2.5 m. The site has sandy clay loam soil. What would be the change in the storage capacity if the same area is used to construct a rectangular pond?

10. Design a pond in an arid region to supply water to a farm having twelve milk cows, six calves, six bulls, and a family of eight. The pond also supplies irrigation to fruits and vegetables grown over 1.0 ha. Assume that seepage and evaporation losses of about 35% in the pond. The average annual rainfall is 450 mm, and the runoff curve number for the farm area is 75.

Multiple Choice Questions

Answer the questions by choosing the correct option.

1. The process of constructing earthen embankments by laying and compacting soil in layers by power rollers under optimum moisture content (OMC) is known as

   (A) hydraulic-fill method   (B) zoned-fill method
   (C) rolled-fill method   (D) OMC method

2. In a diaphragm embankment, the diaphragm is called a thin core when

   (A) its thickness at any elevation is less than 10 m
   (B) its thickness at any elevation is less than the embankment height
   (C) it is constructed of pervious materials compared to the embankment body
   (D) it is constructed of compacted earth fill

3. For constructing an earthen embankment, the most favoured foundation is

   (A) rock foundation   (B) silt and clay foundation
   (C) sand and gravel foundation   (D) non-uniform foundation

4. In an earthen embankment, the flow line above which there is no hydrostatic pressure is known as

    (A) equipotential line   (B) phreatic line
    (C) stream line   (D) seepage line

5. For satisfying the permeability requirement of the filter material in an earthen embankment, the minimum ratio of the 15% finer by weight of the filter material and the base material, D15(F)/D15(B), shall be greater than

    (A) 0.1   (B) 0.2
    (C) 1.0   (D) 5.0

6. When the water standing against an earthen embankment drops down suddenly, then there is an imminent risk of sliding failure to the

    (A) downstream face   (B) foundation
    (C) impervious core   (D) upstream face

7. The ponds preferred in low-lying areas to utilise the shallow water table as the water source are known as

    (A) embankment ponds
    (B) excavated ponds
    (C) dugout ponds
    (D) surface ponds

8. Considering the ease of excavation while constructing a farm pond, the most vital dimension to select is the

    (A) length   (B) width
    (C) side slopes (D) depth

9. A natural colloidal clay, having the characteristics of absorbing water and swelling 8–15 times its dry size, used as an efficient and effective sealing material, is

    (A) Butyl
    (B) Bentonite
    (C) EPDM (Ethylene propylene diene monomer)
    (D) Soil cement

10. For lining a large farm pond having a 3.5 m depth of water storage, the lining material that should be preferred, considering the cost and the efficiency, is

    (A) EPDM (Ethylene propylene diene monomer)
    (B) HDPE (high-density polyethylene)
    (C) LDPE (low-density polyethylene)
    (D) PVC (polyvinyl chloride)

References

Agoramoorthy G, Chaudhary S, Hsu MJ (2008) The check dam route to mitigate India's water shortages. Nat Resour J 48:565–583

Bassel B (1907) Earth dams. Engineering News Publishing Co., New York

Bureau of Indian Standards (BIS) (1995) IS 911532: Construction and maintenance of river embankments (levees)— guidelines. River Training and Diversion Works (WRD 22), Bureau of Indian Standards, New Delhi

Bureau of Indian Standards (BIS) (1999) IS 9429: Drainage system for earth and rockfill dams—code of practice. Water Resources, Dams and Spillways (WRD 9), Bureau of Indian Standards, New Delhi

Bureau of Indian Standards (BIS) (2009) IS 15828: Design and construction of plastic lined farm ponds—code of practice. Bureau of Indian Standards, New Delhi

Casagrande A (1937) Seepage through dams. Contribution to soil mechanics, 1925–1940. Boston Soc Civil Eng, Boston

Clarke D, Smith M, El-Askari K (2001) CropWat for windows: user guide. University of Southampton, Southampton, UK

Critchley W, Siegert K, Chapman C (1991) A manual for the design and construction of water harvesting schemes for plant production. Food and Agriculture Organisation of the United Nations, Rome

Das BM (2020) Advanced soil mechanics, 5th edn. CRC Press, London

Elnashar A, Abbas M, Sobhy H, Shahba M (2021) Crop water requirements and suitability assessment in arid environments: a new approach. Agronomy 11:260

Foster M, Fell R, Spannagle M (2000) The statistics of embankment dam failures and accidents. Can Geotech J 37:1000–1024

Fukuchi T (2020) Accurate empirical calculation system for predicting the seepage discharge and free surface location of earth dam over horizontal impervious foundation. Eng 1:60–95

Justin JD (1932) Earth dams projects. Wiley, New York

Justin JD, Hinds J, Creager WP (1944) Engineering for dams: earth, rockfill, steel and timber dams, vol III. Wiley, New York

Kumar A, Singh R (2010) Plastic lining for water storage structures. Technical Bulletin No. 50. Directorate of Water Management, Bhubaneswar, India

Natural Resources Conservation Service (NRCS) (2005) Earth dams and reservoirs. TR- 60. USDA Natural Resources Conservation Service, Washington, DC

Phansalkar SJ (2006) Livestock-water interaction: status and issues. 5th IWMI-Tata Annual Partners Meet, Gujarat, India

Press Information Bureau (PIB) (2020, March 02). Per Capita Availability of Water [Press release]. Press Information Bureau (PIB), Government of India. https://pib.gov.in/PressReleasePage.aspx?PRID=1604871

Reddy KS, Kumar M, Rao KV, Maruthi V, Reddy BMK, Umesh B, Ganesh Babu R, Srinivasa Reddy K, Vijayalakshmi VB (2012) Farm ponds: a climate resilient technology for rainfed agriculture: planning, design and construction. Central Research Institute for Dryland Agriculture, Hyderabad

Schuyler JD (1905) Reservoirs for irrigation. Wiley, New York

Sharda VN, Juyal GP, Prakash C, Joshi BP (2007) Training Manual: soil conservation and watershed management, Vol-II (soil water conservation engineering). Central Soil and Water Conservation Research and Training Institute (CSWCRTI), Dehradun

Stephens T (2010) Manual on small earth dams: a guide to siting, design and construction. FAO Irrigation and Drainage Paper 64, Food and Agriculture Organisation of the United Nations, Rome

Tamilnadu Agricultural University (TNAU) Agritech Portal (2021). http://agritech.tnau.ac.in/agriculturalengineering/farmand_reservoir.pdf. Accessed 3 September 2021

United States Bureau of Reclamation (USBR) (1987) Design of small dams, 3rd ed. Govt. Printing Office, Washington, DC

United States Bureau of Reclamation (USBR) (2021) Design Standards No. 13, Embankment Dams, Chapter 8: Seepage. U.S. Department of the Interior, Bureau of Reclamation, Technical Service Centre, Denver, Colorado

Venkateswarlu B, Osman M, Padmanabhan MV, Kareemulla K, Mishra PK, Korwar GR, Rao KV (2013) Field manual on watershed management. Central Research Institute for Dryland Agriculture, Hyderabad

Wood AD, Richardson EV (1975) Design of small water storage and erosion control dams. Department of Civil Engineering, Colorado State University, Fort Collins, Colorado

Chapter 13
Remote Sensing and GIS Applications in Soil Conservation

Abstract Information on the extent and spatial distribution of soil erosion is essential for planning and implementing soil conservation measures. Remote sensing (RS) and Geographic Information System (GIS) concepts and tools are widely adopted in natural resources mapping, monitoring, and modelling. This chapter introduces the basic concepts of the RS and GIS and their applications in soil conservation. RS data acquisition and GIS software are particularly emphasised. The chapter includes the RS and GIS applications in land degradation, soil erosion mapping, and parameter estimation of various soil loss estimation models. The site selection for soil conservation structures is also discussed.

13.1 Remote Sensing and Geographic Information Systems

Remote sensing (RS) and Geographical Information System (GIS) have gained considerable importance and application in hydrology and water resources, including soil conservation. With the increasing number of the Earth Observation satellite missions and advancements in GIS technology, integration of remote sensing and GIS have brought revolutionary changes in data availability, analysis, and applications. The following sections present the basic concepts of remote sensing and GIS and their application in soil conservation.

13.2 Remote Sensing Basics

Remote sensing is the science of acquiring information from a distance, i.e., without coming in the physical contact with the target, which may be an object or area. The information is acquired by sensing and measuring the electromagnetic energy transmitted or radiated by the target and subsequently processing, analysing, and applying the information.

Remote sensing is used in our day-to-day life, e.g., while reading the newspaper or viewing the computer monitor's screen. The monitor screen (source) emits or

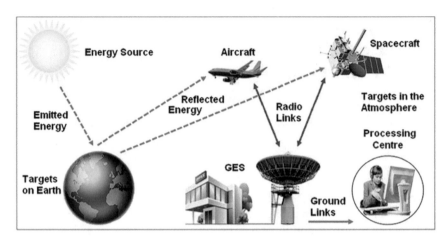

Fig. 13.1 Elements of remote sensing (After Llcev 2019; Reproduced with permission from Springer Nature)

radiates a light that travels over a distance and is captured by our eyes (sensor). Subsequently, the eye sends a signal to the processor (our brain), which analyses the data and converts it into information.

Like the human system, the Earth Observation (EO) system uses sensors to measure the Electromagnetic Radiation (EMR) reflected or emitted by the target. A typical imaging system involves the following seven elements (Fig. 13.1):

1. Energy Source or Illumination: The energy source is the prime mover of the remote sensing process. It emits EMR for illuminating or providing electromagnetic energy to the target of interest. The energy source may be the sun or radar.
2. Radiation and the Atmosphere: The EMR interacts with the atmosphere twice, first travelling from the source to the target and then travelling from the target to the sensor. During the path, EMR properties change due to energy loss and alteration in the wavelength.
3. Interaction with the Target: Once the EMR reaches the target, it interacts with the target in several ways. Based on the properties of the target and its own properties, the EMR may get reflected, absorbed, transmitted, and emitted in varying quantities.
4. Recording of Energy by the Sensor: After interacting with the target, the scattered or emitted EMR is recorded by a sensor based on the nature of the EMR and capability.
5. Transmission, Reception, and Processing: The EMR recorded by the sensor is transmitted, often in an electronic form, to a receiving and processing station where the data are processed into a digital or analogue image.
6. Interpretation and Analysis: The processed image is interpreted visually, digitally, or electronically to extract information about the target.

7. Application: The extracted information is utilised to understand or obtain a new insight or solve a particular problem.

13.2.1 Electromagnetic Radiation (EMR)

EMR is the energy emitted by the energy source for illuminating or providing electromagnetic energy to the target of interest. It is a dynamic form of energy comprising electromagnetic waves that travel through space in the form of periodic disturbances of electric and magnetic fields. The electromagnetic waves are transverse waves with electric and magnetic fields varying perpendicularly to the direction of propagation of the wave.

The energy source governs the frequency and the wavelength of the electromagnetic wave. Electromagnetic waves have a wide range of frequencies and wavelengths. For example, radio waves have a low frequency, while gamma rays have a high frequency.

The wide frequency range of electromagnetic waves is known as the electromagnetic spectrum.

13.2.2 Electromagnetic Spectrum

The electromagnetic spectum represents the wide range of the EMR wavelength (Fig. 13.2). The spectrum is divided into regions or intervals of different wavelengths (called bands). It includes visible light, radio waves, microwaves, infrared light, ultraviolet light, X-rays, and gamma rays. The visible (VIS, wavelength 0.4–0.7 μm), infrared (IR, wavelength 0.7–1000 μm), and microwave (wavelength 0.001–0.10 m) bands are most commonly used in satellite remote sensing. However, it may be noted that bands have no sharp boundaries; thus, adjacent bands may overlap.

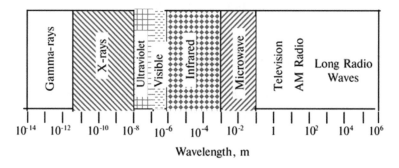

Fig. 13.2 Different regions of the electromagnetic spectrum

The gamma rays (< 0.01 nm) and X-rays (0.01–10 nm) occupy the shorter wavelength end of the spectrum. Typically, their wavelengths are expressed in angstrom (10^{-10} m). Thus, the gamma rays and X-rays have wavelengths of < 0.1 angstrom and (0.1–100 angstrom), respectively. The longer wavelength region beyond 0.1 m belongs to the radio bands, from very high frequency (VHF) to extremely low frequency (ELF). This region is usually expressed in terms of frequency, i.e., in units of Hertz.

As mentioned above, the visible band, which our eyes can see, has a wavelength of 0.4–0.7 μm. Red colour has the longest visible wavelength (0.62–0.7 μm), while violet (0.4– 0.446 μm) has the shortest wavelength. The other colours that we can see are blue (0.446–0.500 μm), green (0.500–0.578 μm), yellow (0.578–0.592 μm), and orange (0.592–0.620 μm). Blue, green, and red are the primary colours, as all other colours can be formed by combining these three colours in various proportions.

The infrared (IR) region, which spans the wavelength range of 0.7–1000 μm, may be divided into reflected IR (0.7–3.0 μm) and thermal IR (3.0–1000 μm).

The IR region is also divided into near-IR (NIR, 0.7–1.3 μm), mid-IR (MIR, 1.3–3 μm), and far-IR (FIR, 3–1000 μm) bands.

13.2.3 Types of Remote Sensing

EMR from the energy source illuminates or provides electromagnetic energy to the target of interest. The detection and differentiation of targets are done based on the reflected or emitted EMR from the target, as different targets reflect or emit varying amounts of energy in different electromagnetic spectrum bands. Typically, a *sensor* (e.g., a camera or a scanner) mounted on a *platform* (e.g., an aircraft or a satellite) detects the reflected or emitted EMR from a target.

Remote sensing may be classified in the following ways:

1. Based on platforms
2. Based on energy sources
3. Based on the range of electromagnetic spectrum
4. Based on the number of bands.

13.2.3.1 Classification Based on Platforms

Based on the earth's surface elevation, remote sensing platforms can be ground based, airborne, or spaceborne.

Ground based

Ground-based platforms include hand-held devices, ladders, tripods, towers, tall buildings, and cranes. These platforms are typically used for sensor calibration, quality control, close-range characterisation of objects, improving existing technologies, or developing new sensors. Permanent ground-based platforms like towers may also be used for long-term monitoring of terrestrial features.

Airborne

Airborne platforms mainly include aircraft, although balloons are sometimes used for carrying disposable probes for meteorological research. Usually, stable wing aircraft are used for remote sensing, though helicopters may also be used occasionally. The choice of aircraft is governed by the altitude and stability requirements of the sensor, i.e., sophisticated aircraft must be used if higher altitude or superior instrument stability is needed. Helicopters are preferred for low altitudes, especially when hovering ability is needed. Nowadays, unmanned aerial vehicles (UAVs), i.e., *drones*, are employed in remote sensing for mapping or real-time monitoring of various activities.

Spaceborne

Satellites are commonly used as space-borne platforms, though space shuttles or space stations may occasionally be used. Remote sensing satellites may be geostationary or polar orbiting.

Geostationary satellites have a rotation period equal to the earth, i.e., 24 h, so these stay over the same location. These are typically located over the equator at about 35,800 km and used for weather forecasting, satellite TV, and communications.

Polar-orbiting satellites constantly circle the earth in an almost north–south orbit, passing close to both poles. Sun-synchronous orbit (SSO) is a special case of polar orbit, and sun-synchronous satellites in SSO pass over the equator at the same local sun time. Thus, sun-synchronous satellites always maintain the same relative position with the sun and ensure illumination.

13.2.3.2 Classification Based on Energy Sources

Based on the EMR source, remote sensing can be classified as passive remote sensing or active remote sensing.

Passive remote sensing

Passive remote sensing systems utilise an external source of energy, e.g., the sun, or naturally emitted energy, while active remote sensing systems have their own source of energy, e.g., RADAR.

Most remote sensing systems are passive as these use solar energy as an EMR source. Solar energy either gets reflected (visible and reflective IR wavelengths) or absorbed and reemitted (thermal infrared wavelengths) by the targets and is recorded using passive sensors on-board airborne or space-borne platforms. Passive remote sensing that detects reflected energy can occur only when the sun is illuminating the earth, i.e., during the daytime. However, the naturally emitted energy by targets can be detected during day or night, provided it is sufficient to be recorded.

Active remote sensing

As mentioned earlier, active remote sensing systems have their own energy source for illumination. The energy is emitted and sent towards the targets. The energy reflected from the targets is recorded using active sensors on-board airborne or space-borne platforms. Most of the microwave remote sensing is done through active remote sensing. Active remote sensing offers the advantage of obtaining data regularly, irrespective of the day or season.

13.2.3.3 Classification Based on Range of Electromagnetic Spectrum

Based on the range of the electromagnetic spectrum, remote sensing can be classified as optical, thermal, or microwave remote sensing.

Optical Remote Sensing

The optical remote sensing devices operate in the electromagnetic spectrum's visible and reflective infrared portion. These devices are sensitive to wavelengths ranging from 0.4 to 3.0 μm. Optical remote sensing sensors usually use the sun as an energy source and record reflected energy from targets.

Thermal Remote Sensing

The thermal remote sensing devices operate in the thermal infrared portion of the electromagnetic spectrum. These devices are sensitive to wavelengths ranging from 3.0 to 1000 μm. Thermal infrared remote sensing sensors usually use the naturally emitted energy from targets.

Microwave Remote Sensing

The microwave remote sensing devices operate in the microwave portion of the electromagnetic spectrum. Microwave remote sensing can be both passive and active. In passive microwave remote sensing, the microwave radiation emitted from the object is detected, while in the active microwave remote sensing, the backscattered microwaves are detected. These devices record the backscattered microwaves in wavelengths of 0.001–0.10 m. However, most microwave sensors are active, having their own energy sources. The active microwave remote sensing has an advantage over others as it is independent of weather, i.e., cloud cover, and solar radiations.

13.2.3.4 Classification Based on Number of Bands

Based on the number of sensitive bands a sensor uses, remote sensing can be classified as panchromatic, multispectral, or hyperspectral remote sensing.

Panchromatic Remote Sensing

In panchromatic remote sensing, images use a single band that combines the information from the visible bands of blue, green, and red. In other words, the band is formed by using the total light energy in the visible spectrum (instead of partitioning it into different spectra). Each pixel of a panchromatic image gives the total intensity of solar radiation reflected by the objects in the pixel and is commonly visualised in a greyscale image.

Multispectral Remote Sensing

In multispectral remote sensing, sensors capture data in multiple bands (usually covering the visible and infrared portions of the electromagnetic spectrum) to produce multispectral images. Usually, the higher the number of spectral bands, the more information the sensor gathers. Multispectral remote sensing may be performed in the optical, thermal, and microwave regions. However, sensors and imaging techniques are different for different regions.

Hyperspectral Remote Sensing

Hyperspectral remote sensing is an extension of the multispectral remote sensing technique. Contrary to the multispectral images with a few relatively broad wavelength bands, hyperspectral images capture data from dozens or hundreds of narrow adjacent spectral bands. Consequently, hyperspectral sensors can provide more than 100 spectral bands compared to less than 15 bands for multispectral sensors. Hyperspectral remote sensing is usually performed in the optical and thermal regions of the electromagnetic spectrum.

13.2.3.5   Interaction of EMR with the Atmosphere

As mentioned earlier, the EMR emitted from the sun interacts with the atmosphere twice, first travelling from the source to the target and then travelling from the target to the sensor. While travelling to the target on the earth, the EMR interacts with the particles and gases in the atmosphere, and consequently, its characteristics are impacted by the following physical processes:

1. Scattering
2. Absorption
3. Refraction.

The processes and their effect on the EMR are discussed below.

13.2.3.6   Scattering

Scattering is the change in the direction of motion of the EMR due to collision with the particles of the large gas molecules present in the atmosphere.

The magnitude of scattering depends on the EMR wavelength and the distance it travels through the atmosphere and the size and the concentration of the particles in the atmosphere.

Based on the EMR wavelength and size of the atmospheric particles involved in the process, scattering can be classified as *Rayleigh* scattering, *Mie* scattering, or *non-selective* scattering.

Rayleigh scattering occurs when the size of the particles causing the scattering is much smaller than the EMR wavelength. The particles responsible may be fine dust particles or oxygen and nitrogen molecules. Rayleigh scattering is the most dominant scattering, and its magnitude is inversely proportional to the fourth power of the EMR wavelength. Consequently, the EMRs with shorter wavelengths are scattered significantly more than EMRs with longer wavelengths. Therefore, blue light is scattered more than red light. Thus, the sky appears blue during the day and red at sunrise and sunset due to Rayleigh's scattering.

Mie scattering occurs when the size of the particles causing scattering is similar in size to the EMR wavelength. The particles responsible may be aerosols like smoke, dust, pollen, and water vapour. Mie scattering influences a larger spectral region than the Rayleigh scattering, i.e., the influence extends to the near-infrared region. It mainly occurs in the lower atmosphere due to the abundance of larger particles and is the dominant scattering process under overcast conditions. Since atmospheric aerosols cause it, it becomes prominent during forest fires, dust storms, or such events that increase the aerosol load in the atmosphere.

Non-selective scattering occurs when the size of the particles causing scattering is much larger than the EMR wavelength. The particles responsible may be large dust particles and water droplets. This scattering process is called non-selective as all wavelengths are scattered almost equally. It is responsible for clouds and fog appearing white as the combination of red, green, and blue in equal quantities produces white.

13.2.3.7   Absorption

Absorption is the mechanism in which the EMR is partially absorbed by atmospheric constituents like ozone, carbon dioxide, and water vapour at specific wavelengths (Fig. 13.3). Ozone absorbs ultraviolet radiation and restricts its transmission to the earth, thus preventing most living things, including human beings, from its harmful effect. Carbon dioxide absorbs radiation strongly in the mid and far-infrared spectrum regions, i.e., the portion of the EMR spectrum responsible for thermal heating. Thus, carbon dioxide traps the heat within the atmosphere and is called a greenhouse gas. On the other hand, water vapour absorbs most thermal infrared and shortwave microwave radiation. Absorption reduces the light reaching our eyes.

Atmospheric Window

Since atmospheric constituents absorb EMR in specific spectrum regions, only a limited electromagnetic spectrum is available for remote sensing. The region of the

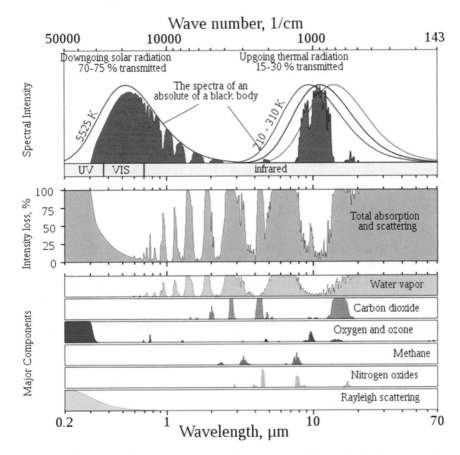

Fig. 13.3 Wavelengths absorbed by atmospheric constituents (Atmospheric Transmission by Cepheiden; Creative Commons CC0 1.0 Universal Public Domain Dedication)

electromagnetic spectrum over which the atmosphere is relatively transparent, i.e., not influenced by absorption, is known as the *atmospheric window*. The visible portion of the spectrum, i.e., 0.4–0.76 μm, belongs to the atmospheric window and has the peak energy level of the sun. Similarly, the thermal IR portion of the spectrum, i.e., around 10 μm, belongs to the atmospheric window, in which the earth emits heat energy. The atmospheric window also includes a large portion of the spectrum beyond 10 mm wavelength, used for microwave remote sensing.

13.2.3.8 Refraction

Refraction is the bending of EMR from its straight-line path due to variation in the density of the atmospheric layers. It occurs as EMR moves from outer space (density ≈ 0), enters the atmosphere (density > 0), and moves through various atmospheric layers of varying density. The angle at which the electromagnetic wave bends is governed by the refractive indices of the two adjoining layers. The refractive index of a layer (or medium) is the ratio of the speed of the EMR in the vacuum (c) to that in the layer. The refractive index of the atmosphere at the sea level is 1.0003, and it decreases with the altitude, depending on the temperature, humidity, and air pressure. Refraction produces mirage, the optical phenomenon that produces a displaced image of distant objects or the sky. It also makes astronomical objects appear higher in the sky than in reality. Besides, the twinkling of the stars or the deformation in the shape of the sun at sunrise or sunset is also caused by refraction.

13.2.4 Spectral Signature

Various materials of the earth's surface have different spectral reflectance characteristics at different wavelengths. The plot of the spectral reflectance as a function of the wavelength is called a *spectral reflectance curve*. Since spectral reflectance varies with the physical and chemical characteristics of the object, it usually has a range of values. Therefore, spectral response patterns are typically averaged to get a unique spectral response pattern for the object. The unique spectral response pattern of an object is referred to as the *spectral signature*. The spectral signature helps distinguish different objects on the earth's surface.

The spectral characteristics of a few select objects are discussed below.

Clear and Turbid Water: A significant component of the EMR incident upon clear water gets absorbed or transmitted, especially the wavelengths in the spectrum's longer visible or near-infrared regions (Davis et al. 1978). Thus, the reflectance of clear water is low. The reflectance of clear water is maximum at the blue end of the spectrum and decreases with the increasing wavelength. Therefore, clear water appears blue or blue-green. On the other hand, turbid water has suspended sediment particles, which results in an increased reflectance in the red end of the spectrum. Therefore, turbid water appears brownish.

The reflectance of a water body depends on its depth and the suspended materials' characteristics.

Dry and Wet Soil: A significant component of the EMR incident upon the soil surface gets reflected or absorbed, and only a limited portion is transmitted (Clark 1999; Zhang et al. 2012). The reflectance of the soil surface depends on its texture, structure, organic matter content, iron content, and moisture content. In the case of dry soil, the reflectance increases monotonically with increasing wavelength. Therefore, dry soil appears yellowish.

The presence of moisture in the soil decreases its reflectance. Thus, wet soil has lower reflectance than dry soil. Consequently, wet soil appears yellowish-red or brownish.

Vegetation: Vegetation has a unique spectral signature that varies with wavelength. The chlorophyll in the plant leaves significantly absorbs radiation in the red and blue wavelengths but reflects green wavelength. Therefore, the reflectance curve of the vegetation has a peak in the green region, which is responsible for its green appearance to the human eye (Clark 1999; Zhang et al. 2012). Also, the reflectance is significantly high in the near-infrared region due to the cellular structure of healthy leaves. The measurement of the vegetation reflectance in the near-infrared region helps determine the health of the vegetation.

13.2.5 Image Processing

13.2.5.1 Image

The earth's features are detected and monitored by measuring the reflected and emitted EMR by cameras or sensors attached to a satellite or an aircraft. The two-dimensional representation of the object on the earth is obtained as an *image*. The image may be *analogue* or *digital*. Aerial photographs are analogue images, while satellite images acquired using electronic sensors are examples of digital images.

An analogue image uses a format in which data are saved continuously, i.e., the information is not broken into several small blocks or pieces. On the other hand, a digital image uses a format that saves data in several separate blocks. Digital images are usually used nowadays.

A digital image includes a two-dimensional array of individual picture elements, called *pixels*. Each pixel has an intensity value, denoted by a digital number, and a location address, referenced by row and column numbers in the two-dimensional image. The column-row address of the pixel corresponds to the geographical coordinates, i.e., the latitude and longitude, of the object on the earth's surface.

13.2.5.2 Image Resolution

A remote sensing image is characterised by its spatial, spectral, temporal, and radiometric resolutions.

Spatial resolution refers to the ability of the remote sensing sensor to detect details, i.e., the instantaneous field of view of the sensor representing the ground area viewed in a given instant in time. It denotes the size of the smallest object that can be resolved on the ground. It is usually expressed in metres. For example, a spatial resolution of 30 m means that one pixel represents an area of 30 m × 30 m on the ground.

A high spatial resolution refers to a small pixel size containing finer details of the object. On the contrary, a low spatial resolution refers to a large pixel size containing coarse details of the object. However, a low-resolution image may include many different objects in one pixel.

Spectral resolution refers to the ability of the remote sensing sensor to resolve or differentiate fine wavelength intervals. In other words, it describes the amount of spectral detail in a band. The higher the spectral resolution, the narrower the wavelength range for a particular band. For example, a panchromatic image has a low spectral resolution, including a single large (wide) band. On the other hand, a hyperspectral image has a high spectral resolution, including many small-width bands. Thus, a high spectral resolution image has higher discriminating power than a low spectral resolution image.

Typically, remote sensing images have a high spatial resolution and a low spectral resolution, or a low spatial resolution and a high spectral resolution.

Radiometric resolution refers to the ability of the remote sensing sensor to detect the slightest change in the reflected or emitted energy. A sensor having a higher radiometric resolution is more sensitive to detecting slight differences in reflected or emitted energy. It is usually measured in bits (a number to the exponential power of 2). For example, if a sensor uses 8-bits to record the data, it will have 2^8 or 256 digital numbers or grey tones, ranging from 0 (black) to 255 (white). On the other hand, if only 4-bits were used, only 2^4 or 16 grey tones would be available, ranging from 0 (black) to 15 (white). The comparison between 8-bits and 4-bits images shows a significant difference in the level of detail included in the images.

Temporal resolution refers to the time elapsed between consecutive images of the same ground location taken by the sensor. It reflects the revisit period of the satellite or the time it takes to complete an orbit cycle, usually expressed in days. For example, suppose the temporal resolution of a satellite is ten days. In that case, it will take 10 days to revisit a particular ground location and provide consecutive images of the location at a 10-day interval.

13.2.5.3   Types of Images

An image may be classified as a multilayer, multispectral, hyperspectral, or panchromatic image.

A *multilayer image* is formed by stacking several images involving various measurements over the exact ground location covered by a pixel. Each measurement, carrying some specific information about the area, forms an image, with each component image acting as a layer in the multilayer image.

Multilayer images may also include images obtained from different sensors and other subsidiary data. For example, a multilayer image may consist of three layers from a SPOT (Satellite Pour l'Observation de la Terre, France) multispectral image, a layer of the synthetic aperture radar image, and perhaps a layer consisting of the digital elevation map of the area being studied.

A *multispectral image* consists of a few image layers of the same scene, with each layer having an image acquired at a particular wavelength band. A multispectral image usually refers to three to ten bands, including the visible, near-infrared, and shortwave infrared bands. Since various objects reflect and absorb differently at different wavelengths, the spectral reflectance curve of the objects helps differentiate among them.

A *hyperspectral image* consists of more than a hundred spectral bands, including the visible, near-infrared, and mid-infrared bands. The bands are much narrower in a hyperspectral image than in a multispectral image. Moreover, the bands are often contiguous and do not have a descriptive name. A hyperspectral image provides detailed information, thus enabling a better identification and characterisation of the object.

A *panchromatic image* consists of a single band that combines all three visible bands, i.e., red, green, and blue, to provide a greater spatial resolution. Since the visible bands get a high amount of solar radiation, panchromatic images are significantly brighter than multispectral images. However, the use of the red–green–blue (RGB) bands amounts to sacrificing colour for brightness as panchromatic images are greyscale images. Also, it does not have any wavelength-specific information.

13.2.5.4 Image Classification and Analysis

Image classification and analysis is a vital process in the digital image analysis that includes digital identification and pixel classification based on spectral information from one or more spectral bands. Each pixel in an image is assigned a particular class or theme, e.g., urban, agricultural, and wetlands, based on the statistical characteristics of the pixel brightness values. The classified image is essentially a *thematic map* of the original image as it comprises a mosaic of pixels, each of which belongs to a particular theme.

Image classification approaches can be broadly classified into two types: supervised classification and unsupervised classification. The critical difference between the supervised and unsupervised classification approaches is whether the classification is based on a priori information about the land cover types at different area locations or entirely on the image data.

Supervised image classification involves a priori knowledge of the type and location of different land cover classes, say, urban, forest, or wetlands, through a training dataset. The training dataset may be obtained through field visits, aerial photographs, previous maps, and the investigator's knowledge. The classification procedure identifies *training sites* having known spectral characteristics on the image to train the classification algorithm. Subsequently, the trained algorithm is applied to the remainder of the image to identify and classify spectrally similar areas. It is also called a hard classification scheme, as each pixel is assigned to only one class. The classification result depends significantly on the training dataset.

The procedure begins by defining the classes of interest based on the intended application. The number of training pixels required for a particular class depends on the distinctive variability of the spectral response. As a thumb rule, the number of sites should be at least ten times the number of spectral bands in the image or 100 per class. Also, the training sites should be spread across the image to account for all possible variations in the image. Typically, statistics-based techniques, like the maximum likelihood method, modified maximum likelihood method, Bayesian's method, decision tree classification, and discriminant functions, are used for the supervised image classification.

Unsupervised image classification does not utilise a priori knowledge of different land cover classes or training datasets. Instead, it relies on the numerical information in the image (data) to group the spectral classes. The grouped spectral classes are subsequently interpreted to identify the land cover classes. Unsupervised classification uses clustering algorithms to identify and assign pixels to a spectral group. Typically, the clustering algorithms utilise the user inputs like the number of classes, the number of bands, and separation distance among the clusters for grouping the spectral classes. The process is iterative as the clusters may be broken down further or combined to obtain the final group of spectral classes. Two standard algorithms used for unsupervised classification are k-means clustering and Iterative Self-Organising Data Analysis Technique (Algorithm) (ISODATA).

13.2.6 Earth Observation Satellites and Missions

The Earth Observation satellites (EOS) usually fly at an altitude of 250–1000 km above the earth's surface and an orbit inclination of about 90°. The orbit is called a *polar orbit* or a Low Earth Orbit (LEO). The sun-synchronous orbit is a special LEO used for all observation systems that are synchronised to always be in the same fixed position relative to the sun and the same surface illumination angle of the sun. The orbit has an altitude of 600–800 km above the earth's surface.

Like the earth, the satellites rotate once per year around the sun, i.e., 360° in 365 days. The orbit is essential for imaging systems like optical cameras. Another orbit, which is used primarily for weather satellites, is the Geostationary Orbit (GEO). The GEO orbit has an altitude of about 36,000 km from the earth's surface. The satellites in GEO have the same angular velocity as the earth, i.e., 360° in 24 h; thus, these occupy the same position relative to the earth's surface and are called *geostationary*.

Many satellite missions collect data related to the land surface, energy, climate elements, water vapour, and other variables. A few of the existing National Aeronautics and Space Administration (NASA) earth science missions are highly relevant to hydrology and water resources, e.g., Global Precipitation Measurement (GPM), Soil Moisture Active Passive (SMAP), Gravity Recovery and Climate Experiment (GRACE), Ice, Cloud, and Land Elevation Satellite-2 (ICESat-2), and Surface Water & Ocean Topography (SWOT).

Similarly, the European Space Agency (ESA) has specific missions that are relevant to hydrology and water resources, e.g., CryoSat-2, European Organisation for the Exploitation of Meteorological Satellites (EUMETSAT), the Soil Moisture and

Ocean Salinity (SMOS), and Sentinel series. Table 13.1 lists the major satellite missions relevant to hydrology and water resources.

In addition to the satellite missions summarised in Table 13.1, several other satellites and sensors have been used significantly in hydrology and water resources. Moreover, the Indian Remote Sensing (IRS) programme of the Indian Space Research Organisation (ISRO) has a series of land observation satellites aimed at the observation and management of the country's natural resources, with particular emphasis on agriculture, hydrology, geology, drought and flood monitoring, marine studies, snow studies, and land use.

Table 13.2 lists the popular satellites and sensors, including the satellites under IRS programme.

Table 13.1 Summary of the satellite missions relevant to hydrology and water resources

Hydrological cycle component	Mission/ sensor[*]	Standard spatial resolution (km)	Standard temporal resolution (days)	Launch year
Precipitation	GPM	10	0.02	2014
Evapotranspiration	Terra MODIS	0.25, 0.5 and 1	1–2	1999
	Aqua MODIS	0.25, 0.5 and 1	1–2	2002
	Landsat 8	15 (Panchromatic) 30 (Multispectral)	16	2013
	Landsat 9	15 (Panchromatic) 30 (Multispectral)	16	2021
Snow and ICE cover	ICESat-2	20	91	2018
	CryoSat-2	0.25	369	2010
Soil moisture	SMOS	35	3	2011
	SMAP	36	3	2015
Streamflow	SWOT	1 (Ocean) 0.05 (Land water)	21	2021
Groundwater	GRACE	300–400	30	2002
Water cycle	WCOM	15–50 (FPIR) 4–50 (PMI) 2–5 (DFPSCAT)	–	–

[*] GPM = Global Precipitation Measurement; MODIS = Moderate Resolution Imaging Spectroradiometer; ICESat Ice, Cloud, and land Elevation Satellite; CryoSat = Earth Explorer Opportunity Mission; SMOS = Soil Moisture and Ocean Salinity; SMAP = Soil Moisture Active Passive; SWOT = Surface Water & Ocean Topography; GRACE = Gravity Recovery and Climate Experiment; WCOM = Water Cycle Observation Mission

Table 13.2 Summary of the satellites and sensors popularly used in hydrology and water resources

Satellite	Sensor	Standard spatial resolution (m)	Standard temporal resolution (days)	Launch year
Landsat 1–3	Multispectral Scanner (MSS)	80	18	1972–1978
Landsat 4–5	Thematic Mapper (TM) + MSS	30 (TM) 80 (MSS)	16	1982–1984
Landsat 7	Enhanced Thematic Mapper Plus (ETM +)	15 (PAN) 30 (MS)	16	1999
Terra	ASTER	15 (VNIR) 30 (SWIR) 90 (TIR)	16	1999
SRTM	SIR-C/X-SAR	90 30	11	2000
IKONOS-2	PAN (Panchromatic) MS (Multispectral)	1 (PAN) 4 (MS)	14	1999
IRS-1A IRS-IB	LISS-I LISS-II	72.5 36	22	1988 1991
IRS-1C IRS-1D	PAN LISS-III WiFS	5.8 (PAN) 23.5 (VNIR) 70 (SWIR) 188.3 (WiFS)	5	1995 1997
IRS-P4 (Oceansat-1) Oceansat-2	OCM MSMR	360	2	1999 2009
IRS-P5 (Cartosat-1)	PAN	2.5	5	2005
IRS-P6 (Resourcesat-1) Resourcesat-2 Resourcesat-2A	LISS-III LISS-IV AWiFS	23.5 5.8 56	24	2003 2011 2016
IRS-P7 (Cartosat-2) Cartosat-2C Cartosat-3	PAN	1 0.6 0.25	4	2007 2016 2019
RISAT-2B RISAT-2BR1	X-band SAR	0.5 0.35	5	2019 2019

ASTER = Advanced Spaceborne Thermal Emission and Reflection Radiometer; SRTM = Shuttle Radar Topography Mission; SIR-C/X-SAR = Spaceborne Imaging Radar-C/X-band Synthetic Aperture Radar; IRS = Indian Remote Sensing Satellite; LISS = Linear Imaging Scanning Sensors; WiFS = Wide Field Sensor; OCM = Ocean colour monitor; MSMR = Multifrequency scanning microwave radiometer; AWiFS = Advanced Wide Field Sensor; RISAT = Radar Imaging Satellite

13.2.7 Spectral Vegetation Indices

One of the significant applications of remote sensing has been the qualitative and quantitative assessment of the vegetative covers, specifically the crop growth, vigour, biomass, and chlorophyll content. The information has been used beneficially for managing agricultural production and crop yield estimation (Mulla 2013).

These successful remote sensing applications in conventional agriculture have even led to the development of a well-established discipline category, precision agriculture. Using spectral measurements, researchers developed the vegetation index (VI) concept for evaluating vegetative covers (Bannari et al. 1995). A vegetation index, also called a vegetative index, is a single number that quantifies vegetation biomass and plant vigour for each pixel in a remote sensing image. Several vegetation indices have been developed since the spectral response of vegetated areas is impacted by vegetation types, soil moisture and colour, and spatio-temporal atmospheric variations. The vegetation indices focus on different vegetation properties and provide information about the plant's biomass, leaf area, and health based on different electromagnetic spectrum bands (Hatfield et al. 2019).

The VI concept is based on the relative amounts of light reflected, transmitted, and absorbed by a surface as a function of the wavelength. Consequently, for analysing the vegetation characteristics, the red part of the visible spectrum (0.62–0.7 μm) and near-infrared (NIR, 0.7–1.3 μm) are used as these contain about 90% of the information (Baret et al. 1989a). The visible radiation (0.4–0.7 μm) is typically absorbed by chlorophyll in plant leaves, while the leaf cellular structures strongly reflect NIR radiation. Therefore, vegetative covers can be distinguished by their spectral behaviour with respect to the ground cover and soils by measuring and quantifying the difference between these two spectral bands (Bannari et al. 1995). Table 13.3 presents a few popular vegetation indices.

13.3 GIS Basics

13.3.1 GIS Definition

A Geographic Information System (GIS) is defined in various ways to elucidate its capabilities. The GIS definition has changed over time to highlight its broad applications. A few example definitions are as follows:

- 'Set of tools for collecting, storing, retrieving at will, transforming and displaying spatial data from the real world for a particular set of purposes'. (Burrough 1986)
- 'Computer-based system that provides four sets of capabilities to handle georeferenced data, viz., input data management (data storage and retrieval), manipulation and analysis, and output'. (Aronoff 1989) and retrieval of all forms of spatially indexed and related descriptive data'. (Jackson 1992)

Table 13.3 Summary of the popular vegetation indices

Index	Abbreviation	Equation	Reference
Ratio vegetation index	RVI	$\frac{R}{NIR}$	Pearson and Miller (1972)
Vegetation index number	VIN	$\frac{NIR}{R}$	Pearson and Miller (1972)
Normalised difference vegetation index	NDVI	$\frac{NIR-R}{NIR+R}$	Rouse et al. (1974)
Difference vegetation index	DVI	$NIR - R$	Richardson and Wiegand (1977)
Soil-adjusted vegetation index	SAVI	$\frac{NIR-R}{NIR+R+L}(1+L)$	Huete (1988)
Modified SAVI	MSAVI	$0.5\left[2NIR + 1 - \sqrt{(2NIR+1)^2 - 8(NIR-R)}\right]$	Qi et al. (1994)
Transformed SAVI	TSAVI	$\frac{[a(NIR-aR-b)]}{[(R+aNIR-ab+X(1+a^2))]}$	Baret et al. (1989b)
Optimised SAVI	OSAVI	$\frac{NIR-R}{NIR+R+X}$	Rondeaux et al. (1996)
Vegetation index_green[a]	VI_{green}	$\frac{G-R}{G+R}$	Gitelson et al. (2002)
Visible atmospheric resistant Index[a]	VARI	$\frac{G-R}{G+R-B}$	Gitelson et al. (2002)
Red green blue vegetation Index[a]	RGBVI	$\frac{(G)^2-(B+R)}{(G)^2+(B+R)}$	Bendig et al. (2015)

[a] Vegetation indices based on UAV remote sensing in the visible spectral region; R = reflectance in the red spectrum range; NIR = reflectance in the near-infrared spectrum; L = soil conditioning index; X = adjustment factor to minimise soil brightness effects; G = reflectance in the green range of the spectrum; B = reflectance in the blue range of the spectrum

- 'Automated systems for the capture, storage and retrieval of spatial data'. (Clarke 1997)
- 'Everyone has their own favourite definition of a GIS, and there are many to choose from'. These include GIS as follows: 'a container of maps in digital form...a computerised tool for solving geographical problems...a spatial decision support system...a mechanised inventory of geographically distributed features and facilities...a tool for revealing what is otherwise invisible in geographic information...a tool for performing operations on geographic data that are too tedious or expensive or inaccurate if performed by hand...'. (Longley et al. 2005)
- 'System that creates, manages, analyses, and maps all types of data'. (Environmental Systems Research Institute, Inc., ESRI 2022)

These definitions show that GIS has evolved from a computerised system for a specific application to a more generalised system comprising hardware and software tools to facilitate creating, managing, analysing, and displaying spatial information (Wieczorek and Delmerico 2009).

13.3.2   GIS Components

GIS concepts and technologies have evolved from various fields, and GIS has become a generic term referring to all automated systems used primarily to manage maps and geographic data. A GIS has five key components: hardware, software, data, people, and methods (Fig. 13.4).

Fig. 13.4 GIS components

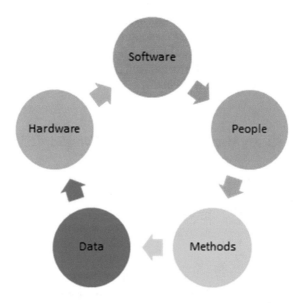

Hardware is the computer system on which GIS software runs. The computer system may range from a cloud-based or centralised server to a personal computer with an advanced processor.

The hardware may also include input and output devices like scanners, digitisers, printers, and plotters.

The software provides tools to query, edit, run, and display GIS data. It may include a graphical user interface (GUI) for operational ease, a relational database management system (RDBMS) to store the data, and tools to analyse, store, and display spatial information. A few popular GIS software are ArcGIS, ArcView, QGIS, Google Earth Pro, and MapInfo Pro.

People form a vital component of the successful operation of the GIS. People may include technical experts who design and maintain the system and users who run the GIS software to carry out their assigned tasks.

Methods govern the GIS operations based on the predefined plan and rules unique to an individual organisation. Accordingly, the models and techniques to carry out a specific activity will be chosen.

Data are the most critical component of a GIS. A digital map forms the primary input data for GIS. The digital data may also have tabular data related to the map objects attached to it. A GIS integrates spatial data with other data resources and can even use a database management system (DBMS), which most organisations use to maintain their data and manage spatial data. Most GIS software includes an RDBMS to create and maintain a database for organising and managing data.

A GIS typically stores and organises data in layers, with each layer representing a particular theme linked precisely to the geographical location (Fig. 13.5). Once registered within a standardised reference system, these thematic layers help us simultaneously compare and analyse information stored in different layers. Thus, this simple but powerful thematic approach simplifies the organisation and representation of the information and paves the way for solving real-world planning and decision-making problems.

13.3.3 Map Projections and Coordinate Systems

The earth has a near-spherical shape, called a spheroid, and consequently, spherical globes are often used to represent the three-dimensional earth. A globe typically has horizontal lines representing the parallels of latitude and vertical lines representing the meridians of the longitude. The network of parallels and meridians, called the graticule, facilitates the drawing of maps. However, globes have several drawbacks. For example, they are too big and bulky compared to a map, and their small scale limits their practical use as the geographic details are sacrificed. Moreover, the latitude–longitude–spherical coordinate system specifies positions on the globe, which can

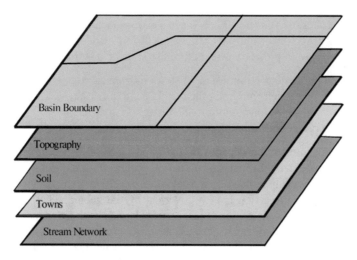

Fig. 13.5 Map layers representing individual themes

only be used to measure angles, not distances or areas. These drawbacks may be overcome by transforming the spherical three-dimensional earth into a two-dimensional planar surface like a flat piece of paper or a digital display screen. The mathematical models used to transform the spherical coordinates, i.e., latitude and longitude, into two-dimensional planar coordinates, i.e., X and Y, constitute the *map projection*. Alternatively, the map projection is the process of drawing the graticule on a flat surface.

To illustrate the concept of a map projection, let us imagine that a light bulb is placed in the centre of a translucent globe. When the light bulb is switched on, the outline of the graticule will be projected as shadows on the wall or any other nearby surface. This projection of the graticule demonstrates the map projection.

Typically three surfaces, a cylinder, a cone or a plane, are used to project the graticule's shadows. These surfaces, known as *developable surfaces,* are placed outside the globe in the light bulb experiment such that shadows are cast upon them. Subsequently, the developable surfaces may be opened flat to display the geographical features on a flat surface. Based on the nature of the developable surface, the map projections are categorised as *cylindrical, conical,* or *zenithal* (or *planar* or *azimuthal*) projections.

These projections are further classified as *equatorial* (or *normal or standard*), *transverse,* or *oblique* based on the position of the plane touching the globe. If the developable surface touches the globe at the equator, i.e., tangential to the equator, it is called the equatorial projection. If the developable surface is tangential to the pole, it is called the transverse projection. All other orientations of the developable surface between the pole and the equator are called oblique projections.

Several distortions are introduced in the shape, distance, direction, and area during the map projection. The distortions in the specified features vary depending on the map projection, map scale, and the mapped spatial extent. The objective of reducing the amount of distortion governs the choice of a map projection. However, the minimisation of a particular distortion may increase the distortion of one or more of the other features. The types of projections that minimise specific distortions are as follows:

- *Conformal* projections minimise the distortion in shape.
- *Equidistant* projections minimise the distortion in the distance.
- *True-direction* projections minimise the distortion in direction.
- *Equal-area* projections minimise the distortion in area.

The choice of projection depends on the importance of a specific feature for a project. For example, conformal projections are preferred for navigational purposes as maintaining a bearing or heading is vital when travelling great distances.

The choice of a projection is also governed by the geographical object being mapped and the map scale. For example, the North and South Poles maps typically use planar or azimuthal projections. In contrast, the conical projections are best suited for the mid-latitude areas of the earth. On the other hand, the standard cylindrical projection represents the features that stretch east–west well, e.g., Russia. In contrast, the transverse projection better represents north–south countries, e.g., Chile and Norway.

As evident from the above discussion, theoretically, an infinite number of map projections exist.

13.3.3.1   Coordinate Systems

Coordinate systems define the unique location of points in space. The *Cartesian* system, named after the French mathematician and philosopher *René Descartes,* is the most commonly adopted coordinate system in a two-dimensional plane. In the Cartesian system, points are labelled by their distance along a horizontal (x) and a vertical (y) axis from the origin, a reference point with coordinates (0, 0). Similarly, for identifying the locations on the three-dimensional curved surface of the earth, a *geographic coordinate system* (GCS) is used. In GCS, the locations are expressed in angular units from the earth's centre relative to two planes, the planes defined by the equator, and the prime meridian that passes through Greenwich, London. Therefore, a point is located by two values, i.e., a latitude and a longitude. Latitude is measured relative to the equator at 0°, with a maximum of ninety degrees south (90° N) at the North Pole or ninety degrees south (90° S) at the South Pole. On the other hand, longitude is measured relative to the prime meridian at zero degrees, with a maximum of 180° west or 180° east. The latitudes and longitudes may be expressed in degrees-minutes-seconds (DMS) or decimal degrees (DD). When expressed in decimal degrees, latitudes above the equator, i.e., north, and longitudes east of the prime meridian are positive, and latitudes below the equator and longitudes west of the prime meridian are negative. For example, if a point is located at 45°45' N (45.75° N) latitude and 65°15' W (65.25° W) longitude, in decimal degrees, the location will be +45.75° and −65.25°.

An ellipsoid, geoid, and datum typically define a GCS.

Ellipsoid, Spheroid, and Geoid

The earth's shape is best represented by an *ellipsoid* having a semi-major axis (the equatorial radius) and a semi-minor axis (the polar radius). However, the earth's equatorial radius (6378 km) and the polar radius (6357 km) do not vary significantly. Therefore, the ellipsoid shape of the earth closely resembles a sphere and is often considered a *spheroid*. It may be noted that though the spheroid assumption works well for small-scale maps, the ellipsoid model of the earth is essential for accuracy, especially for large scales.

The ellipsoid models assume that the earth's surface is smooth and homogenous. However, in reality, the earth's surface is not a smooth ellipsoid; instead, it is quite uneven and rugged. The undulations on the earth's surface are due to variations in the gravitational pull across its surface. Thus, the ellipsoid models of the earth's surface have a drawback that they do not take topography into account and may introduce errors in elevation estimations. Therefore, we use geoid models to overcome this drawback of the ellipsoid models.

Geoid models use mean sea level (MSL) to account for the earth's surface elevation variations. Mean sea level represents the earth's sea-level surface elevation if the water was completely flat, i.e., without tides and currents. Since water on the earth's surface responds to its gravitational pull, a high MSL reflects a stronger gravitational force and indicates more mass beneath the surface, i.e., a mountain. On the contrary, a low MSL indicates dips in elevation on the ocean floor. The geoid is also called the *surface of the equal gravitational potential.*

Therefore, the need to adopt a simple mathematical model of the earth's surface while genuinely representing the undulating nature of the earth's surface requires aligning the ellipsoid or the spheroid with the geoid. The alignment can be local in which the ellipsoid surface closely fits the geoid at a particular location on the earth's surface or geocentric, where the ellipsoid is aligned with the earth's centre. This alignment of the ellipsoid and the geoid defines a datum.

A *datum* is typically a spheroid or an ellipsoid model that specifies the orientation and origins of the lines of latitude and longitude relative to the centre of the earth. The GCS, which is used for specifying the latitude and longitude of a point on the earth's surface, also requires specifying the datum. Since different models (spheroid or ellipsoid) define the critical parameters like the point of origin, and equatorial and polar axes of the ellipsoid, differently, changing the datum alters the latitude and longitude of a place. Therefore, while describing a place by its latitude and longitude, it is essential to specify the datum. The Everest ellipsoid is used as the datum for the Survey of India maps in India.

Figure 13.6 shows the earth's surface along with the ellipsoid and the geoid. The *geoid height* represents the vertical distance between the geoid and the ellipsoid, while the *ellipsoid height* represents the difference between the ellipsoid and the coordinate on the earth's surface.

As mentioned earlier, a datum can be local or geocentric. The local and geocentric data are described below.

Fig. 13.6 Earth's surface
along with ellipsoid and
geoid

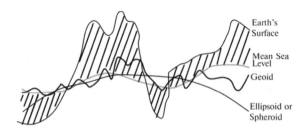

A projected coordinate system (PCS) is a reference system used for identifying the locations and measuring features on a flat (map) surface. It consists of lines that intersect at right angles, forming a grid. PCS is based on the Cartesian coordinates, and the unit of measurement is linear, usually in metres. Conversion from a GCS to a PCS requires a mathematical transformation. As described earlier, the projections can be planar, cylindrical, and conical.

Local Datum

A local datum aligns a geoid with the ellipsoid at a particular location on the earth's surface. There are several local data, a few of which are old, while others are recently defined. A particular datum is chosen based on the location. For example, the North American Datum of 1927 (NAD27) works well in the USA but fails in other parts of the world. Similarly, the European Datum of 1950 (ED50) suits well for Europe. Table 13.4 presents a few popular local datum.

Geocentric Datum

A geocentric datum aligns a geoid with the ellipsoid at the earth's centre. Most modern datum uses a geocentric alignment, e.g., the popular World Geodetic Survey for 1984 (WGS84) or the North American Datum of 1983 (NAD83). Most popular geocentric datum uses the WGS84 ellipsoid or the Geodetic Reference System 1980

Table 13.4 Popular local datum used around the world

Name of the local datum	Acronym	Most suitable region for application	Remarks
North American datum of 1927	NAD27	Continental US	An old datum, but still prevalent because of the wide use of older maps
European datum of 1950	ED50	Western Europe	Developed after World War II and is still quite popular. However, not used in the UK
World geodetic system 1972	WGS72	Global	Defined and maintained by the United States National Geospatial-Intelligence Agency (NGA)

Table 13.5 Popular geocentric datum used around the world

Name of the geocentric datum	Acronym	Most suitable region for application	remarks
North American Datum of 1983	NAD83	Continental US	One of the most popular modern datum for the contiguous USA
European terrestrial reference system 1989	ETRS89	Western Europe	One of the most popular modern datum for much of Europe
World geodetic system 1984	WGS84	Global	Used by the global positioning system

(GRS80) ellipsoid. These two ellipsoids share nearly identical semi-major and semi-minor axes: 6,378,137 and 6,356,752 m, respectively. Table 13.5 presents a few popular geocentric datum.

Projected Coordinate Systems

The planar projection is popularly used for mapping Polar Regions but may be used for other locations on the earth's surface. However, in such cases, it is called an oblique planar projection. Three examples of polar projections are orthographic, gnomonic, and equidistant projections.

One of the widespread cylindrical projections is the *Transverse Mercator projection* (named after the Belgian cartographer Gerardus Mercator). In this projection, parallels and meridians are rendered as straight lines spaced to produce a correct ratio of latitude to longitude at any point. A particular subset of the Transverse Mercator is the *Universal Transverse Mercator* (UTM), commonly used by scientists and federal organisations. UTM is based on a series of 60 Transverse Mercator projections, in which the globe is divided into 60 6° wide zones between 84° S and 84° N. Each UTM zone has its central meridian and spans 3° west and 3° east from the centre of the zone. Within each zone, a local coordinate system is defined. The X-origin is located 500,000 m west of the central meridian, and the Y-origin is the South Pole or the equator, depending on the hemisphere. For zones in the northern hemisphere, the X-origin is at 500,000 m west of the central meridian, and the Y-origin is the equator (the South Pole for the southern hemisphere). The false easting or northing values are applied to eliminate negative coordinates. UTM coordinate system is commonly used with NAD27 and NAD83 datum in the USA. A WGS84-based UTM coordinate system also exists

Three widespread conical projections are Albers equal area (preserves area), equidistant (preserves distance), and conformal (preserves shape). Equidistant Conic and Albers Equal Area Conic (area-preserving) are popularly used in the contiguous USA. In contrast, Europe Albers Equal Area Conic and Europe Lambert Conformal Conic projections are used in the European maps. Different countries have adopted different standard projections for mapping. In India, the Survey of India (SOI) commonly uses the polyconic projection for its toposheets.

13.3.4   GIS Data

A GIS utilises two kinds of data: *spatial data* and *attribute data*. The spatial data describe the coordinate location of the geographical feature on the earth's surface, while the attribute data describe the characteristics of the geographical feature. For example, the coordinate location of a historical monument would be spatial data, while its name, the year it was built and by whom, and such other information would form the attribute data.

13.3.4.1   Spatial Data

Spatial data represent features that have a known location on the earth. Features may include natural objects, e.g., rivers, mountains, vegetation, or man-made constructions, e.g., highways, streets, buildings, pipelines, and classifications, e.g., cities, countries, land use patterns.

Spatial data have the following four specific features that distinguish them from other digital data:

1. Spatial data always have the associated attribute data. For example, if a point feature is highlighted on a map, the attribute data describing the feature may also be highlighted in an associated table, called an attribute table. The geographical features and their attributes are usually dynamically linked in a GIS environment, i.e., if the features are edited on the map, the changes are automatically updated in the attribute table or vice versa.
2. Spatial data are georeferenced, i.e., they are linked to a GCS.
3. Spatial data may be categorised based on the features represented. Thus, spatial data may reflect point, line, or area features and are stored separately.
4. Spatial data are typically arranged in different thematic layers, with each layer representing a specific feature. For example, land use, soil type, and river network may be stored as separate layers. This thematic layer approach facilitates efficient data storage, management, and analysis.

Spatial data may also include photographs, videos, historical information, public amenities, and land records.

13.3.4.2   Attribute Data

Attribute data are descriptive data linked to the geographical features (spatial data) represented by a point, line, or area. For points, attribute data may include the location's name, elevation, and other features. For lines, attribute data may include the name of a highway or a canal and other associated information. Similarly, for polygons, attribute data may include the name of a state, its area and population, the area under forest in the state, and other features. Attribute data are typically kept in tabular form in attribute tables and are tagged to spatial data. Attribute data are also called the Z value in a GIS. The X and Y coordinates describe the geographic location in a GIS database, while the Z value specifies an attribute. It may be noted that in an elevation or terrain model, the Z value represents elevation, while in other kinds of surface models, it represents the density or quantity of a particular attribute. For example, Mount Everest has X and Y coordinates of 27.9881° N latitude and 86.9250° E longitude and a Z value of 8849 m as elevation.

13.3.5 Data Models

Spatial data models are approaches for storing spatial data in a database. Data models help represent geographical features as discrete objects in GIS. The two primary data models are vector and raster.

13.3.5.1 Vector Data Model

Vector data models represent geographical features as points, lines, or polygons (Fig. 13.7). A point representing features like cities or towns is usually defined by Cartesian coordinates (X, Y) or geographical coordinates (latitude, longitude). A line representing features like roads, streams, or canals is defined by a series of sequential points having Cartesian or geographical coordinates, with the two endpoints of the line identified. The endpoints and the intermediate points of the line are called nodes and vertices, respectively. A polygon, representing area features like districts, lakes, or agro-ecological zones, is defined by combining a series of lines signifying the boundaries of an enclosed area. Usually, two types of data structures are used in vector data models: topologic data structure and the non-topologic or computer-aided drafting (CAD) data structure.

Topologic Data Structure

The topologic data structure, the most popular vector data model, utilises the mathematical concept of topology. Topology is based on feature adjacency and connectivity principles, and it uses absolute location and relative location of the geographical features to establish spatial relationships. Absolute location is the location of a geographic feature, while relative location is the feature's location relative to other features. For example, the absolute location of the Taj Mahal is 27°10′26.0076″ N latitude and 78°2′31.4448″ E longitude, while its relative location is in the city of Agra, south of Yamuna river and 180 km south-east of New Delhi. Topology helps distinguish a GIS from other graphics or cartography systems as topologic data structure facilitates the easy tracing of spatial relationships between geographic features.

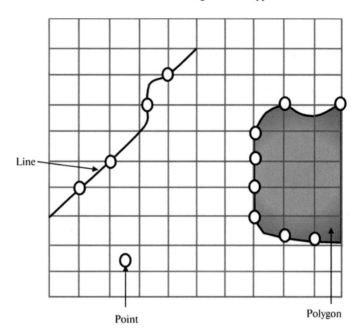

Fig. 13.7 Vector data model showing point, line, and polygon features

Non-Topologic Data Structure

The computer-aided drafting (CAD) data structure is the non-topologic data structure that defines geographic features as elements using points, lines, or polygons, defined by vertices. Thus, features, e.g., polygons, are independent, without any spatial relationship with other features. Consequently, feature adjacency and connectivity, two vital attributes of the topologic data structure, are absent in the CAD data structure.

13.3.5.2   Raster Data Formats

A raster data model uses a matrix of grid cells or pixels to store data, with grid cells or pixels representing the geographic area organised into rows and columns (Fig. 13.8). The raster data resolution and the appearance of the geological features are governed by the pixel or grid cell size. A smaller pixel or grid cell size will contain more feature details than a larger pixel or grid cell size. However, a small pixel or grid cell size will result in many cells, thus, requiring longer data processing time and more storage space.

Figure 13.9 presents the point, line, and area representation using the grid cells or pixels. Raster data formats are conveniently used for representing remote sensing imageries, aerial photographs, and scanned paper maps.

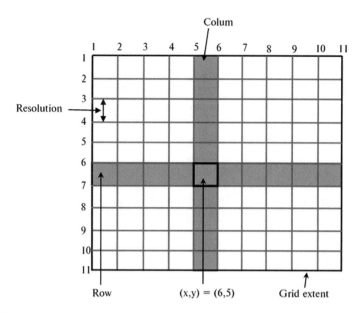

Fig. 13.8 Raster data model showing matrix of grid cells

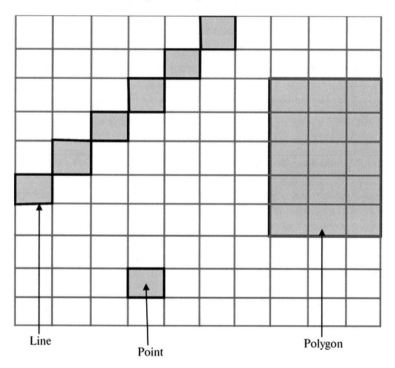

Fig. 13.9 Point, line, and polygon representation in a raster data model

There are three commonly used raster data structures: cell-by-cell encoding, run-length encoding, and quadtree.

Cell-by-Cell Encoding Data Structure

The cell-by-cell encoding data structure is the simplest raster data structure. It uses the conventional grid cells or pixels matrix to store data, with each grid cell or pixel assigned a value. Digital elevation models (DEM) are the best example of the cell-by-cell encoding data structure. This data structure can also be used in satellite images.

However, a pixel in a satellite image may have more than one value as most satellite images contain data from multispectral bands. Thus, for storing data in a multiband/multispectral imagery, Band Sequential (BSQ), Band Interleaved by Lines (BIL), or Band Interleaved by Pixels (BIP) formats are used.

Run-Length Encoding Data Structure

The run-length encoding (RLE) data structure helps overcome the problem of redundant or missing data in a grid. In the RLE method, adjacent cells along a row with the same value are treated as a group and called a run. Thus, instead of repeatedly storing the same value for each cell, the value is stored only once, along with the number of the cells included in the run. The attribute data for a run remains the same.

Quadtree Data Structure

The quadtree data structure is used to compress the data and save space. A quadtree works by partitioning a two-dimensional space by recursively subdividing it into four quadrants or regions. The regions may be square or rectangular or may have an arbitrary shape. The attribute data for all the pixels of a quadrant remain the same even if it is divided.

13.3.5.3 Comparison of Vector and Raster Data Models

Both vector and raster data models have their advantages and disadvantages. However, most modern GIS packages can handle both vector and raster data. Table 13.6 compares the vector and raster data models.

13.3.5.4 Attribute Data Models

Attribute data are stored and maintained using a separate data model. These data models may be inbuilt into the GIS software or available through external database management system (DBMS). The commonly used attribute data models are hierarchical, network, relational, and object oriented.

A brief description of these models is as follows:

Table 13.6 Comparison of vector and raster data models

Properties	Vector	Raster
Data structure	Complex; stored as (x, y) coordinate; useful for data having discrete boundaries, such as country borders, land parcels, and streets	Simple; stored as pixels or array of cells representing a digital number; helpful in storing continuously varying data, such as aerial photographs or satellite images
Topology	Easier to represent topology among graphical objects	Extremely difficult to represent topology among graphical objects with pixels
Storage	Relatively less space is needed as only point coordinates are stored. It can also store attribute information	Relatively huge space is needed as data include all pixels. It has no attribute information
Data conversion	Fast	Slow
Access	It does not require high computer processing power	Requires high computer processing power
Analysis	Complex as data must be converted into a topological structure; processing intensive; difficult to perform spatial analysis	Easy due to the nature of the data storage technique; easy to programme and quick to perform; simple to perform spatial analysis
Accuracy	Produces high spatial accuracy as the data are discrete; representation of features is accurate	No accurate representation of more than one feature covered under a cell; accuracy depends on the size of the cell; high spatial variability like elevation is well represented
Visualisation and output	Poor for continuous data but suitable for various kinds of maps; pleasing to the eye; not suitable for 3D display	Good for images but poor for discrete data; less pleasing to the eye; suitable for 3D display

Hierarchical Model

The hierarchical model organises data into a tree-like structure with a single root. The hierarchy begins from the root, and it expands like a tree with child nodes for further expansion. A child node can only have a single parent node, but a parent can have multiple child nodes. This model is not very popular and is used only with stable data, where primary relationships among data do not change substantially.

Network Model

Data is organised in a network or plex structure in the network model. Like the tree-like structure in the hierarchical model, the network or plex structure can also be described in terms of parents and children. However, in this model, children may have more than one parent. This model is also not adaptive towards changes, i.e., even a minor modification in the system requires changes in the whole system. Also, the system is complex as all records are connected through some pointers, which makes maintenance challenging.

Relational Model

In the relational data model, data are organised in tables, with each table having a unique name. The tables have multiple rows and columns, with columns representing a specific attribute and rows representing records. Tables are referenced to each other by common columns or relational fields, which have an identification number for a specific geographic feature. The identification number acts as the primary key for the table. The core of the relational model is its ability to join various tables using the common column. Usually, Structured Query Language (SQL) is used for retrieving data. The relational model is the most widely used database model for the attributes of geographic data. Most GIS software has an internal relational data model, besides provision for using external DBMS.

Object-Oriented Model

In the object-oriented model, data are managed through objects. The data model deals with object structure, object class, and object identity. The object structure denotes the object properties that are as attributes. Thus, an object structure may also be defined as a real-world entity with specific attributes. Each object belongs to a class, a group of all objects having the same properties and behaviour. Individual objects of a class are also called instances of the class. Thus, each instance contains its graphical location, characteristics, and other associated data in a GIS. The object-oriented model offers the advantage of simplified and fast queries, besides operational benefits for geographic data processing. Consequently, it is gaining popularity in the GIS fraternity.

In addition to vector and raster data models, spatial data may also be presented using a Triangulated Irregular Network (TIN) model described below.

Triangulated Irregular Network (TIN)

In the TIN model, a surface is represented as a set of connecting, non-overlapping triangles. The triangles are drawn by joining irregularly spaced points, called mass points. Within a triangle, the surface is represented by a plane. The accuracy of the surface model is governed by the mass points representing a significant variation in the shape of the surface, e.g., the peak of a mountain, the floor of a valley, or the edges of a cliff. Heterogeneous surfaces that vary sharply in some areas and less in others can be modelled accurately. TIN provides the flexibility of placing many mass points for an irregular surface and fewer points for a flat surface. TIN also provides a more efficient method to store data.

13.3.6 Data Input

GIS input data are derived from various sources and by using various procedures. The primary spatial data sources are analogue maps, aerial photographs, remotely sensed imageries, point data collected during sample surveys, Global Positioning System (GPS) data, and existing digital data files. The popular procedures for spatial data input in a GIS include manual digitisation, automatic scanning, converting existing digital data, and GPS data processing.

13.3.7 Spatial Analysis

In a GIS, both location and attribute information are stored together. For vector data, locations of vector map objects are stored in a spatial table, and the corresponding attribute data are stored in an attribute table. The vector data may or may not have a topology structure. For raster data, location is stored in the raster cells' position, and the raster cell's value is the attribute data. The availability of topological and attribute data together helps visualise raw data spatially, besides facilitating various raw data analyses to create new data or information.

Spatial analysis refers to various operations to analyse existing data for creating new data or information. The spatial analysis aims at answering questions that may not be explicitly answered using the stored data, thus making GIS a valuable tool for decision-making. It performs well if the data are error free and follow a topological structure. Most GIS provide a standard set of spatial analysis tools for analysing or manipulating existing data to generate new information.

The usually adopted spatial analysis functions are queries, classifications and generalisation, neighbourhood operations, measurements, network analysis, and overlay analysis.

Overlay analysis is frequently used in a GIS. Overlay operations combine multiple spatial data layers representing different themes for analysing or identifying relationships among them. Two input data layers are chosen in a typical overlay operation, assuming they overlap in the study area and are georeferenced in the same system. During the spatial overlay, the characteristics of the same location in both data layers are compared to produce an output data layer. Overlay operations may be logical or mathematical. With raster data, the characteristics are compared for a cell chosen from each input raster. Raster overlay operations use map algebra, i.e., algebraic operations are performed on input data layers to generate a new map. With vector data, the characteristics are compared between the point, line, and polygon features and their associated attributes from each input layer. Theoretically, any vector features can be overlaid with other vector feature types. Boolean algebra that involves binary logical operations linking two spatial selection criteria using the Boolean operators AND, OR, NOT, and XOR is used in vector overlay operations.

13.3.8   GIS and Remote Sensing

Although remote sensing and GIS developed quasi-independently, they are increasingly used together in an interactive and complementary way. Presently, GIS software can analyse and display remote sensing images, while image processing software can analyse ancillary geospatial data. Ehlers (1990) envisaged three levels of possible remote sensing-GIS integration. The first-level integration refers to data exchange between separate RS and GIS systems, while the second-level integration includes combined raster-vector processing through a common user interface. The third-level integration aims at a single, integrated RS-GIS system. Since GPS is a critical element of contemporary RS-GIS integration, Ehlers' three-level classification must include GPS (Gao 2002). The published literature shows that the first two levels of integration envisaged by Ehler have been successfully achieved. An example of first-level integration of remote sensing and GIS is the overlay of a digital image with a cartographic dataset derived from a GIS (e.g., roads selected from a digital line graph). The integration produces a merged product that allows the visualisation of information derived from both (Merchant and Narumalani 2009). Similarly, digital image classification employing multisource data is an example of second-level integration. The digital image classification procedures use multiple sources of data (ancillary data and multiple types of imagery) for extracting information from images through image interpretation while invoking the reasoning and logic employed in the visual interpretation of images (Merchant and Narumalani 2009).

Remote sensing is the prime source of various thematic data critical to GIS analyses. A simple example could be DEMs, which form a broad GIS application. Satellite imageries acquired by IKONOS, ASTER, or Light Detection and Ranging (LIDAR) are often used for generating DEMs due to their high resolution and accuracy. Similarly, satellite RS and GPS technologies have revolutionised the GIS application. Consequently, GPS–GIS–remote sensing is increasingly employed in precision farming, emergency response, and mobile mapping.

13.3.9   Data Sources

Presently, a wide variety of satellite-based sensors are providing operational raster remotely sensed data for GIS applications. These data differ in spatial, temporal, and spectral characteristics based on a satellite mission's nature and specific goals. Table 13.7 presents the popular data sources along with the spatial resolution of data for a typical water resources modelling project. Since scores of remote sensing platforms and hundreds of sensors exist, Table 13.7 provides limited information. A more comprehensive list is provided in the Manual of Remote Sensing series, Volume 1.1: Earth Observing Platforms and Sensors, published in 2009 by the American Society for Photogrammetry and Remote Sensing (ASPRS).

Table 13.7 Popular data sources relevant to hydrology and water resources modelling

Data	Source	Spatial resolution
Digital elevation model (DEMs)	Shuttle radar topography mission (SRTM)	30 × 30 m
	ALASKA satellite facility	12.5 × 12.5 m
Land use and land cover (LULC)	Food and agriculture organisation	1 × 1 km
Soil datasets	Food and agriculture organisation	1 × 1 km
Rainfall datasets	India meteorological department	0.25° × 0.25°
Temperature datasets	India meteorological department	1° × 1°
Streamflow and river stage data	Central water commission	Station-based data
Groundwater level data	Central groundwater commission	Station-based data
Climate datasets	GCM	250–600 km
	RCM Coordinated regional climate downscaling experiment (CORDEX)[a]	50 km
Vegetation indices (NDVI and EVI)	MODIS	250 m–1 km
Soil moisture datasets	Global soil moisture data	0.25° × 0.25°
Evapotranspiration datasets	MOD16A2	500 m
Water quality datasets	Central pollution control board	
	India water resources information system	

[a] Available for 14 domains, i.e., South America, North America, Central America, Europe, Africa, South Asia, East Asia, Central Asia, Australia, Antarctica, Arctic, Mediterranean, Middle East North Africa, and South-East Asia

13.3.9.1 GIS and Remote Sensing Software

Table 13.8 presents a few popular GIS and remote sensing software. For example, ArcGIS, MapInfo, and Intergraph are mainly GIS software, while Earth Resources Data Analysis System (ERDAS) and Environment for Visualising Images (ENVI) are mainly remote sensing software. On the contrary, Geographic Resources Analysis Support System (GRASS), an open-source GIS and IDRISI Selva are software that include GIS, remote sensing, and spatial analysis tools. Several open-source software are also available for GIS and RS analyses, e.g., QGIS and SAGA GIS. Also, the statistical software package, R, or the mathematical package, MATLAB, provide tools for spatial analysis and image processing.

Table 13.8 List of commonly used GIS and remote sensing software

Software	Vendor name	Website	Cost (100,000 INR)
ArcGIS	ESRI	http://www.esri.com/products	11.5 (standard) 21.5 (advanced)
IDRISI	Clark Labs	http://clarklabs.org/	1.0 (general) 0.55 (academic)
MapInfo Pro	Precisely	https://www.precisely.com/product/precisely-mapinfo/mapinfo-pro	0.65 (standard) 1.0 (Premium)
GRASS GIS	OSGeo	http://grass.osgeo.org/	Open source
PIX 4D	PIX4D	https://www.pix4d.com/pricing	1.5 (educational)
Global mapper	Blue marble geographics	https://www.bluemarblegeo.com/global-mapper/	0.5 (standard) 1.0 (professional)
SURFER	Golden Software	http://www.goldensoftware.com/products/surfer/surfer.shtml	0.76
ENVI	Exelis	http://www.exelisvis.com/ProductsServices/ENVI.aspx	0.3/month
ERDAS	Hexagon	https://www.hexagongeospatial.com/products/power-portfolio/erdas-imagine	16 (academic concession may be available)
Manifold GIS	Manifold	https://manifold.net/	0.11
AUTO CAD MAP 3D	AUTO desk	https://www.autodesk.com/products/autocad/included-toolsets/autocad-map-3d	1.4/year
TNTmips	MicroImages	https://www.microimages.com/products/tntmips.htm	3.8
QGIS	qgis	https://qgis.org/en/site/index.html	Open source
gvSIG	gvSIG association	http://www.gvsig.com/en/products/gvsig-desktop	Open source
ILWIS	ITC, University of Twente	https://www.itc.nl/ilwis/	Open source
SAGA GIS	SAGA	gttps://saga-gis.sourceforge.io/en/index.html	Open source
GeoDa	The University of Chicago	https://geodacenter.github.io/download.html	Open source
MapWindow	MAPWINDOW	https://www.mapwindow.org/#download	Open source

(continued)

Table 13.8 (continued)

Software	Vendor name	Website	Cost (100,000 INR)
ORFEO Toolbox	Orfeo ToolBox	https://www.orfeo-too lbox.org/	Open source

13.4 Remote Sensing and GIS Applications

13.4.1 Soil Erosion Mapping

Remote sensing and GIS applications are well established in soil erosion mapping due to the availability of up-to-date data at various resolutions. RS and GIS techniques have been extensively applied for soil erosion mapping (Sepuru and Dube 2018; Senanayake et al. 2020), for identifying and delineating eroded and deposition areas (Sarapatka and Netopil 2010), and for identifying erosion features like rills and gullies (Martinez-Casasnovas 2003). RS and GIS techniques have also been used to develop GIS-enabled databases, such as the Global and National Soils and Terrain Digital Database (SOTER) (van Engelen and Wen 1995). SOTER provides detailed information on natural resources that can readily be accessed and analysed for specified purposes, including soil conservation needs.

The continued advancements in remote sensing technologies facilitate the assessment of the nature, extent, and magnitude of soil erosion, which is critical for planning and executing soil conservation measures to ensure food security (Seutloali et al. 2017; Sepuru and Dube 2018). In the past, visual interpretation of aerial photographs was used for soil erosion mapping (Frazier et al. 1983). However, with the launch of the first operational Earth Observation satellite, Landsat 1 in 1972, researchers started using satellite data extensively for mapping soil erosion (Stephens and Cilhar 1982). Subsequently, the availability of satellite data with improved spatial and spectral resolutions has enhanced its application for timely, affordable, and accurate investigation of soil erosion at larger spatial scales.

Landsat has remained the most popular source of remotely sensed data for soil erosion mapping (Luleva et al. 2012). With seven spectral bands recording energy in the visible, near-infrared, mid-infrared, and thermal infrared regions of the electromagnetic spectrum, Landsat TM has been widely used for mapping eroded areas, including gullies (Hochschild et al. 2003; Jensen 2005). Besides Landsat TM, SPOT data have also been used to map eroded areas (Wentzel 2002; Servenay and Prat 2003).

Subsequently, IKONOS 2 (launched in 1999), Quikbird (launched in 2001), and SPOT 5 (launched in 2002) satellites, providing high spatial resolution imagery, also found favour in soil mapping studies (Vrieling 2006; Luleva et al. 2012; Sepuru and Dube 2018). However, their utility remained limited due to high data cost (Vrieling et al. 2008) and low spectral resolution (Seutloali et al. 2016). Recently launched satellites, such as Landsat 8 (launched in 2013), Sentinel 2 (launched in 2015), and

Landsat 9 (launched in 2021), provide free multitemporal data at higher spatial, spectral, and radiometric resolutions. Thus, these offer an enhanced opportunity for soil erosion mapping, especially in developing countries.

Furthermore, satellite data vegetation indices, such as NDVI, have been used extensively for soil erosion mapping (Taruvinga 2009; Seutloali et al. 2016; Puente et al. 2019). NDVI has been derived from various satellite imageries, like SPOT (Mathieu et al. 1997), multitemporal NOAA/AVHRR (Symeonakis and Drake 2004), and IRS-1B LISS-II (Vaidyanathan et al. 2002) and applied in the soil erosion research. However, the NDVI values are sensitive to the non-vegetation factors, such as soil colour, soil brightness, atmosphere, and cloud cover (Xue and Su 2017). Consequently, the Soil-Adjusted Vegetation Index (SAVI), Transformed SAVI (TSAVI), and the soil and atmospherically resistant vegetation index (SARVI) have been derived and used in the soil erosion mapping (Gandhi et al. 2015; Sonawane and Bhagat 2017).

13.4.2   Soil Erosion Estimation

RS and GIS technologies have been widely applied in soil erosion studies due to new tools and software with enhanced capabilities (Seutloali et al. 2017). RS and GIS techniques have been often integrated with various soil erosion models to assess soil erosion (Pandey et al. 2007; Dabral et al. 2008; Jahun et al. 2015; Seutloali et al. 2017; Sepuru and Dube 2018). Although several physically based models have been developed for the soil loss estimation, such as Water Erosion Prediction Project (WEPP) and European Soil Erosion Model (EUROSEM), the empirical models, like Universal Soil Loss Equation (USLE) and RUSLE (Revised USLE), are extensively used globally because of their simplicity and wider acceptability (Sects. 13.3.1–13.3.2, Chap. 3). The USLE and RUSLE predict soil loss through sheet and rill erosion. The empirical equation integrates various factors, like rainfall erosivity, soil erodibility, slope length, slope steepness, vegetative cover, and conservation measures, to estimate the soil loss.

RS and GIS techniques have been used extensively for extracting the inputs to USLE and RUSLE. However, RUSLE has been a popular choice because of its integration with GIS. RS data have frequently derived the land and cover management factor, C, of USLE. For example, Pradhan et al. (2012) utilised SPOT 5 and Landsat ETM + data for LULC classification to generate the USLE crop cover (C) factor. The USLE C factor was also derived using IRS P-6 LISS-III data (Biswas 2012) and Spot 5 (Rizeei et al. 2016).

In RUSLE applications, the conventional rainfall and soil data have been used to develop rasterised rainfall erosivity and soil erodibility maps to obtain the rainfall erosivity and soil erodibility factor (Kouli et al. 2009). Nowadays, with GIS technology, many researchers have adopted using the DEM for generating topography data (Prasannakumar et al. 2011; Mhangara et al. 2012). The RUSLE C factor has been derived using NDVI derived from Landsat ETM + (Kouli et al. 2009), Landsat

TM (Fathizad et al. 2014), and MODIS (Yan et al. 2020). *C* factor has also been derived from LULC maps of IRS 1-D LISS-III data (Ganasri and Ramesh 2016) and SPOT 5 (Nampak et al. 2018).

Unmanned aerial vehicles (UAV) derived data are gaining popularity as it offers the advantages of simple operation, affordability, and fast imaging speed (Anderson and Gaston 2013; Valavanis and Vachtsevanos 2015). The UAV-based LIDAR and hyperspectral images provide precise estimates of the land surface processes, which are helpful in soil erosion estimation (Hutton and Brazier 2012). UAV data can also generate DTM, land cover map, or NDVI (Lelong et al. 2008; Sankey et al. 2017).

Many researchers have used RS and GIS tools for soil erosion assessment and forecasting. For example, Borrelli et al. (2020) have modelled potential global soil erosion using SSP-RCP (Shared Socioeconomic Pathway-Representative Concentration Pathway) scenarios of the Intergovernmental Panel on Climate Change (IPCC). Thus, it is evident that RS and GIS technologies have established themselves as an integral part of the soil erosion assessment projects.

13.4.3 Watershed Prioritisation

Watershed prioritisation refers to ranking watersheds based on the erosion severity to set the soil conservation treatment priorities (Kumar and Kumar 2011). It involves the identification and mapping of eroded lands for setting priorities. RS data and GIS have been effectively used to identify and map eroded lands and their severity. Visual interpretation of aerial photographs was used earlier for identifying and mapping eroded areas, especially gullies and ravines (Kamphorst and Iyer 1972; Servenay and Prat 2003). Later on, visual interpretation has been extended to satellite data like false colour composite (FCC) of medium resolution sensors such as Landsat MSS or TM for identifying areas affected by sheet erosion (Krishna et al. 2009), rill erosion (Sujatha et al. 2000), and gully erosion (Sujatha et al. 2000; Krishna et al. 2009). Researchers also used the fusion of multispectral data and other relevant information, including high-resolution panchromatic (PAN) data, for identifying eroded lands, e.g., multitemporal IRS 1C PAN data (Singh et al. 1998) and IRS-1C LISS-III merged with PAN data (Krishna et al. 2009). High-resolution QuickBird data have also been used for gully classification (Vrieling et al. 2007).

Supervised and unsupervised classification techniques have been used as an alternative to visual interpretation for identifying eroded lands. Supervised classification has been used with Landsat TM and SPOT data (Dwivedi et al. 1997) and ASTER and MODIS data (Vrieling et al. 2007), while unsupervised classification has been used with SPOT data (Servenay and Prat 2003) to distinguish various erosion classes.

Many researchers attempted morphometric analysis using RS and GIS techniques and applied for watershed characterisation and prioritisation (Singh and Singh 1997; Ali and Singh 2002; Pandey et al. 2011). These studies have typically used Landsat and IRS 1-D LISS III data and combined them with the data obtained from traditional sources to create various thematic layers. The watersheds are prioritised based on their soil loss potential, and the highest-ranked watershed is given top priority for employing the conservation measures (Khare et al. 2014).

13.4.4 Site Selection for Water Harvesting Structures

RS and GIS techniques are used extensively to identify suitable sites for the water harvesting structures in watersheds (De Winnaar et al. 2007; Kumar et al. 2008; Jasrotia et al. 2009; Elewa et al. 2012). Typically, check dams, ponds, terraces, contour strips, percolations tanks, and nala bunds are the preferred water harvesting structures in a watershed management programme (Oweis et al. 2012). Among these structures, check dams are the most versatile as they perform multiple functions of water harvesting, erosion control, and flow and sediment diversion (Ammar et al. 2016). FAO (2003) has recommended guidelines based on six criteria: climate, hydrology, topography, agronomy, soils, and socio-economics, to identify potential water harvesting sites (Kahinda et al. 2009). Before FAO guidelines, studies considered only the biophysical criteria, i.e., rainfall, slope, soil type, drainage network, and land use (Prinz et al. 1998). FAO guidelines introduced the socio-economic factors in the site selection studies (De Winnaar et al. 2007).

RS and GIS tools have been used for the site selection of water harvesting structures, either in isolation (Prinz et al. 1998) or in combination with hydrological models (Durbude and Venkatesh 2004; De Winnaar et al. 2007). A few studies have also combined multicriteria analysis (MCA) and hydrological models with RS and GIS (Elewa et al. 2012; Rahmati et al. 2019). These studies have employed data products from the Landsat MSS, TM, ETM+ , IRS LISS-II and III, IKONOS, MODIS, SRTM, and AVHRR to extract relevant information for various thematic layers (Kumar and Rashmidevi 2013).

Questions

1. Describe a digital image. Enumerate the advantages of a digital image over a traditional hard copy image.
2. What is an unsupervised classification? What are its advantages and disadvantages?
3. What is NDVI? Explain the process of determining it and its applications.
4. A raster data structure is more suitable for analytical operations. Explain its advantages and disadvantages
5. What are map projections? Explain various types of projections.
6. What is a geographic coordinate system? How is it different from the projected coordinate system?
7. Explain the differences between local and geocentric datum. Is WGS84 a local or geocentric datum?

8. To know the extent of area affected by a flood event, which spatial analysis tool will be used and why?
9. What are the differences between the topological and spaghetti vector models? What are the advantages and disadvantages of using each one?
10. Suppose groundwater level data are available at 50 different locations in a watershed. What is the best data model to store this information, and why?

Multiple Choice Questions
Answer the questions by choosing the correct option.

1. The near-infrared portion of EMR lies between

    (A) 0.4–0.7 μm
    (B) 0.5 mm–1.0 μm
    (C) 0.7–1.3 μm
    (D) 0.7–14 μm

2. The system that uses the sun as the source of energy and records the naturally radiated and reemitted energy from an object is called

    (A) Active remote sensing
    (B) Passive remote sensing
    (C) Global Positioning System
    (D) Geographical Information System

3. The scattering that occurs when the size of the particles causing scattering is much larger than the EMR wavelength is called

    (A) Rayleigh scattering
    (B) Mie scattering
    (C) Selective scattering
    (D) Non-selective scattering

4. The ratio of the total solar radiant energy returned by a planetary body to the total radiant energy incident on the body is

    (A) Absorptance
    (B) Emittance
    (C) Transmittance
    (D) Albedo

5. The image resolution that refers to the ability of the remote sensing sensor to differentiate fine wavelengths in the EMR spectrum is called

    (A) Radiometric resolution
    (B) Spatial resolution
    (C) Spectral resolution
    (D) Temporal resolution

6. The vegetation index based on the UAV remote sensing in the visible spectral region, Vegetation Index_Green, is expressed as (G and R represent the reflectance in the green and red spectrum range, respectively)

   (A) $\frac{G+R}{G-R}$
   (B) $\frac{G-R}{G+R}$
   (C) $\frac{G-R}{GR}$
   (D) $\frac{G+R}{GR}$

7. The surface of the equal gravitational potential is popularly known as

   (A) Ellipsoid
   (B) Geoid
   (C) Spheroid
   (D) Datum

8. A raster data structure that helps overcome the problem of redundant or missing data in a grid is called

   (A) Cell-by-cell encoding
   (B) Run-length encoding
   (C) Quadtree
   (D) TIGER

9. The geoid height represents the

   (A) Vertical distance between geoid and the earth's surface
   (B) Vertical distance between geoid and mean sea level
   (C) Difference between geoid and the coordinates on the earth's surface
   (D) Vertical distance between geoid and ellipsoid

10. The attribute data model, in which data is organised in tables, with each table having a unique name, is known as

    (A) Hierarchical data model
    (B) Network data model
    (C) Relational data model
    (D) Object-oriented data model

References

Ali S, Singh R (2002) Morphological and hydrological investigation in Hirakud catchment for watershed management planning. J Soil Water Conserv 1:246–256

Ammar A, Riksen M, Ouessar M, Ritsema C (2016) Identification of suitable sites for rainwater harvesting structures in arid and semi-arid regions: a review. Int Soil Water Conserv Res 4:108–120

Anderson K, Gaston KJ (2013) Lightweight unmanned aerial vehicles will revolutionize spatial ecology. Front Ecol Environ 11:138–146

Aronoff S (1989) Geographic information systems: a management perspective. WDL Publications, Ottawa, Ontario, Canada

Bannari A, Morin D, Bonn F, Huete AR (1995) A review of vegetation indices. Remote Sens Rev 13:95–120

Baret F, Guyot G, Major DJ (1989a) Crop biomass evaluation using radiometric measurements. Photogrammetria (PRS) 43:241–256

Baret E, Guyot G, Major DJ (1989b) TSAVI: a vegetation index which minimizes soil brightness effects on LAI and APAR estimation. In: Proceedings 12th Canadian symposium on remote sensing, Vancouver

Bendig J, Yu K, Aasen H, Bolten A, Bennertz S, Broscheit J, Gnyp ML, Bareth G (2015) Combining UAV-based plant height from crop surface models, visible, and near infrared vegetation indices for biomass monitoring in barley. Int J Appl Earth Obs Geoinf 39:79–87

Biswas S (2012) Estimation of soil erosion using remote sensing and GIS and prioritization of catchments. Int J Emerg Technol Adv Eng 2:124

Borrelli P, Robinson DA, Panagos P, Lugato E, Yang JE, Alewell C, Wuepper D, Montanarella L, Ballabio C (2020) Land use and climate change impacts on global soil erosion by water (2015–2070). Proc Natl Acad Sci USA 117:21994–22001

Burrough PA (1986) Principles of geographical information systems for land resources assessment. Geocarto Int 1:54

Clark RN (1999) Spectroscopy of rocks and minerals, and principles of spectroscopy. In: Rencz AN (ed) Remote sensing for earth sciences. Wiley, New York

Clarke KC (1997) Getting started with geographic information systems. Pearson Prentice Hall, New Jersey

Dabral PP, Baithuri N, Pandey A (2008) Soil erosion assessment in a hilly catchment of North Eastern India using USLE, GIS and remote sensing. Water Resour Manag 22:1783e1798

Davis SM, Landgrebe DA, Phillips TL, Swain PH, Hoffer RM, Lindenlaub JC, Silva LF (1978) Remote sensing: the quantitative approach. McGraw-Hill International, New York

De Winnaar G, Jewitt GPW, Horan M (2007) A GIS-based approach for identifying potential runoff harvesting sites in the Thukela river basin, South Africa. Phys Chem Earth 32:1058–1067

Durbude DG, Venkatesh B (2004) Site suitability analysis for soil and water conservation structures. J Indian Soc Remote Sens 32:399–405

Dwivedi RS, Kumar AB, Tiwari KN (1997) The utility of multi-sensor data for mapping eroded lands. Int J Remote Sens 18:2303–2318

Ehlers M (1990) Remote sensing and geographic information systems: towards integrated spatial information processing. IEEE Trans Geosci Remote Sens 28:763–766

Elewa HH, Qaddah AA, El-Feel AA (2012) Determining potential sites for runoff water harvesting using remote sensing and geographic information systems-based modeling in Sinai. Am J Environ Sci 8:42–55

ESRI (2022) What is GIS? Overview. https://www.esri.com/en-us/what-is-gis/overview. Acccessed on 28 March 2022

FAO (2003) Planning of water harvesting schemes, unit 22: training course on RWH (CDROM). Land and water digital media series 26. Food and Agriculture Organization of the United Nations, FAO, Rome

Fathizad H, Karimi H, Alibakhshi SM (2014) The estimation of erosion and sediment by using the RUSLE model and RS and GIS techniques (Case study: arid and semi-arid regions of Doviraj, Ilam province, Iran). Int J Agric Crop Sci 7:303

Frazier BE, Mc Cool DK, Engle CF (1983) Soil erosion in the Palouse: an aerial perspective. J Soil Water Conserv 38:70–74

Gandhi MG, Parthiban S, Thummalu N, Christy A (2015) Vegetation change detection using remote sensing and GIS—a case study of Vellore District. Procedia Comput Sci 57:1199–1210

Ganasri BP, Ramesh H (2016) Assessment of soil erosion by RUSLE model using remote sensing and GIS-A case study of Nethravathi Basin. Geosci Front 7:953–961

Gao J (2002) Integration of GPS with remote sensing and GIS: reality and prospects. Photogram Eng Remote Sens 68:447–453

Gitelson AA, Kaufman YJ, Stark R, Rundquist D (2002) Novel algorithms for remote estimation of vegetation fraction. Remote Sens Environ 80:76–87

Hatfield JL, Prueger JH, Sauer TJ, Dold C, O'Brien P, Wacha K (2019) Applications of vegetative indices from remote sensing to agriculture: past and future. Inventions 4:71

Hochschild V, Märker M, Rodolfi G, Staudenrausch H (2003) Delineation of erosion classes in semiarid southern African grasslands using vegetation indices from optical remote sensing data. Hydrol Process 17:917–928

Huete AR (1988) A soil-adjusted vegetation index (SAVI). Remote Sens Environ 25:295–309

Hutton C, Brazier R (2012) Quantifying riparian zone structure from airborne LiDAR: vegetation filtering, anisotropic interpolation, and uncertainty propagation. J Hydrol 442–443:36–45

Jackson MJ (1992) Integrated geographical information systems. Int J Remote Sens 13:1343–1351

Jahun BG, Ibrahim R, Dlamini NS, Musa SM (2015) Review of soil erosion assessment using RUSLE model and GIS. J Biol Agric Healthcare 9:36–47

Jasrotia AS, Majhi A, Singh S (2009) Water balance approach for rainwater harvesting using remote sensing and GIS techniques, Jammu Himalaya, India. Water Resour Manag 23:3035–3055

Jensen JR (2005) In: Introductory digital image processing, 3rd edn. Pearson Prentice Hall, New Jersey, USA

Kahinda JM, Taigbenu AE, Sejamoholo BBP, Lillie ESB, Boroto RJ (2009) A GIS-based decision support system for rainwater harvesting (RHADESS). Phys Chem Earth, Parts a/b/c 34:767–775

Kamphorst A, Iyer HS (1972) Application of aerial photo-interpretation to ravine surveys in India. In: Proceedings 12th Congress International Society Photogram Engineering, Ottawa

Khare D, Mondal A, Mishra PK, Kundu S, Meena PK (2014) Morphometric analysis for prioritization using remote sensing and GIS techniques in a hilly catchment in the state of Uttarakhand, India. Indian J Sci Technol 7:1650–1662

Kouli M, Soupios P, Vallianatos F (2009) Soil erosion prediction using the revised universal soil loss equation (RUSLE) in a GIS framework, Chania, Northwestern Crete, Greece. Environ Geol 57:483–497

Krishna G, Kushwaha SPS, Velmurugan A (2009) Land degradation mapping in the upper catchment of river Tons. J Indian Soc Remote Sens 37:119–128

Kumar B, Kumar U (2011) Micro watershed characterization and prioritization using geomatics technology for natural resources management. Int J Geomat Geosci 1:789–802

Kumar DN, Reshmidevi TV (2013) Remote sensing applications in water resources. J Indian Inst Sci 93:163–187

Kumar GM, Agarwal AK, Bali R (2008) Delineation of potential sites for water harvesting structures using remote sensing and GIS. J Indian Soc Remote Sens 36:323–334

Lelong CCD, Burger P, Jubelin G, Roux B, Labbé S, Baret F (2008) Assessment of unmanned aerial vehicles imagery for quantitative monitoring of wheat crop in small plots. Sensors 8:3557–3585

Llcev SD (2019) Global satellite meteorological observation (GSMO) applications, vol 2. Springer, Cham, Switzerland

Longley PA, Goodchild MF, Maguire DJ, Rhind DW (2005) Geographic information systems and science. Wiley, West Sussex, England

Luleva MI, van de Werff H, van der Meer F, Jetten V (2012) Gaps and opportunities in the use of remote sensing for soil erosion assessment. Chem : Bulg J Sci Educ 21:748–764

Martinez-Casasnovas JA (2003) A spatial information technology approach for the mapping and quantification of gully erosion. CATENA 50:293–308

Mathieu R, King C, Le Bissonnais Y (1997) Contribution of multi-temporal SPOT data to the mapping of a soil erosion index. The case of the loamy plateau of Northern France. Soil Technol 10:99–110

Merchant JW, Narumalani S (2009) Integrating remote sensing and geographic information systems. SAGE Publications Ltd., London

Mhangara P, Kakembo V, Lim KJ (2012) Soil erosion risk assessment of the Keiskamma catchment, South Africa using GIS and remote sensing. Environ Earth Sci 65:2087–2102

Mulla DJ (2013) Twenty five years of remote sensing in precision agriculture: key advances and remaining knowledge gaps. Biosyst Eng 114:358–371

Nampak H, Pradhan B, Mojaddadi Rizeei H, Park HJ (2018) Assessment of land cover and land use change impact on soil loss in a tropical catchment by using multitemporal SPOT-5 satellite images and revised universal soil loss equation model. Land Degrad Dev 29:3440–3455

Oweis TY, Prinz D, Hachum AY (2012) Rain water harvesting for agriculture in the dry areas. CRC Press, London

Pandey A, Behra S, Pandey RP, Singh RP (2011) Application of GIS for watershed prioritization and management: a case study. Int J Environ Sci Dev Monit 2:25–42

Pandey A, Chowdary VM, Mal BC (2007) Identification of critical erosion prone areas in the small agricultural watershed using USLE, GIS and remote sensing. Water Resour Manag 21:729–746

Pearson RL, Miller LD (1972) Remote mapping of standing crop biomass for estimation of the productivity of the short-grass prairie. In: Asrar G (ed) Proceedings of the 8th international symposium on remote sensing of environment. Pawnee National Grasslands, Colorado, USA

Pradhan B, Chaudhari A, Adinarayana J, Buchroithner MF (2012) Soil erosion assessment and its correlation with landslide events using remote sensing data and GIS: a case study at Penang Island, Malaysia. Environ Monit Assess 184:715–727

Prasannakumar V, Shiny R, Geetha N, Vijith H (2011) Spatial prediction of soil erosion risk by remote sensing, GIS and RUSLE approach: a case study of Siruvani river watershed in Attapady Valley, Kerala, India. Environ Earth Sci 64:965–972

Prinz D, Oweis T, Oberle A (1998) Rainwater harvesting for dry land agriculture-developing a methodology based on remote sensing and GIS. In: Proceedings of XIII international congress agricultural engineering, ANAFD, Rabat, Morocco

Puente C, Olague G, Trabucchi M, Arjona-Villicaña PD, Soubervielle-Montalvo C (2019) Synthesis of vegetation indices using genetic programming for soil erosion estimation. Remote Sens 11:156

Qi J, Chehbouni A, Huete AR, Kerr YH, Sorooshian S (1994) A modified soil adjusted vegetation index. Remote Sens Environ 48:119–126

Rahmati O, Kalantari Z, Samadi M, Uuemaa E, Moghaddam DD, Nalivan OA, Destouni G, Tien Bui D (2019) GIS-based site selection for check dams in watersheds: considering geomorphometric and topo-hydrological factors. Sustainability 11:5639

Richardson AJ, Weigand C (1977) Distinguishing vegetation from soil background information. Photogram Eng Remote Sens 43:1541–1552

Rizeei HM, Saharkhiz MA, Pradhan B, Ahmad N (2016) Soil erosion prediction based on land cover dynamics at the Semenyih watershed in Malaysia using LTM and USLE models. Geocarto Int 31:1158–1177

Rondeaux G, Steven M, Baret F (1996) Optimization of soil adjusted vegetation indices. Remote Sens Environ 55:95–107

Rouse JW, Haas RH, Schell JA, Deering DW (1974) Monitoring vegetation systems in the great plains with ERTS. In: Proceedings of the third earth resources technology satellite-1 symposium, Greenbelt, NASA SP-351

Sankey T, Donager J, McVay J, Sankey JB (2017) UAV Lidar and hyperspectral fusion for forest monitoring in the Southwestern USA. Remote Sens Environ 195:30–43

Sarapatka B, Netopil P (2010) Erosion processes on intensively farmed land in the Czech Republic: comparison of alternative research methods. In: 2010 19th World congress of soil science, soil solutions for a changing world, Brisbane, Australia

Senanayake S, Pradhan B, Huete A, Brennan J (2020) A review on assessing and mapping soil erosion hazard using geo-informatics technology for farming system management. Remote Sens 12:4063

Sepuru TK, Dube T (2018) An appraisal on the progress of remote sensing applications in soil erosion mapping and monitoring. Remote Sens Appl: Soc Environ 9:1–9

Servenay A, Prat C (2003) Erosion extension of indurated volcanic soils of Mexico by aerial photographs and remote sensing analysis. Geoderma 117:367–375

Seutloali KE, Dube T, Mutanga O (2017) Assessing and mapping the severity of soil erosion using the 30-m Landsat multispectral satellite data in the former South African homelands of Transkei. Phys Chem Earth 100:296–304

Seutloali KE, Beckedahl HR, Dube T, Sibanda M (2016) An assessment of gully erosion along major armoured roads in south-eastern region of South Africa: a remote sensing and GIS approach. Geocarto Ints 31:225–239

Singh S, Singh MC (1997) Morphometric analysis of Kanhar river basin. Natl Geogr J India 43:31–43

Singh AN, Sharma YK, Singh S (1998) Evaluation of IRS 1C PAN data for monitoring gullied and Ravinous lands of western UP. In: Remote sensing and geographic information system for natural resources management, Indian Society of Remote Sensing and NNRMS, Bangalore

Sonawane KR, Bhagat VS (2017) Improved change detection of forests using Landsat TM and ETM+ data. Remote Sens Land 1:18–40

Stephens PR, Cihlar J (1982) Mapping erosion in New Zealand and Canada. In: Johanmen CJ, Sanders JL (eds) Remote sensing for resource management. Soil & Water Conservation Society, Iowa

Sujatha G, Dwivedi RS, Sreenivas K, Venkataratnam L (2000) Mapping and monitoring of degraded lands in part of Jaunpur district of Uttar Pradesh using temporal space borne multispectral data. Int J Remote Sens 21:519–531

Symeonakis E, Drake N (2004) Monitoring desertification and land degradation over sub-Saharan Africa. Int J Remote Sens 25:573–592

Taruvinga K (2009) Gully mapping using remote sensing: case study in KwaZulu-Natal, South Africa. Dissertation, University of Waterloo

Vaidyanathan NS, Sharama G, Sinha R, Dikshit O (2002) Mapping of erosion intensity in the Garhwali Himalaya. Int J Remote Sens 23:4125–4129

Valavanis KP, Vachtsevanos GJ (2015) Handbook of unmanned aerial vehicles. Springer, Dodlerk, The Netherlands

van Engelen VWP, Wen TT (eds) (1995) Global and national soils and terrain digltal databases (SOTER). Procedures manual, revised. Wageningen, ISRIC

Vrieling A, De Jong SM, Sterk G, Rodrigues SC (2008) Timing of erosion and satellite data: a multi-resolution approach to soil erosion risk mapping. Int J Appl Earth Obs Geoinf 10:267–281

Vrieling A, Rodrigues SC, Bartholomeus H (2007) Automatic identification of erosion gullies with ASTER imagery in the Brazilian Cerrados. Int J Remote Sens 28:2723–2738

Vrieling A (2006) Satellite remote sensing for water erosion assessment: a review. CATENA 65:2–18

Wentzel K (2002) Determination of the overall soil erosion potential in the Nsikazi district (Mpumalanga Province, South Africa) using remote sensing and GIS. Can J Remote Sens 28:322–327

Wieczorek WF, Delmerico AM (2009) Geographic information systems. Wiley Interdiscip Rev Comput Stat 1:167–186

Xue J, Su B (2017) Significant remote sensing vegetation indices: a review of developments and applications. J Sens 1353691

Yan H, Wang L, Wang TW, Wang Z, Shi ZH (2020) A synthesized approach for estimating the *C*-factor of RUSLE for a mixed-landscape watershed: a case study in the Gongshui watershed, southern China. Agric Ecosyst Environ 301:107009

Zhang F, Tiyip T, Ding J, Sawut M, Tashpolat N, Kung H, Han G, Gui D (2012) Spectral reflectance properties of major objects in desert oasis: a case study of the Weigan-Kuqa river delta oasis in Xinjiang, China. Environ Monit Assess 184:5105–5119

Chapter 14
Impact of Climate and Land Use Land Cover Changes on Soil Erosion

Abstract Climate change and land use land cover (LULC) changes are recognised as two of the most significant causes of environmental change. Climate change and LULC changes are related to one another. Land use change may drive climate change, and a changing climate may result in land cover changes. Climate change and LULC changes are believed to influence soil erosion. This chapter analyses the impacts of climate and LULC changes on soil erosion. The causes and effects of climate change on precipitation, temperature, solar radiation, atmospheric CO_2 concentrations, and radiative forcing are discussed. The chapter includes the impacts of climate change on soil characteristics, vegetation cover, runoff, floods, and droughts and extends the impacts of these changes on water and wind erosion. The chapter explores the human alterations of LULC changes in terms of changes in the forest cover, alterations in agricultural lands, increase in urban areas, and decrease in wetland areas. The influence of the LULC changes on soil erosion and sediment production processes is discussed. Also, the combined impact of climate and LULC changes on soil erosion is explored, and mitigation strategies like sustainable land management practices and appropriate policy incentives to conserve soil are discussed.

14.1 Climate Change

Intergovernmental Panel on Climate Change (IPCC) defines *climate* as the 'average weather, in a narrow sense, or more rigorously, as the statistical description in terms of the mean and variability of relevant quantities over a period of time ranging from months to thousands or millions of years'. The temperature, precipitation, and wind are the relevant quantities, typically averaged over 30 years, as recommended by the World Meteorological Organisation (IPCC 2021).

However, the *climate* is often wrongly used to represent *the weather*, which represents precipitation, temperature, wind speed and direction, and sunshine hours over a short period (up to several days). Thus, the climate and weather can be differentiated based on the *measure of time*, i.e., weather refers to the atmospheric conditions over a short period, whilst climate represents the atmosphere behaviour over a relatively long period.

© The Author(s), under exclusive license to Springer Nature Singapore Pte Ltd. 2023   415
R. Singh, *Soil and Water Conservation Structures Design*, Water Science and Technology
Library 123, https://doi.org/10.1007/978-981-19-8665-9_14

IPCC defines climate change as 'a change in the state of the climate that can be identified (e.g., by using statistical tests) by changes in the mean and/or the variability of its properties and that persists for an extended period, typically decades or longer'. The change may be due to internal variability or natural causes such as solar variations, ocean currents, volcanic eruptions, or human interventions like burning of fossil fuels and land use changes (IPCC 2021).

The United Nations Framework Convention on Climate Change (UNFCCC), on the other hand, defines climate change as 'a change of climate which is attributed directly or indirectly to human activity that alters the composition of the global atmosphere and which is in addition to natural climate variability observed over comparable time periods'. The UNFCCC thus distinguishes between climate change caused due to human activities and natural climate variability (UNFCCC 2015).

Even if the greenhouse gas emissions were strongly reduced today, changes in the climate system will continue due to the inertia in the geophysical and socioeconomic systems. The unavoidable long-term (centuries to millennia) change in the climate system due to the effect of past greenhouse gas emissions and slow response of some elements of climate system is termed as climate change commitment (IPCC 2021). Climate change commitment focuses on limiting global warming to avoid extreme fluctuations in weather phenomena and large-scale changes in the hydrological variables and sea level (IPCC 2021).

14.1.1 Evidence of Climate Change

Climate change has been observed and analysed independently by several groups of climate scientists worldwide. These studies unequivocally pointed to changes in various climate system components that have led to rapid warming since the industrial revolution (IPCC 2021).

The consensus statement of the Geological Society of America (GSA 2020) on climate change science summed up the climate change assessment by various groups of climate scientists. The statement read, 'The Geological Society of America (GSA) concurs with assessments by the National Academies of Sciences, Engineering and Medicine (2019), the National Research Council (2011), the US Global Change Research Programme (2017, 2018), and the Intergovernmental Panel on Climate Change (IPCC 2014) that global climate has warmed recently in response to increasing concentrations of greenhouse gases, especially carbon dioxide (CO_2), and that human activities (mainly greenhouse gas emissions) are the dominant cause of rapid warming since the middle 1900s, whilst other natural factors contribute, at most, only marginally'.

IPCC has reported the impact of human-induced climate change on weather and climate extremes around the globe (IPCC 2021). Changes have been observed in the extremes of precipitation, droughts, heatwaves, and tropical cyclones since the fifth assessment report of IPCC, AR5 (IPCC 2013).

The global surface temperature has likely increased by 1.07 °C (0.8 °C to 1.3 °C) from 1850–1900 to 2010–2019. The global surface temperature increases since 1970 is the fastest compared to any previous 50-year period in the past 2000 years (IPCC 2021).

Besides temperature, the globally averaged land precipitation has also been increasing since 1950, with the period after 1980 exhibiting a faster rate of increase. Due to climate change, heavy precipitation events have also occurred frequently since the 1950s. Also, land evapotranspiration has increased, leading to agricultural and ecological droughts in some regions (IPCC 2021).

IPCC has also reported a rise in the concentration of greenhouse gases (GHG) since around 1750. With an annual average concentration of 410 ppm for CO_2, the atmospheric CO_2 concentrations in 2019 were reportedly the highest in the past 2 million years. The increased CO_2 emissions have caused global ocean surface acidification, leading to a drop in oxygen levels and warming of the global upper ocean (0–700 m) (IPCC 2021).

14.2 Climate Change Causes and Effects

14.2.1 Climate Change Drivers

Climate change drivers can be broadly classified into two categories: natural drivers and anthropogenic drivers. The natural drivers include the changes in solar irradiance, volcanic eruptions, and the El Nino-Southern Oscillation (ENSO). The anthropogenic drivers comprise aerosols, greenhouse gases, and land use land cover changes.

Natural Drivers

Solar Irradiance

The radiant energy from the sun, referred to as solar irradiance, is the fundamental source of energy to the earth. Therefore, the total solar irradiance (TSI) changes may considerably influence the earth's temperature. It is well-established that the total solar irradiance follows an 11-year solar cycle, becoming brighter than average at solar maximum and dimmer at solar minimum (Solanki et al. 2013).

Satellite-based measurements of solar irradiance have been available since 1978, in addition to the proxy indicators of solar cycles, available since the early 1600s. The variations in the solar irradiance during the industrial era (1760 onward) have been estimated to account for about 0.05 W m^{-2} of the radiative forcing, which represents the amount of warming or cooling produced by the changes in the global energy balance (IPCC 2021). However, the analysis of the post-1978 satellite-based measurements shows a fluctuation of about 0.1% in the solar irradiance over the 11-year solar cycles, resulting in less than 0.1 °C fluctuations in the global mean temperature. Therefore, recent years' observed global warming trend suggests a considerable decrease in the relative influence of the total solar irradiance changes on global warming (Mendoza 2005; IPCC 2021).

Volcanoes

Volcanic eruptions influence climate change by releasing vast amounts of volcanic gases and dust particles into the atmosphere. Though volcanic CO_2 can cause global warming, the sulphate aerosols formed due to the conversion of sulphur dioxide to sulphuric acid in the stratosphere can cause global cooling. The sulphate aerosols reflect the solar radiation into space and cool the earth's lower atmosphere or troposphere. However, since 1900, the average amount and variability of volcanic aerosols have remained at the same level as recorded over the past 2500 years (IPCC 2021). Moreover, the decline in global surface temperature due to volcanic aerosols lasts only 2–5 years.

El Nino-Southern Oscillation (ENSO)

ENSO, an influential climate phenomenon, considerably impacts the global climate and influences the seasonal-to-interannual climate prediction. However, global warming has influenced the ENSO characteristics and its impacts in the twenty-first century compared to the twentieth century. In the twenty-first century, ENSO has tended to occur with the warmest sea-surface temperature (SST) anomalies in the tropical central Pacific instead of the eastern Pacific in the twentieth century. Although the climate model projections confirm the changes in the ENSO in the warming future, the future ENSO conditions and their impact on temperature and precipitation are uncertain (Yang et al. 2018).

Anthropogenic Drivers

Greenhouse Gases

Increased concentration of atmospheric greenhouse gases (GHG) has reportedly dominated the climate change drivers since 1750 (IPCC 2021). CO_2, methane (CH_4), and nitrous oxide (N_2O) are the primary greenhouse gases. Recent IPCC report suggests that concentrations of CO_2, CH_4, and N_2O have touched the highest levels in at least 800,000 years, and current CO_2 concentrations are the highest over at least 2 million years. Consequently, the total anthropogenic effective radiative forcing (ERF) has likely risen at a faster rate since the 1970s. With reference to 1750, ERF was 2.72 W m^{-2} in 2019 (IPCC 2021).

Figure 14.1 presents CO_2 concentrations in the atmosphere from hundreds of thousands of years ago through 2019 (EPA 2021). Over 1750–2019, CO_2 has increased by 131.6 ± 2.9 ppm (47.3%) and reached 409.9 (±0.3) ppm in 2019. The combustion of fossil fuels like coal, oil and gas, and deforestation are the primary CO_2 emission sources.

Similarly, since 1750, concentrations of CH_4 increased by 1137 ± 10 ppb (157.8%) to reach 1866.3 (±3.3) ppb in 2019. N_2O increased by 62.0 ± 6.0 ppb since 1750 and reached 332.1 (± 0.4) ppb in 2019. The N_2O concentration primarily

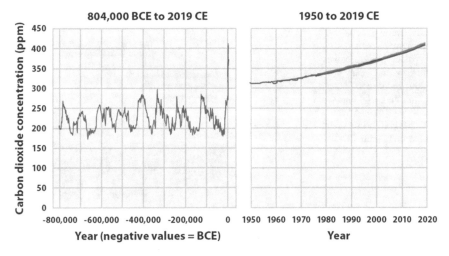

Fig. 14.1 Global atmospheric concentrations of carbon dioxide over time (After EPA 2021)

increased due to a 30% rise in emissions caused by worldwide agricultural expansion since 1980 (IPCC 2021).

Aerosols

Atmospheric aerosols primarily have a cooling effect on the climate system as they reflect incoming solar radiation to outer space. Aerosols have diverse anthropogenic and natural sources and are categorised by composition: sulphate, black carbon, organic, nitrate, dust, and sea salt. Referring to 1750, total aerosol ERF was -1.1 W m^{-2} in 2019 (IPCC 2021).

In addition to the changes in the GHG and aerosols, tropospheric ozone has been rising since 1750 due to emissions of hydrocarbons, including N_2O and CH_4. The ozone depletion may increase the ultraviolet (UV) radiation reaching earth's surface and enhance skin cancer, eye cataracts, and genetic and immune system damage.

Land Use and Land Cover Change (LULCC)

Land use and land cover (LULC) combines two terms, separate but often used interchangeably, i.e., land use and land cover. Land use describes the way humans use the land for economic and cultural activities, for example, agricultural, residential, industrial, mining, and recreational uses. On the other hand, land cover refers to the biophysical characteristics of the earth's surface that integrate and reflect the climate, geology, soils, and available biota over decades or longer.

LULCC is a significant driver of climate change as it exerts biophysical and biogeochemical effects (Liping et al. 2018). Recent IPCC report suggests that since 1750, the increase in global albedo has cooled the climate due to biophysical effects of land use change. On the contrary, the biogeochemical effects due to GHG emissions or sinks have resulted in net warming. ERF due to land use and land cover changes stands at -0.2 W m^{-2} (IPCC 2021).

Fig. 14.2a Change in decadal average global temperature as recreated from paleoclimate archives (solid grey line, years 1–2000) and direct observations (solid black line, 1850–2020) (After IPCC 2021; Reprinted with permission from IPCC)

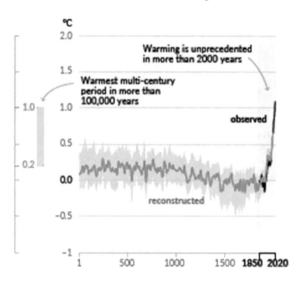

14.2.2   Increase in Temperature

Since the Industrial Revolution, the global average surface temperature has risen overall by approximately 1 °C. The total human-induced global surface temperature has increased by about 1.07 °C during 1850–1900 and 2010–2019. However, global surface temperature increase since 1970 has been faster than any previous 50-year period in the past 2000 years. Moreover, 2011–2020 has been even warmer than the earlier recorded warm period around 6500 years ago (IPCC 2021).

Worldwide, nine of the ten warmest years since 1880 have occurred since 2005, and the five warmest years recorded ever have happened since 2015. Additionally, 2016 and 2020 have been the warmest and the second-warmest years on record, with 2011–2020 being the warmest decade.

Figures 14.2a and 14.2b present the changes in the global surface temperature relative to 1850–1900 (IPCC 2021). The left vertical bar in Fig. 14.2a shows the temperature estimates for the warmest multicentury period. The grey shading presents the likely range of the reconstructed temperature. Similarly, the shadings in Fig. 14.2b show the very likely range of simulations.

14.2.3   Alterations in Precipitation Patterns

As mean temperatures have risen, mean precipitation has also increased due to increased evaporation. Since 1901, global precipitation has increased at an average rate of 2.5 mm per decade. Also, heavy precipitation events have occurred frequently since the 1950s and at an accelerated rate since the 1980s.

Fig. 14.2b Change in global observed and simulated annual average temperature considering human and natural and only natural factors (both 1850–2020) (After IPCC 2021; Reprinted with permission from IPCC)

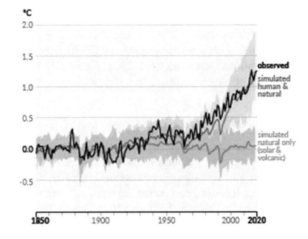

The precipitation patterns are likely to vary worldwide, as precipitations tend to increase due to warming from GHG emissions and decrease due to cooling from aerosol emissions. For example, monsoon precipitation over South Asia and East Asia has increased during the past four decades. In the coming decades, global precipitation during the monsoon season will likely increase, over Asia, especially the East, South and Southeast Asia, and West Africa. However, the monsoon is likely to onset late over North and South America and West Africa (IPCC 2021).

14.2.4 Changes in CO_2 Concentration and Radiative Forcing

In nature, CO_2 exchange occurs continuously between the atmosphere and plants (through photosynthesis), the atmosphere and animals (through respiration) and between the atmosphere and ocean (through gas exchange).

However, the CO_2 level in 2019 has been 40% higher than in the nineteenth century, with most of this increase taking place since 1970. The CO_2 increase has been due to fossil fuel burning and deforestation, disturbing the carbon cycle balance. In 2019, annual average CO_2 concentrations reached 410 parts per million (ppm), the highest level in at least 2 million years.

Radiative forcing represents the amount of warming or cooling produced by the changes in the global energy balance. The global energy balance refers to the balance between the energy coming from the sun and the energy returned to space by the earth's radiative emission. The energy balance changes resulting in warming are called *positive* forcing, whilst those resulting in cooling are called *negative* forcing. The earth's average surface temperature varies due to the imbalance between the positive and negative forcing.

Radiative forcing depends on the changes in the atmospheric greenhouse gas concentrations. The recent report of IPCC suggests that relative to 1750, radiative forcing reached 2.72 W m^{-2} in 2019, causing warming of the climate system due to increased atmospheric GHG concentrations. The radiative forcing has increased by 0.43 W m^{-2} (19%) relative to the IPCC fifth assessment report (2013), with GHG concentrations increasing since 2011, accounting for 0.34 W m^{-2} (IPCC 2021).

14.2.5 Changes in Sea Levels and Glacier Retreat

Due to increased greenhouse gas concentrations, global warming has resulted in the global mean sea-level increase of 0.20 m between 1901 and 2018. The average rate of sea-level rise has increased from 1.3 mm year^{-1} between 1901 and 1971 to 1.9 mm year^{-1} between 1971 and 2006 and further to 3.7 mm year^{-1} between 2006 and 2018. Moreover, the global ocean has warmed faster over the past century since the end of the last deglacial transition (around 11,000 years ago) (IPCC 2021).

Since the 1990s, human interventions have substantially influenced the global glaciers retreat. The Arctic sea ice area has reduced by around 40% and 10% in September and March, respectively, during 1979–1988 and 2010–2019. Similarly, human influence has contributed to the decrease in Northern Hemisphere spring snow cover since 1950 and the Greenland Ice Sheet since 2000. The annual average ice area of the Arctic sea was at its lowest level during 2011–2020 since at least 1850. Also, the Arctic sea ice area recorded in the late summer was the smallest in at least the past 1000 years.

14.3 Climate Change Impacts

14.3.1 Runoff

Climate change is likely to affect the hydrology and water cycle considerably, and the quantitative assessment of these effects is essential to deal with its consequences. As discussed in Sect. 14.2.2, it is established that the global surface temperature has increased during 1850–1900 and 2010–2019 by about 1.07 °C. Since the temperature impacts the hydrological cycle components like precipitation, evapotranspiration, and interception, rising temperatures may influence the river runoff and regional water resources (Bates et al. 2008; Huo and Li 2013). Since river runoff plays a vital role in water supply, transport of nutrients and sediment, drainage and flooding in an area, its accurate assessment is essential for effective water management.

A large number of studies have confirmed an increase in the global mean annual runoff due to rise in global surface temperature (Zhang et al. 2014; Lehner et al. 2019), but the results vary regionally (Chen et al. 2017; Cook et al. 2020). Lehner et al. (2019) reported that Coupled Model Intercomparison Project 5 (CMIP5) models displayed a large variability in the runoff to precipitation ratio for the present and future climate.

In an independent study, Ghiggi et al. (2019) introduced a global gridded monthly runoff reconstructed from 1902 to 2014 (Fig. 14.3). The long-term mean runoff rates differed by three orders of magnitude across the globe, with the highest rates in the tropics and extensive mountain ranges and lowest rates in the extratropics and major world deserts such as the Sahara (Fig. 14.3a). Figures (14.3b) and (14.3c) present the months with the minimum and maximum of the long-term mean runoff. As evident, the snow-dominated regions in the Northern Hemisphere have the lowest runoff in winter and runoff peaks towards the end of spring due to spring snowmelt. On the other hand, maximum runoff occurs during the rainy season in the tropics. In the humid mid-latitudes, the lowest and the highest runoff occurred during the summer and winter months.

14.3.2 Floods

As discussed in Sect. 14.2.3, besides an average 2.5 mm per decade increase in global precipitation since 1901, heavy precipitation events have occurred frequently since the 1950s and at an accelerated rate since the 1980s. The increased frequency and intensity of heavy precipitation are likely to increase the intensity and frequency of flooding. Since floods severely impact human societies by causing disruption and loss of life, besides hampering the aquatic ecosystems and the economy, they are considered one of the significant impacts of climate change. Global-scale flood assessments have shown decreases and increases in the future floods under global warming (Tabari 2020).

Besides precipitation amount and intensity, many other factors affect floods, e.g., the antecedent soil moisture, land use and land cover characteristics, and stream morphology. In cold regions, snow water-equivalent and snowmelt timing may significantly affect floods. However, extreme precipitation may have a high correlation with flash floods.

The earlier reports by SREX (Managing the Risks of Extreme Events and Disasters to Advance Climate Change Adaptation) (Seneviratne et al. 2012) and IPCC (AR5; Hartmann et al. 2013) found low confidence in the observed changes in the flood frequency or magnitudes at the global scale. However, SR15 (IPCC Special Report: Global Warming of 1.5 °C) reported an increase in the magnitude and frequency of floods in some regions but a decrease in others (Hoegh-Guldberg et al. 2018). Similarly, peak flow trends have shown high regional variability worldwide with no statistically significant increase or decrease (Do et al. 2017).

In recent times, flash floods in urban areas have been rising. This rise in flash floods may be due to increased frequency and intensity of heavy precipitation occurring over impervious surfaces in urban areas (Hettiarachchi et al. 2018). However, the design of stormwater drainage systems and the overland flow rate in urban areas may play a

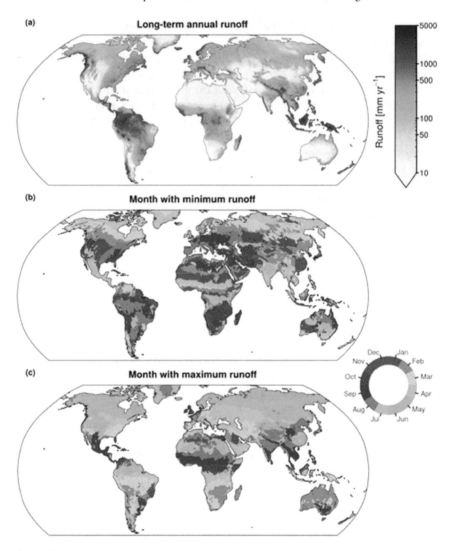

Fig. 14.3 Runoff variation over 1902–2014 **a** long-term mean annual runoff rates, **b** month with the minimum, and **c** the maximum long-term mean monthly runoff (After Ghiggi et al. 2019; Creative Commons Attribution 4.0)

crucial role in the occurrences of such floods (Maksimovic et al. 2009). Flash floods may also occur due to glacier lake outbursts or dam breaks (Schwanghart et al. 2016).

14.3.3 Drought

Droughts are defined as a period of unusually dry weather that lasts long enough to cause a shortage in a region's water supply. Drought usually occurs in areas where rainfall is below the long-term average. Drought may be classified as meteorological (precipitation deficits), agricultural (soil moisture deficits, leading to crop yield reductions or failure), hydrological (water deficits in streams or lakes, reservoirs, and groundwater), or ecological (plant water stress that may lead to tree mortality).

Droughts are usually assessed using indices that quantify their duration, magnitude, and frequency. The standardised precipitation index (SPI) and the standardised precipitation evapotranspiration index (SPEI) are preferred due to simplicity. As discussed in Sects. 14.2 and 14.3, climate change will result in temperature rise, besides changing the precipitation patterns and temporal and spatial runoff. Consequently, the drought characteristics of a region will also change due to changes in the water cycle components and increased water demand.

Following the 2015 Paris Agreement of the United Nations Framework Convention on Climate Change (UNFCCC), several studies have shown significant impacts of a 0.5 °C warming on drought (Su et al. 2018). The Paris agreement aimed at limiting the global mean temperature (GMT) increases below 2.0 °C above preindustrial (PI) levels whilst targeting to keep the increase to 1.5 °C above PI levels (UNFCCC 2015). The recent IPCC report suggests that increasing greenhouse gas concentrations impact various drought types differently. Agricultural and ecological droughts primarily occur due to insufficient soil moisture and increased atmospheric evaporative demand (AED). On the other hand, hydrological drought is caused by a lack of runoff and surface water. Agricultural and ecological droughts have reportedly increased worldwide. However, the frequency of hydrological droughts has increased only in fewer regions (IPCC 2021).

14.3.4 Agriculture

The agricultural sector shares a unique relationship with climate change. On the one hand, the agriculture sector is directly affected by climate change due to altered weather conditions like warming, changing precipitation patterns, and increased frequency of extreme events. On the other hand, the agriculture sector is responsible for climate change by contributing a substantial portion of the greenhouse gas emissions, i.e., 17% through agricultural practices and 7–14% through inevitable land use changes. The agriculture sector generates about 50% of anthropogenic methane emissions through ruminant animals and rice cultivation; and about 75% of anthropogenic nitrous oxide emissions resulting from cultivated soils and fertiliser and manure applications. Both methane and nitrous oxide pose a more significant threat to global warming potential than CO_2.

Climate change has already affected food security worldwide. Studies have shown a decline in the yield of crops like maize and wheat in several lower-latitude regions. On the contrary, yields of crops like maize, wheat, and sugar beets have increased in higher-latitude regions. Since many countries in the world are agriculture dependent; therefore, climate change may lead to poverty in many developing countries. For example, in subSaharan Africa, the agricultural sector employs almost 52% of the population and contributes about 14% to the gross domestic product (GDP).

Besides impacting the stability of the food supply, increased atmospheric CO_2 levels are likely to reduce the nutritional content of crops (IPCC 2019). Simulation results of global crop and economic models under the Shared Socioeconomic Pathway 2, i.e., SSP2, show a median increase of 7.6% in cereal prices in 2050, leading to higher food prices and increased risk of food insecurity and hunger. Also, by 2050, droughts and their associated consequences are projected to impact about 178, 220 and 277 million people in dry regions at 1.5, 2 and 3 °C warmings (IPCC 2019). The SSPs are described later in Sect. 14.7.

The agricultural sector can contribute significantly to climate change adaptation and mitigation by improved fertiliser and crop management, increasing soil organic matter, and using heat and drought-tolerant crop varieties. Moreover, better land management for grazing, improved manure management, and higher-quality feed for livestock can also contribute. Improved harvesting, on-farm storage and food supply-chain management can reduce the reported loss of about 25–30% of the total food produced and consequently reduce total anthropogenic GHG emissions by 8–10% (IPCC 2019).

14.3.5   Landslides

Landslides denote the down-slope movement of a mass of soil, rock, or debris under the direct influence of gravity (Hungr et al. 2013). Landslides involve falling, flowing, toppling, sliding, and spreading, or a combination of different types of movements. Landslides occur worldwide and influence landscape evolution. In many areas, they pose serious geological hazards (Petley 2012).

Landslides cause significant loss of lives and economic damages globally and account for the 17% of fatalities attributed to natural hazards worldwide. Landslides occur mainly due to slope failures influenced by different phenomena, e.g., precipitation, snowmelt, temperature rise, earthquakes, volcanic eruptions, and human actions. Since climate change influences several of these phenomena, especially precipitation and temperature, it is expected that climate change will significantly impact slope failures and landslides (Seneviratne et al. 2012).

With the rise in temperature due to human-induced climate change, the frequency of the intense storm will go up. Consequently, shallow landslides, including debris flows, debris avalanches, and rockfalls, will increase in high mountain areas (Stoffel et al. 2014). These areas include the Alps, the American Cordillera, the Himalayas, the Atlas Mountains in northwestern Africa, East Africa's Rift Valley mountains and hills in southwestern Africa, the Arabian Peninsula, the Appalachians in eastern North America, and the Carpathians in Eastern Europe.

14.4 Climate Change and Soil Erosion

14.4.1 Water Erosion

Soil erosion poses a universal threat to food and water security and the environment (Panagos et al. 2015; Li and Fang 2016). Soil erosion by water is mainly affected by rainfall, topography, soil, land use, and land cover (Panagos et al. 2015). Climate change will have a direct impact on soil erosion through variations in rainfall amount, intensity and spatio-temporal distribution patterns, and indirectly through changes in land cover and soil moisture due to temperature rise (Li and Fang 2016; Duulatov et al. 2019).

The precipitation and temperature changes combined will also impact the crop planting and harvesting dates and cultivars (Nunes and Nearing 2011; Parajuli et al. 2016). An average annual soil erosion rate of $1.0\,t\,ha^{-1}\,year^{-1}$ is considered tolerable, and an average annual soil erosion rate of more than $1.0\,t\,ha^{-1}\,year^{-1}$ is considered irreversible over 50–100 years (Jones et al. 2004; Verheijen et al. 2009). Yang et al. (2003) estimated the global scale average soil erosion rate as $10.2\,t\,ha^{-1}\,year^{-1}$, with even the lowest rate of $3.0\,t\,ha^{-1}\,year^{-1}$ amongst continents for Australia exceeding the tolerable erosion rate. Figure 14.4 presents the spatial distribution of soil erosion worldwide (Li and Fang 2016). As evident, soil erosion is severe in Australia, Africa, China, India, Europe, South America, and the USA and may worsen due to climate change.

As mentioned earlier, changes in precipitation amount, intensity, and spatio-temporal distribution are likely to impact soil erosion directly. Research studies have shown that increased rainfall will likely increase runoff and soil erosion (Zabaleta et al. 2014). A precipitation variation of 1% may lead to a 1.3–2% variation in runoff and a 1.7–2% change in soil erosion (Pruski and Nearing 2002; Lu et al. 2013). Soil erosion is also likely to be affected by rainfall intensity. Several studies have reported increased soil erosion due to the increased frequency of extreme rainfall events (Plangoen et al. 2013; Simonneaux et al. 2015).

Li and Fang (2016) reviewed about 205 research papers dealing with climate change impact on soil erosion. Out of these, 136 studies from Austria, Brazil, China, Germany, India, Portugal, the UK and parts of the USA concluded that erosion rates would increase over 1.2%–16.14% (nearly 17 times) due to climate change. On the contrary, 55 studies from Brazil, India, the UK, parts of the USA and Vietnam showed decreased soil erosion. Most studies reported erosion rates higher than the tolerable threshold soil loss of $1.0\,t\,ha^{-1}\,yr^{-1}$. Soil erosion is likely to be more intense under future climate change. However, future soil erosion projections show high spatial variability. Soil erosion rates may vary based on geographic locations, topographic

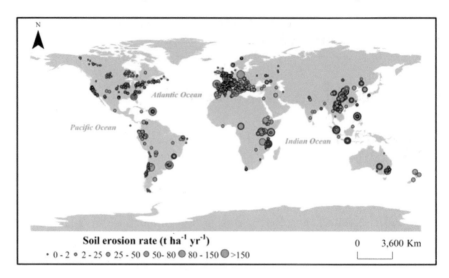

Fig. 14.4 Spatial distribution of soil erosion around the world (After Li and Fang 2016; Reprinted with permission from Elsevier)

conditions, precipitation patterns, and land management practices. The measures that can protect soils include afforestation, no-till or conservation tillage, and sowing drought-resistant cultivars (Routschek et al. 2014; Mondal et al. 2015; Serpa et al. 2015).

14.4.2 Wind Erosion

Wind erosion is a global environmental issue that impacts around one-third of the world's land (Pimentel et al. 1995; Shen et al. 2018). It is the leading cause of concern in arid, semi-arid, and even subhumid regions worldwide. Climate change impacts wind erosion by affecting wind speed, precipitation, and temperature (Du et al. 2017; Chi et al. 2019). The wind is the dominant climatic factor influencing wind erosion as wind speed governs sediment transport. Moreover, near-surface wind speeds can strongly influence the wind erosion process. Temperature and precipitation contribute to wind erosion by influencing the soil water content and crop residue cover (Zhang et al. 2018).

Several researchers have analysed the impact of climate change on wind erosion worldwide. Gao et al. (2002) found a 25% increase in the wind erosion of grassland due to a temperature rise by 2 °C in Asia. Liddicoat et al. (2012) reported that a 5–20% drop in precipitation, 1.5 °C rise in temperature and 390 to 480 ppm rise in CO_2 concentration by 2030 might lead to reduced biomass cover in agricultural lands of Australia and enhance wind erosion. Bohner et al. (2004) assessed the impact of climate change on wind erosion in Europe using the Wind Erosion on European

Light Soils (WEELS) model. They predicted extended periods of erosion during spring in the future. The European Environment Agency (2012) also reported that fine-textured soils across central Europe might face enhanced wind erosion due to extreme wind speeds and increased aridity.

The recent IPCC report (IPCC 2021) suggests that central and southern Australia may face increased wind erosion and dust emission due to altered vegetation owing to hydrological drought and a decrease in severe winds (Webb et al. 2020). Similarly, Central Asia may face higher risks of desertification, dust storms, and wildfire due to higher temperatures and less precipitation (Hoegh-Guldberg et al. 2018).

14.5 Land Use and Land Cover (LULC) Changes

LULC changes play a critical role in invoking sustainability issues, including food and water security, amidst global challenges like climate change. Therefore, it is essential to analyse how LULC changes occur across space and time for understanding their climatic and environmental impact. LULC change has affected about 32% of land across the globe during 1960–2019 (Winkler et al. 2021). Major LULC changes can be categorised as changes in the forest cover (deforestation and afforestation), alternation in the agricultural lands (agricultural land loss and agricultural land expansion), decrease in wetland areas (conversion and degradation), and increase in urban areas (urbanisation).

Figure 14.5 presents the land transitions in percentage terms relative to the total global area that changed over 1992–2018 (Radwan et al. 2021). As evident, the most significant land transition occurred from forest to agriculture, followed by natural vegetation converting to the forest.

Overall, forests, agricultural lands, and naturally vegetated areas accounted for about 57% of all global land cover transitions during the period. It is also evident from the figure that the urbanisation of the land ends the land transition as an urbanised land did not change further. The urban development mainly occurred on the agricultural land, amounting to 68% of the global urban growth.

14.5.1 Changes in Forest Cover

Forests cover about 4.06 billion hectares or 31% of the global land area (Winkler et al. 2021). About 50% of the forest area is intact, with primary forests free from human activities, accounting for more than one-third area. However, the remainder of the forest area has suffered from deforestation at alarming rates.

Deforestation accounts for 38% of all land transitions. Winkler et al. (2021) recounted a global net loss of forest area of 80 million hectares during 1960–2019. FAO (2021) estimates show that since 1990, about 420 million hectares of forest have been converted to other land uses. Agricultural expansion has been the key motive

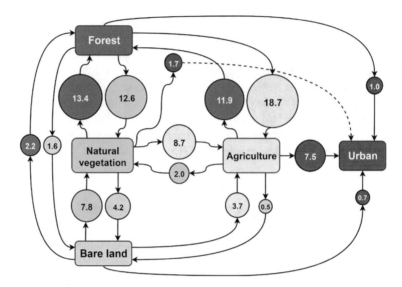

Fig. 14.5 Schematic representation of the global land transitions between 1992 and 2018 (After Radwan et al. 2021; Creative Commons Attribution Licence 4.0 (CC BY))

behind deforestation, accounting for about 48% of the total deforested area. However, in the past three decades, the deforestation rate has slowed down to 10 million hectares per year (2015–2020) compared to 16 million hectares per year in the 1990s. Moreover, large-scale afforestation has occurred in several countries around the world. China heads the list of countries with the largest area under forest by 2015 with 79 million hectares, followed by the USA (26.3 million hectares), Russia (19.8 million hectares), Canada (15.8 million hectares), Sweden (13.7 million hectares), and India (12 million hectares) (FAO 2015).

14.5.2 Alterations in Agriculture Lands

The total land used for agriculture, i.e., the sum of the cropland and grazing land, accounts for about 4.9 billion hectares or 37 per cent of the global land area. Out of this, about 1.6 billion hectares or 12 per cent of the global land area are used for cropland and 3.3 billion hectares for grazing (Ritchie and Roser 2018). The current area under cropland is only about 36 per cent of the total land estimated to be fit for crop cultivation worldwide.

The changes in the agricultural land use, i.e., land transitions from one crop type to the other, e.g., subsistence to cash crops, account for 53% of all land transitions. The agricultural land use changes frequently occur in the developed countries of Europe, Australia, and the USA and fast-growing economies like India due to agricultural intensification.

The global land use changes, in general, and the agricultural land use changes, in particular, depict two distinct temporal phases: an acceleration phase during 1960–2004 and a decelerating phase during 2005–2018 (Radwan et al. 2021). The market globalisation and enhanced international trade were responsible for the accelerated land use changes, especially the agricultural land use changes during the 1990s. On the other hand, climate change-induced extreme events like floods and droughts have a dominant role in the deceleration phase. For example, droughts have affected agricultural land use in East and West Africa during the 2000s.

14.5.3 Decrease in Wetland Areas

Wetlands cover about 1.2 billion hectares of the global land area and provide vital services like climate change mitigation, flood control, and water purification for humankind. Wetlands provide fresh water for human consumption and support almost 40% of the world's species for living and breeding. Wetlands also act as a carbon sink and help deliver global commitments on reducing the carbon footprint. However, wetlands are disappearing fast, with about a 35% decline in wetland areas during 1970–2015 (Ramsar Convention 2018).

The wetland area has declined as natural wetlands have been drained and converted into agricultural, industrial, or urban lands worldwide. The conversion of wetlands into agricultural lands and the resultant increased pollutant load due to fertilisers, pesticides, and animal faeces lead to algal bloom and subsequent ecosystem damages. The urbanisation of the coastal areas and river deltas and changing food habits has further added to the problem.

14.5.4 Increase in Urban Areas

Though there is no universal definition of *urban*, urbanisation represents a process involving conversion of forest cover and agricultural lands to accommodate the rising numbers of the *urban workforce*, i.e., people employed in the non-agricultural sector like manufacturing. Therefore, urbanisation essentially involves the spatial expansion of built-up areas. As per the 2018 UN World Urbanisation Prospects, about 4.2 billion people, or 55% of the global population, lived in urban areas compared to 1.0 billion in 1960 (UN 2019). The urban population grew steadily and surpassed the rural population in 2007 (Ritchie and Roser 2018).

LULC changes due to urbanisation significantly affect the quality of air, soil, and water (Haas and Ban 2014; Liu et al. 2019). The increase in the built-up area increases imperviousness, which, in turn, enhances the surface runoff and facilitates drainage of the diffuse pollutants into the surface water bodies (Buffleben et al. 2002). Figure 14.5 shows that once a land is urbanised, it cannot be used for any other purpose. It is also evident from the figure that almost 68% of the urban area results from the conversion of agricultural land, followed by 16% from the natural vegetation and 9% from forest cover.

14.6 LULC Changes and Soil Erosion

LULC changes, including deforestation and alterations in agricultural practices, typically enhance soil erosion. The global soil erosion from arable lands worldwide was estimated as 75 billion tonnes (Pg) per year in 1993 (Myers 1993). However, these estimates were based on static observations and did not consider the effect of LULC changes. Borrelli et al. (2013) carried out the RUSLE-based modelling of approximately 125 million km^2 area, representing about 84.1% of the earth's land surface belonging to 202 countries, and revised the soil erosion estimates to 35 Pg year^{-1} or 2.8 Mg ha^{-1} year^{-1} for 2001. Also, the soil erosion rates were estimated as 35.9 Pg year^{-1} for 2012, showing a 2.5% increase in soil erosion due to LULC changes. During 2001–2012, around 4.0 million km^2 underwent a land use change, of which 2.9 million km^2 resulted in an increase of 1.74 Pg year^{-1} in soil erosion and the remainder 1.1 million km^2 showed a decline of 0.88 Pg year^{-1}. Thus, the soil erosion from areas with LULC change increased by 0.86 Pg year^{-1}.

The LULC change dynamics during 2001–2012 showed a 1.65 million km^2 decline in the forest cover and an increase of 0.22 million km^2 in croplands and 1.43 million km^2 in the semi-natural vegetation (grassland, scrubland and transition forest) (Fig. 14.6). About 2.26 million km^2 of forest cover was lost during the period, and 0.61 million km^2 was gained, accounting for a negligible decrease of 0.02 Pg year^{-1} in soil erosion. About 0.55 million km^2 of cropland was abandoned, whilst 0.76 million km^2 of new cropland was established. The increased cropland resulted in an increase of 0.69 Pg year^{-1} in soil erosion; thus contributing to about 80% of the overall increase in soil erosion.

Croplands resulted in about 50% of the estimated overall soil erosion from about 11% of the land surface at an average soil erosion rate of 12.7 Mg ha^{-1} year^{-1}. On the contrary, forests contributed only 1.7% to the soil erosion from 28% of the land surface at a considerably lower average rate of 0.6 Mg ha^{-1} year^{-1}. At the continental scale, Europe, North America, and Oceania had significantly lower soil erosion rates of 0.92, 2.23, and 0.9 Mg ha^{-1} year^{-1}, respectively, in 2001. On the contrary, South America had the highest average soil erosion rate of 3.53 Mg ha^{-1} year^{-1} in 2001, whilst Africa (3.51 Mg ha^{-1} year^{-1}) was second, followed by Asia (3.47 Mg ha^{-1} year^{-1}). However, with a 10% increase in the average soil erosion rate, Africa had the highest average soil erosion rate (3.88 Mg ha^{-1} year^{-1}) in 2012. South America and Asia showed an 8% and 1% increase in the average soil erosion rate. Interestingly, despite being the two most populous countries globally, China and India showed a decrease in soil erosion estimates, primarily due to emphasis on afforestation.

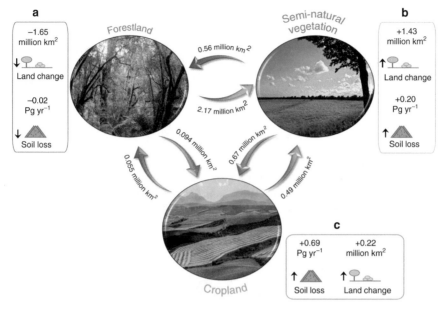

Fig. 14.6 Land use change dynamics and the resultant soil erosion during 2001–2012. Insets a–c demonstrate the net change in the area (million km^2) and soil erosion (Pg yr^{-1}) (After Borrelli et al. 2013: Creative Commons Attribution Licence 4.0 (CC BY))

14.7 Combined Impact of Climate and LULC Changes on Soil Erosion

As discussed in Sects. 14.4 and 14.6, both climate and LULC changes influence soil erosion. Moreover, climate and LULC changes are interdependent as LULC change is a significant driver of climate change (Sect. 14.2.1), whilst climate change drives changes in land use, especially impacting agriculture (Sect. 14.3.4). Since anthropogenic activities are likely to impact both climate and land use, the effects of climate and LULC changes should be analysed in an integrated manner to evaluate the future changes in soil erosion.

Yang et al. (2003) initiated the trend of analysing the possible changes in soil erosion due to global changes in climate and LULC. They used GIS-based RUSLE and the available global datasets at a 0.5° grid. Their analysis showed that anthropogenic activities induced almost 60% of soil erosion. Also, climate change might influence soil erosion more significantly than LULC change in the future from a global perspective. Similarly, Ito (2007) analysed the impacts of climate and LULC changes on soil erosion by water. They carried the analysis over 1901–1990 based on historical data, and 1991–2100 utilising the climate projections from the IPCC Fourth Assessment Report. They concluded that the conversion of forests into croplands significantly influenced soil erosion during the historical period. However, climate change-induced rainfall and land use changes influenced soil erosion during the latter phase.

Since 2010, researchers have joined forces to develop the Shared Socioeconomic Pathway (SSP)–Representative Concentration Pathway (RCP) framework (Moss et al. 2010). Though the details of the RCPs and SSPs were known earlier, these were formally published in 2011 (van Vuuren et al. 2011) and 2017 (Riahi et al. 2017), respectively. The SSP–RCP framework helps carry out climate change analysis in an integrated manner by combining the concurrently developed climate and societal futures.

The Shared Socioeconomic Pathways (SSPs) present five different development pathways of the world, considering socioeconomic factors such as population, economic growth, education, urbanisation, and the rate of technological development (O'Neill et al. 2016). The SSPs describe a range of possible worlds without climate policy whilst considering climate change mitigation and adaptation challenges. The specific features of individual SSPs are as follows (Riahi et al. 2017):

(1) SSP1 Sustainability—Taking the Green Road (low challenges to mitigation and adaptation)
(2) SSP2 Middle of the Road—Development following historical patterns (medium challenges to mitigation and adaptation)
(3) SSP3 Regional Rivalry—A Rocky Road (high challenges to mitigation and adaptation)
(4) SSP4 Inequality—A Road Divided (low challenges to mitigation, high challenges to adaptation)
(5) SSP5 Fossil-fuelled Development—Taking the Highway (high challenges to mitigation, low challenges to adaptation)

SSP1 and SSP5 present relatively optimistic human development trends, with a booming economy leading to significant investments in health and education. However, SSP1 assumes an increasing shift towards sustainable practices, whilst SSP5 assumes an energy-intensive, fossil fuel-based economy. SSP2 presents a *middle of the road* approach, assuming the continuity of the historical development patterns in the twenty-first century. SSP3 and SSP4 present relatively pessimistic trends in the future social and economic development, with increasing inequalities in underdeveloped countries amidst negligible investment in education and health coupled with unabated population growth.

The Representative Concentration Pathways (RCPs), on the other hand, describe different pathways for greenhouse gases and aerosol concentrations, together with land use change in the future. The pathways are categorised by the radiative forcing, defined earlier in Sect. 14.2.4. There are four pathways representing the radiative forcing of 2.6, 4.5, 6.0, and 8.5 W m^{-2}. RCP2.6 is an ambitious pathway representing a mitigation scenario in which emissions fall due to the active removal of greenhouse gases. On the contrary, RCP8.5 is a high baseline emission scenario representing a failure to reduce emissions. RCP4.5 and RCP6 are two medium stabilisation scenarios in which total radiative forcing is stabilised shortly after 2100, without overshooting the specific radiative forcing target levels.

Borrelli et al. (2020) used the SSP-RCP pathways for analysing the combined impact of climate and land use projections on soil erosion during 2015–2070. They used global land use conditions based on three SSP-RCP pathways, namely SSP1–RCP2.6, SSP2–RCP4.5, and SSP5–RCP8.5. Based on the description of SSPs and RCPs given above, the SSP1–RCP2.6 scenario represents low emission and radiative forcing, characterised by moderate population growth and high economic growth and technological improvements, including agricultural productivity. On the contrary, SSP5–RCP 8.5 scenario represents high emission and radiative forcing due to high fossil fuel use, almost 20% expansion in global cropland, and up to 300% increase in greenhouse gas emission. SSP2–RCP4.5 is the 'middle of the road' scenario, with continuing historical development patterns throughout the twenty-first century.

Figure 14.7 presents the simulation results in terms of overall soil erosion, the contribution of the projected climate change to soil erosion and the projected land use changes and their contribution to soil erosion for SSP1–RCP2.6 (insert a), SSP2–RCP4.5 (insert b), and SSP5–RCP8.5 (insert c).

The results show that overall soil erosion will increase substantially compared to the 2015 baseline, with SSP1–RCP2.6, SSP2–RCP4.5, and SSP5–RCP8.5 scenarios increasing soil erosion to 13.08, 21.78, and 28.58 Pg yr^{-1}, respectively. The results further show that although land use changes will have a considerable effect on the overall soil erosion process, climate change will be the major driver of future changes in soil erosion.

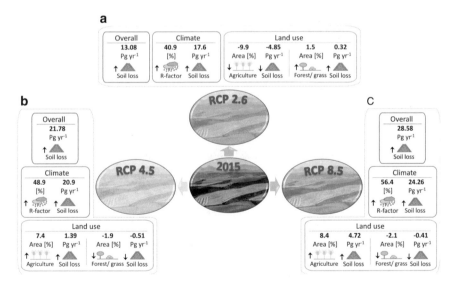

Fig. 14.7 Impact of the projected land use and climate changes on soil erosion for SSP1–RCP2.6 (insert a), SSP2–RCP4.5 (insert b), and SSP5–RCP8.5 (insert c) (After Borrelli et al. 2020; Creative Commons Attribution Licence 4.0 (CC BY))

Questions

1. Discuss the drivers of climate change. Explain the role of the drivers in the warming over the past few decades.
2. How do the increase in anthropogenic carbon dioxide and other greenhouse gases affect climate? How does agriculture affect climate change?
3. What is global warming? Describe various impacts of global warming.
4. Will the world succeed in preventing a greater than 2 °C temperature rise? What role can civil society play in addressing climate change? What role can an individual play in mitigating climate change?
5. What is an El Nino? Discuss its role in altering global weather patterns. Will it help forecast long-term weather more accurately?
6. What are the possible implications of a warming planet in various fields? Describe these implications in terms of their impacts on the environment, economy, and human beings.
7. Describe various land use land cover (LULC) changes over the past decades, along with their climatic and environmental impact.
8. Discuss the possible implications of deforestation and alterations in agricultural practices in enhancing soil erosion.
9. What are the Shared Socioeconomic Pathways (SSPs)? Describe the specific features of the individual SSPs.
10. What will be the combined impact of climate and land use projections on soil erosion during the twenty-first century?

Multiple Choice Questions

Answer the questions by choosing the correct option.

1. Since the Industrial Revolution, the global average temperature has
   (A) cooled down by 0.1°C   (B) warmed up 0.1°C
   (C) warmed up by more than 1°C   (D) warmed up by about 2°C

2. The most significant contribution to positive (warming) radiating forcing of the climate is by

   (A) aerosols   (B) ozone
   (C) greenhouse gases   (D) solar irradiance

3. Which of the following is the primary cause of sea-level rise?

   (A) Melting sea ice
   (B) Rivers accelerating
   (C) Rocks and soil washing into the sea
   (D) Seawater expanding as it gets warmer

4. The greenhouse gases usually trap

    (A) shortwave radiation
    (B) longwave radiation
    (C) microwave electromagnetic radiation
    (D) radio wave electromagnetic radiation

5. Out of the Shared Socioeconomic Pathways (SSPs), the one considering high challenges to mitigation and adaptation is

    (A) SSP1 Sustainability
    (B) SSP2 Middle of the Road
    (C) SSP3 Regional Rivalry
    (D) SSP5 Fossil-fuelled Development

6. As per the 2018 UN World Urbanisation Prospects, the per cent of the global population living in urban areas is about

    (A) 35   (B) 45
    (C) 55   (D) 65

7. The most significant land transition in percentage terms relative to the total global area that has reportedly occurred over the past three decades is from

    (A) agriculture to urban   (B) forest to agriculture
    (C) forest to natural vegetation   (D) natural vegetation to forest

8. Climate change will impact soil erosion indirectly through changes in

    (A) rainfall amount
    (B) rainfall intensity
    (C) spatio-temporal distribution patterns of rainfall
    (D) soil moisture

9. Based on the recent estimates of soil erosion at the continental scale, the continent having the highest average soil erosion rate is

    (A) Africa   (B) Europe
    (C) North America   (D) South America

10. Considering the contributions of the projected climate and land use changes to soil erosion, the scenario having the highest soil erosion rate is

    (A) SSP1–RCP2.6   (B) SSP2–RCP4.5
    (C) SSP4–RCP6.0   (D) SSP5–RCP8.5

References

Bates BC, Kundzewicz ZW, Wu S, Palutikof JP (eds) (2008) Climate change and water. Technical paper of the intergovernmental panel on climate change. IPCC Secretariat, Geneva

Bohner J, Gross J, Riksen M (2004) Impact of land use and climate change on wind erosion: prediction of wind erosion activity for various land use and climate scenarios using the WEELS wind erosion model. In: Goossens D, Riksen M (eds) Wind erosion and dust dynamics: observations, simulations, modelling. ESW Publications, Wageningen

Borrelli P, Robinson DA, Panagos P, Lugato E et al (2020) Land use and climate change impacts on global soil erosion by water (2015–2070). Proc Nat Acad Sci 117:21994–22001

Borrelli P, Robinson DA, Fleischer LR, Lugato E et al (2013) An assessment of the global impact of 21st century land use change on soil erosion. Nat Commun 8:1–13

Buffleben MS, Zayeed K, Kimbrough D, Stenstrom MK, Suffet IH (2002) Evaluation of urban non-point source runoff of hazardous metals entering Santa Monica Bay, California. Water Sci Tech 45:263–268

Chen J, Brissette FP, Liu P, Xia J (2017) Using raw regional climate model outputs for quantifying climate change impacts on hydrology. Hydrol Process 31:4398–4414

Chi W, Zhao Y, Kuang W, He H (2019) Impacts of anthropogenic land use/cover changes on soil wind erosion in China. Sci Total Environ 668:204–215

Cook BI, Mankin JS, Marvel K, Williams AP, Smerdon JE, Anchukaitis KJ (2020) Twenty-first century drought projections in the CMIP6 forcing scenarios. Earth's Future 8: e2019EF001461

Do HX, Westra S, Leonard M (2017) A global-scale investigation of trends in annual maximum streamflow. J Hydrol 552:28–43

Du H, Wang T, Xue X (2017) Potential wind erosion rate response to climate and land use changes in the watershed of the Ningxia-Inner Mongolia reach of the Yellow River, China, 1986–2013. Earth Surf Process Landf 42:1923–1937

Duulatov E, Chen X, Amanambu AC, Ochege FU, Orozbaev R, Issanova G, Omurakunova G (2019) Projected rainfall erosivity over Central Asia based on CMIP5 climate models. Water 11:897

European Environment Agency (EPA) (2021) Climate change indicators: atmospheric concentrations of greenhouse gases. US Environmental Protection Agency https://www.epa.gov/climate-indicators/climate-change-indicators-atmospheric-concentrations-greenhouse-gases. Accessed 14 Jan 2022

European Environment Agency (EPA) (2012) Climate change, impacts and vulnerability in Europe 2012: An indicator-based report. Report no.12, Copenhagen

FAO (2021) State of the world's forests 2020. https://www.fao.org/state-of-forests/en/ Accessed 28 Dec 2021

FAO (2015) Global forest assessment 2015. Food and Agriculture Organization of the United Nations, Rome

Gao Q, Ci L, Yu M (2002) Modeling wind and water erosion in northern China under climate and land use changes. J Soil Water Conserv 57:46–55

Ghiggi G, Seneviratne SI, Humphrey V, Gudmundsson L (2019) GRUN: an observation-based global gridded runoff dataset from 1902 to 2014. Earth Syst Sci Data 11:1655–1674

GSA (2020) Geological Society of America: Climate change position statement. https://www.geosociety.org/gsa/positions/position10.aspx Accessed 14 Nov 2021

Haas J, Ban YF (2014) Urban growth and environmental impacts in Jing-Jin-Ji, the Yangtze River delta and the Pearl River delta. Int J App Earth Obs Geoinf 30:42–55

Hartmann DL, Klein Tank AMG, Rusticucci M, Alexander L, Brönnimann S et al. (2013) Observations: atmosphere and surface supplementary material. In: Stocker TF, Qin D, Plattner GK, Tignor M, Allen SK, Boschung J, Nauels A, Xia Y, Bex V, Midgley PM (eds) Climate Change 2013: The Physical Science Basis. Contribution of Working Group I to the Fifth Assessment Report of the Intergovernmental Panel on Climate Change, Cambridge University Press, Cambridge, United Kingdom and New York, NY, USA

Hettiarachchi S, Wasko C, Sharma A (2018) Increase in flood risk resulting from climate change in a developed urban watershed—the role of storm temporal patterns Hydrol Earth Syst Sci 22: 2041–2056

Hoegh-Guldberg O, Jacob D, Taylor M, Bindi M et al. (2018) Impacts of $1.5°C$ global warming on natural and human systems. Masson-Delmotte V, Zhai P, Pörtner H, Roberts D, Skea J, Shukla PR, Pirani A, Moufouma-Okia W, Péan C, Pidcock R, Connors SL, Matthews JBR, Chen Y, Zhou X, Gomis MI, Lonnoy E, Maycock T, Tignor M, Waterfield T (eds)An IPCC Special Report on the impacts of global warming of 1.5 °C above pre-industrial levels and related global greenhouse gas emission pathways, in the context of strengthening the global response to the threat of climate change, World Meteorological Organization Technical Document, Geneva, Switzerland

Hungr O, Leroueil S, Picarelli L (2013) The Varnes classification of landslide types, an update. Landslides 11:167–194

Huo A, Li H (2013) Assessment of climate change impact on the stream-flow in a typical debris flow watershed of Jianzhuangcuan catchment in Shaanxi Province, China. Environ Earth Sci 69:1931–1938

IPCC (2021) Climate change 2021: the physical science basis.In: Masson-Delmotte V, Zhai P, Pirani A, Connors SL, Péan C, Berger S, Caud N, Chen Y, Goldfarb L, Gomis MI, Huang M, Leitzell K, Lonnoy E, Matthews JBR, Maycock TK, Waterfield T, Yelekçi O, Yu R, Zhou B (eds) Contribution of Working Group I to the Sixth Assessment Report of the Intergovernmental Panel on Climate Change. Cambridge University Press. In Press

IPCC (2019) Climate change and land: An IPCC special report on climate change, desertification, land degradation, sustainable land management, food security, and greenhouse gas fluxes in terrestrial ecosystems. https://www.ipcc.ch/srccl/download/ Accessed 14 Nov 2021

IPCC (2014) Climate Change 2014: Impacts, adaptation, and vulnerability. Part A: Global and sectoral aspects. In: Field CB, Barros VR, Dokken DJ, Mach KJ, Mastrandrea MD, Bilir TE, Chatterjee M, Ebi KL, Estrada YO, Genova RC, Girma B, Kissel ES, Levy AN, MacCracken S, Mastrandrea PR, White LL (eds) Contribution of Working Group II to the Fifth Assessment Report of the Intergovernmental Panel on Climate Change. Cambridge University Press, Cambridge, UK and New York

IPCC (2013) Climate Change 2013: Tthe physical science basis. In: Stocker TF, Qin D, Plattner GK, Tignor M, Allen SK, Boschung J, Nauels A, Xia Y, Bex V, Midgley PM (eds) Contribution of Working Group I to the Fifth Assessment Report of the Intergovernmental Panel on Climate Change. Cambridge University Press, Cambridge, United Kingdom and New York

Ito A (2007) Simulated impacts of climate and land cover change on soil erosion and implication for the carbon cycle, 1901 to 2100. Geophys Res Lett 34:L09403

Jones RJ, Le Bissonnais Y, Bazzoffi P, Sanchez DJ, Düwel O et al. (2004) Nature and extent of soil erosion in Europe. Reports of the Task Group 2. In: Van-Camp L, Bujarrabal B, Gentile AR, Jones RJA, Montanarella L, Olazabal C, Selvaradjou SK (eds) Reports of the Technical Working Groups Established under the Thematic Strategy for Soil Protection. Office for Official Publications of the European Communities, Luxembourg

Lehner F, Wood AW, Vano JA, Lawrence DM, Clark MP, Mankin JS (2019) The potential to reduce uncertainty in regional runoff projections from climate models. Nature Cli Change 9:926–933

Li Z, Fang H (2016) Impacts of climate change on water erosion: a review. Earth Sci Rev 163:94–117

Liddicoat C, Hayman P, Alexander B, Rowland J et al. (2012) Climate change, wheat production and erosion risk in South Australia's cropping zone: linking crop simulation modelling to soil landscape mapping. Government of South Australia, Department of Environment, Water and Natural Resources

Liping C, Yujun S, Saeed S (2018) Monitoring and predicting land use and land cover changes using remote sensing and GIS techniques—a case study of a hilly area, Jiangle China. PLoS ONE 13:e0200493

Liu F, Zhang Z, Zhao X, Wang X, Zuo L et al (2019) Chinese cropland losses due to urban expansion in the past four decades. Sci Total Environ 650:847–857

Lu XX, Ran LS, Liu S, Jiang T, Zhang SR, Wang JJ (2013) Sediment loads response to climate change: a preliminary study of eight large Chinese rivers. Int J Sediment Res 28:1–14

Maksimovic C, Prodanović D, Boonya-Aroonnet S, Leitão JP (2009) Overland flow and pathway analysis for modelling of urban pluvial flooding. J Hydraul Res 47:512–523

Mendoza B (2005) Total solar irradiance and climate. Adv Space Res 35:882–890

Mondal A, Khare D, Kundu S, Meena PK, Mishra PK, Shukla R (2015) Impact of climate change on future soil erosion in different slope, land use, and soil-type conditions in a part of the Narmada River Basin India. J Hydrol Eng 20:C5014003

Moss RH, Edmonds JA, Hibbard KA et al (2010) The next generation of scenarios for climate change research and assessment. Nature 463:747

Myers N (1993) Gaia: an atlas of planet management. J Acad Librariansh 19:200

National Academies of Sciences, Engineering and Medicine (NASEM) (2019) Negative emissions technologies and reliable sequestration: A research agenda: https://www.nap.edu/catalog/25259

National Research Council (2011) America's climate choices. The National Academies Press, Washington DC

Nunes J, Nearing MA (2011) Modelling impacts of climate change: case studies using the new generation of erosion models. In: Morgan RP, Nearing MA (eds) Handbook of erosion modelling. Wiley-Blackwell, Chichester, UK

O'Neill BC, Tebaldi C, van Vuuren DP, Eyring V, Friedlingstein P, Hurtt G, Knutti R, Kriegler E, Lamarque JF, Lowe J, Meehl GA, Moss R, Riahi K, Sanderson BM (2016) The Scenario Model Intercomparison Project (ScenarioMIP) for CMIP6. Geosci Model Dev 9:3461–3482

Panagos P, Borrelli P, Poesen J, Ballabio C, Lugato E, Meusburger K, Montanarella L, Alewell C (2015) The new assessment of soil loss by water erosion in Europe Environ Sci. Policy 54:438–447

Parajuli PB, Jayakody P, Sassenrath GF, Ouyang Y (2016) Assessing the impacts of climate change and tillage practices on stream flow, crop and sediment yields from the Mississippi River basin. Agric Water Manag 168:112–124

Petley D (2012) Global patterns of loss of life from landslides. Geology 40:927–930

Pimentel D, Harvey C, Resosudarmo P, Sinclair K, Kurz D et al (1995) Environmental and economic costs of soil erosion and conservation benefits. Science 267:1117–1123

Plangoen P, Babel M, Clemente R, Shrestha S, Tripathi N (2013) Simulating the impact of future land use and climate change on soil erosion and deposition in the Mae Nam Nan sub-catchment, Thailand. Sustainability 5:3244–3274

Pruski FF, Nearing MA (2002) Runoff and soil-loss responses to changes in precipitation: a computer simulation study. J Soil Water Conserv 57:7–16

Radwan TM, Blackburn GA, Whyatt JD, Atkinson PM (2021) Global land cover trajectories and transitions. Sci Rep 11:1–16

Ramsar Convention (2018) Global wetland outlook: State of the world's wetland and their services to people. Secretariat of the Convention on Wetlands, Gland, Switzerland

Ramsar Convention (2016) In: An introduction to the Ramsar convention on wetlands, 7th edn. (previously The Ramsar Convention Manual). Ramsar Convention Secretariat, Gland, Switzerland

Riahi K, van Vuuren DP, Kriegler E, Edmonds J, O'Neill BC et al (2017) The shared socioeconomic pathways and their energy, land use, and greenhouse gas emissions implications: an overview. Glob Environ Change 42:153–168

Ritchie H, Roser M (2018) Urbanization. Our world in data. https://ourworldindata.org/urbanization Accessed 14 Nov 2021

Routschek A, Schmidt J, Kreienkamp F (2014) Impact of climate change on soil erosion—a high-resolution projection on catchment scale until 2100 in Saxony/Germany. CATENA 121:99–109

Schwanghart W, Worni R, Huggel C, Stoffel M, Korup O (2016) Uncertainty in the Himalayan energy–water nexus: estimating regional exposure to glacial lake outburst floods. Environ Res Lett 11:074005

Seneviratne SI, Nicholls N, Easterling D, Goodess CM et al. (2012) Changes in climate extremes and their impacts on the natural physical environment. In: Field CB, Barros V, Stocker TF, Qin D, Dokken DJ, Ebi KL, Mastrandrea MD, Mach KJ, Plattner KG, Allen SK, Tignor M, Midgley PM (eds) Managing the risks of extreme events and disasters to advance climate change adaptation, A Special Report of Working Groups I and II of the Intergovernmental Panel on Climate Change (IPCC). Cambridge University Press, Cambridge, UK, and New York

Serpa D, Nunes JP, Santos J, Sampaio E, Jacinto R et al (2015) Impacts of climate and land use changes on the hydrological and erosion processes of two contrasting Mediterranean catchments. Sci Total Environ 538:64–77

Shen Y, Zhang C, Wang X, Zou X, Kang L (2018) Statistical characteristics of wind erosion events in the erosion area of Northern China. CATENA 167:399–410

Simonneaux V, Cheggour A, Deschamps C, Mouillot F, Cerdan O, Bissonnais Y (2015) Land use and climate change effects on soil erosion in a semi-arid mountainous watershed (High Atlas, Morocco). J Arid Environ 122:64–75

Solanki SK, Krivova NA, Haigh JD (2013) Solar irradiance variability and climate. Annu Rev Astron Astrophys 51:311–351

Stoffel M, Tiranti D, Huggel C (2014) Climate change impacts on mass movements- case studies from the European Alps. Sci Total Environ 493:1255–1266

Su B, Huang J, Fischer T, Wang Y, Kundzewicz ZW et al (2018) Drought losses in China might double between the 1.5°C and 2.0°C warming. Proc Natl Acad Sci USA 115:10600–10605

Tabari H (2020) Climate change impact on flood and extreme precipitation increases with water availability. Sci Rep 10:13768

UN (2019) United Nations Department of Economic and Social Affairs, Population Division. World Urbanization Prospects: The 2018 Revision (ST/ESA/SER.A/420). United Nations, New York

UNFCCC (2015) Report of the conference of the parties on its twenty-first session, held in Paris from 30 November to 13 December 2015. Part Two: Action taken by the conference of the parties at its twenty-first session. United Nations. https://undocs.org/en/FCCC/CP/2015/10/Add.1

US Global Change Research Program (USGCRP) (2017) Fourth national climate assessment Volume I: Climate science special reports. https://science2017.globalchange.gov

US Global Change Research Program (USGCRP) (2018) Fourth national climate assessment Volume II: Impacts, risks, and adaptation in the United States. https://nca2018.globalchange.gov

van Vuuren DP, Edmonds J, Kainuma M et al (2011) The representative concentration pathways: an overview. Clim Chang 109:5–31

Verheijen FGA, Jones RJA, Rickson RJ, Smith CJ (2009) Tolerable versus actual soil erosion rates in Europe. Earth Sci Rev 94:23–38

Webb NP, Kachergis E, Miller SW, McCord SE et al (2020) Indicators and benchmarks for wind erosion monitoring, assessment and management. Ecol Indicators 110:105881

Winkler K, Fuchs R, Rounevell M, Herold M (2021) Global land use changes are four times greater than previously estimated. Nat Commun 12:2501

Yang DW, Kanae S, Oki T, Koike T, Musiake K (2003) Global potential soil erosion with reference to land use and climate changes. Hydrol Process 17:2913–2928

Yang S, Li Z, Yu J, Hu X, Dong W, He S (2018) El Nino Southern Oscillation and its impact in the changing climate. Natl Sci Rev 5:840–857

Zabaleta A, Meaurio M, Ruiz E, Antigüedad I (2014) Simulation climate change impact on runoff and sediment yield in a small watershed in the Basque Country, northern Spain. J Environ Qual 43:235–245

Zhang HY, Fan J, Cao W, Harris W, Li Y, Chi W, Wang S (2018) Response of wind erosion dynamics to climate change and human activity in Inner Mongolia, China during 1990 to 2015. Sci Total Environ 639:1038–1050

Zhang X, Tang Q, Zhang X, Lettenmaier DP (2014) Runoff sensitivity to global mean temperature change in the CMIP5 models. Geophys Res Lett 41:5492–5498

Answers to Multiple Choice Questions

Chapter 2

Q. 1 B	Q. 2 C	Q. 3 A	Q. 4 A	Q. 5 B
Q. 6 C	Q. 7 C	Q. 8 D	Q. 9 D	Q. 10 A

Chapter 3

Q. 1 A	Q. 2 C	Q. 3 A	Q. 4 A	Q. 5 C
Q. 6 D	Q. 7 C	Q. 8 B	Q. 9 D	Q. 10 B

Chapter 4

Q. 1 D	Q. 2 A	Q. 3 B	Q. 4 C	Q. 5 B
Q. 6 A	Q. 7 C	Q. 8 B	Q. 9 A	Q. 10 B

Chapter 5

Q. 1 D	Q. 2 A	Q. 3 C	Q. 4 C	Q. 5 D
Q. 6 D	Q. 7 B	Q. 8 B	Q. 9 D	Q. 10 B

Chapter 6

Q. 1 A	Q. 2 C	Q. 3 B	Q. 4 A	Q. 5 C
Q. 6 D	Q. 7 B	Q. 8 C	Q. 9 A	Q. 10 C

Chapter 7

Q. 1 B	Q. 2 C	Q. 3 D	Q. 4 D	Q. 5 B
Q. 6 C	Q. 7 A	Q. 8 D	Q. 9 B	Q. 10 A

Chapter 8

| Q. 1 C | Q. 2 B | Q. 3 D | Q. 4 C | Q. 5 D |
| Q. 6 A | Q. 7 B | Q. 8 C | Q. 9 B | Q. 10 A |

Chapter 9

| Q. 1 C | Q. 2 B | Q. 3 A | Q. 4 C | Q. 5 B |
| Q. 6 A | Q. 7 C | Q. 8 B | Q. 9 D | Q. 10 B |

Chapter 10

| Q. 1 C | Q. 2 B | Q. 3 D | Q. 4 C | Q. 5 D |
| Q. 6 A | Q. 7 C | Q. 8 C | Q. 9 B | Q. 10 B |

Chapter 11

| Q. 1 A | Q. 2 C | Q. 3 C | Q. 4 A | Q. 5 B |
| Q. 6 D | Q. 7 B | Q. 8 C | Q. 9 C | Q. 10 C |

Chapter 12

| Q. 1 C | Q. 2 D | Q. 3 A | Q. 4 B | Q. 5 D |
| Q. 6 D | Q. 7 C | Q. 8 D | Q. 9 B | Q. 10 B |

Chapter 13

| Q. 1 C | Q. 2 B | Q. 3 D | Q. 4 D | Q. 5 C |
| Q. 6 B | Q. 7 D | Q. 8 B | Q. 9 D | Q. 10 C |

Chapter 14

| Q. 1 C | Q. 2 C | Q. 3 D | Q. 4 B | Q. 5 C |
| Q. 6 C | Q. 7 B | Q. 8 D | Q. 9 D | Q. 10 D |

Index

Printed in the United States
by Baker & Taylor Publisher Services